Discrete Fractional Calculus

Applications in Control and Image Processing

Series in Computer Vision

ISSN: 2010-2143

Series Editor: C H Chen *(University of Massachusetts Dartmouth, USA)*

Series in Computer Vision - Vol. 4

Discrete Fractional Calculus

Applications in Control and Image Processing

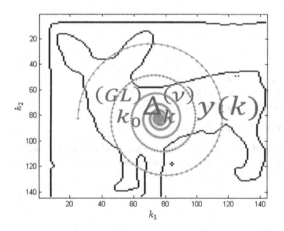

Piotr Ostalczyk

Lodz University of Technology, Poland

World Scientific

NEW JERSEY · LONDON · SINGAPORE · BEIJING · SHANGHAI · HONG KONG · TAIPEI · CHENNAI · TOKYO

Published by

World Scientific Publishing Co. Pte. Ltd.

5 Toh Tuck Link, Singapore 596224

USA office: 27 Warren Street, Suite 401-402, Hackensack, NJ 07601

UK office: 57 Shelton Street, Covent Garden, London WC2H 9HE

British Library Cataloguing-in-Publication Data
A catalogue record for this book is available from the British Library.

Series in Computer Vision — Vol. 4
DISCRETE FRACTIONAL CALCULUS
Applications in Control and Image Processing

Copyright © 2016 by World Scientific Publishing Co. Pte. Ltd.

ISBN 978-981-4725-66-8

Printed in Singapore

Rien n'est parfait.

Antoine de Saint-Exupéry, 'Le Petit Prince'

Preface

The statement of professor Katsuyuki Nishimoto expressed in his book entitled "An Essence of Nishimoto's Fractional Calculus, Calculus of the 21st Century: Integrations and Differentiations of Arbitrary Order" [Nishimoto (1991b)] dated 1991 appeard as the prohecy of the end of XX century. This apt opinion confirms the Science Watch of Thomson Reuters stating in October, 2009 that the fractional calculus nowadays is an 'Emerging Research Front'.

Since the mid-twentieth century, the fractional calculus consequently has been introduced at 'the mathematical and technical scientific fora' of the world by scientists out of whom there should be mentioned Oldham, K. B., Jerome Spannier [Oldham and Spannier (1974)], Katsuiuki Nishimoto [Nishimoto (1984, 1986, 1989)], Stefan Samko, Anatoly A. Kilbas and Oleg Marichev [Samko at al. (1986)], Kenneth S. Miller and Bertram Ross [Miller and Ross (1993)]. It is worth noting the great contribution of professors: Alain Oustaloup [Oustaloup (1983, 1991, 1995)], Virginia Kiryakova [Kiryakova (1994)], Igor Podlubny [Podlubny (1999b)] and their teams contributing to the development of the fractional calculus.

According to professor Katsuyuki Nishimoto's forecast the beginning of twenty-first century brought a rapid growth of works devoted to the theoretical and applied fractional calculus. Among many important monographs one should mention the works of professors: Shantanu Das [Das (2009); Das and Pan (2012)], Tenreiro Machado, Ivo Petráš [Petráš (2011)], YangQuan Chen [Monje et al. (2010)], Dumitru Baleanu [Baleanu et al. (2011)] and Alain Oustaloup [Oustaloup (2014)].

The precise historical review is contained in the articles by Professors: J. Tenreiro Machado, Virginia Kiryakova, Francesco Mainardi [Machado at al. (2011)] and J. Tenreiro Machado, Alexandra M. Galhado and Juan J. Truijllo [Machado at al. (2013)].

Now, also in Poland, the fractional calculus is the subject of research in many scientific centers, among which one should list the following professors and their teams: Małgorzata Klimek at Częstochowa University of Technology (Częstochowa) [Klimek (2009)], Tadeusz Kaczorek [Kaczorek (2011)] and recently deceased Professor Mikołaj Busłowicz at Bialystok University of Technology, Jerzy Klamka at Silesian University of Technology (Gliwice), Andrzej Dzieliński at Warsaw

University of Technology, Wojciech Mitkowski at The AGH University of Science and Technology (Kraków), Stefan Domek [Domek (2013)] at West Pomeranian University of Technology (Szczecin), Krzysztof Latawiec and Rafał Stanisławski at Opole University of Technology.

At this point the Author would like to apologise for not menitoning the remaining prominent scientists working on the fractional calculus developement and applications in different scientific areas.

The fractional calculus originally concerned continuous-variable functions. Such functions describing the so-called analog signals of real world are continuous functions of the temporal variable t. One of the first fractional-order derivatives application appeared in the anomalous diffusion models. In the early sixties of the last century there was proposed the fractional model of the ultra-capacitor. In mechanics, the viscoelasticity phenomenon [Freed *et al.* (2002); Meral *et al.* (2010)] particularly accurately describes mathematical models based on the fractional calculus.

The inerposition of the digital computers to signal processing which can deal with immense quantities of information expressed by numbers, not signals, forced a conversion of the analog signal to a sequence of samples expressed as a set of digital words. This sampling process is usually performed in a digital-to-analog converter (under the well-known restriction expressed by the Shannon theorem). Therefore, one establishes a relation between the continuous variable function and its discrete-variable counterpart. In a discrete version of the fractional calculus the continuous-variable functions are substituted by discrete-variable ones, the fractional-order derivatives are replaced by fractional-order differences and the fractional-order integerals by fractional-order sums. One should admit, that operating on fractional-order differences and sums is more complicated in comparison with the integer-order case. The complications are related to longer signal processing time and larger computer memory needed. The huge development of computers, converters and the memory size compensates for this inconvenience.

This book presents selected applications of the discrete fractional calculus in the discrete system control theory and discrete image processing. In the discrete system identification, analysis and synthesis one can consider integer or fractional models. Usually fractional models are more simple, with lower number of parameters including the fractional orders. In the closed-loop system analysis and synthesis one can consider fractional order controllers or compensators with the integer order plants or more general case where both the regulator and the controlled plant are described by the fractional-order difference equations. The classical problems in the control theory ranges from transient and frequency response analysis of dynamic elements and systems, fractional-oredr PID controllers, fractional-order systems' stability, fractional-order systems' robustness, the optimal control of the fractional-order systems including the variational problem. The classical digital filters can be generalised to the fractional-order case, where backward differences are generalised to the fractional-order case. Also in the digital image processing one can succesfully generalise classical methods to the fractional ones. Here, one should mention early works of the CRONE team at the University of Bordeaux on image edge detection.

The monograph is organised as follows. Basing upon the discrete calculus given in Chapters 1 and 2 fundamental properties of the fractional-order backward-diference and sum are presented. Althought as a fundametal form of the fractional-order backward difference/sum the Grünwald-Letnikov one is used, its five equivalents: Horner, Riemann-Liouville, Caputo, Polynomial-Like Matrix and Laguerre-based are introduced in Chapter 3. The simple graphical interpretation of the Grünwald-Letnikov form is proposed in Chapter 4. Chapter 5 contains the fundamental proprties of the fractional-order backward difference/sum. In the equations only FO backward differences without forward shifts are considered. An application of the discrete fractional calculus starts in Chapter 6 where the non-linear fractional-order dynamical system linearisation procedure is presented. The linearised system descriptions by: the fractional-order difference equation, the state-space equations, discrete fractional transfer function and polynoimial like matrix are given. The linear time-invariant FOE solution $y(k)$ at the discrete time instant k depends only on antecedent solution values $y(k-1), y(k-2), \ldots$. This creates the problem of initial conditions denoted as $y_{k_0-1}, y_{k_0-2}, \ldots$. No forward shift is addmitted. For the four equivalent forms mentioned above homogenous and forced solutions are studied in Chapter 7. Numerous transient responses are presented. This chapter contains also the FOS frequency characterisitcs analysis represented by the discrete Nyquist plots, amplitude and phase discrete characterisitcs and related discrete Bode plots. The FOS reachability and observability supplement the FOS dynamic properties analysis. The FOS stability tests based on the frequency characteristics finish the selected applications of the discrete fractional calculus in the linear discrete system analysis. In Chapter 8 fundamental dynamical elements are analysed. They follow those known in the clasical linear discrete-time: summator, differentiator, lag and oscillation ones. The connections considered in Chapter 9: parallel, serial and with the negative feedback enable building more complicated dynamical strucutres. The closed-loop system structure with the fractional order PID controllers is considered in Chapter 10. Three last chapters presents applications of the discrete fractional calculus in image processing. Considering an image as two variable discrete functions, in Chapters 11, 12, 13 one applies fractional order backward difference and sum to build an image of the fractional potential, detect an image edge and to filter an image.

The book is dedicated to students and engineers working on automatic control, dynamic system identification and image processing. Like in any monograph, there are inevitable errors. Therefore the Author encourages all readers to send any information about them and suggestions improving the topics dealt with in the book.

Here, it's a honour to thank everyone without whom this book would not have appeared. First of all, the Author is very grateful to Professor Chi Hau Chen at the University of Massachssetts Dartmouth (Dartmouth, Massachussetts, USA) for enabling the results presentation the Author have gained for over twenty years of work.

Next, the Author would like to thank Professor Dominik Sankowski, the head of The Institute of Applied Computer Science, Lodz University of Technology for his permanent spiritual encouragement in the Author's research activities and building extremely convenient conditions for scientific activities and especially for this monograph's preparing.

The Author is also honored to thank the inner reviewers. First of all there are many thanks to Professor Tadeusz Kaczorek at The Bialystok University of Technology, a regular member of the Polish Academy of Sciences for his precious remarks concerning the crutial pages of the monograph. It is also the pleasure the thank Professor Jerzy Klamka at The Silesian University of Technology and a regular member of the Polish Academy of Sciences for sections concerning the controllability and reachability of the fractional-order systems.

The Author would also like to acknowledge his PhD students: Piotr Duch (now Ph doctor), Dariusz Brzeziński and Anna Zebrowska at The Institute of Applied Computer Science, Lodz University of Technology who for more then three years have been quite helpful in pointing out problems concerning the fractional calculus applications.

The Author thank the Editor, Steven Patt, at the World Scientific Publishing, Co. Pte. Ltd. Singapure Office for his patience in numerous instructions during the monograph edition process.

I was honoured to work with the thoroughly professional editorial and production staff of the World Scientific (Singapore) represented by Janice Sim. I thank all of of them.

Finally, it is moment to thank Marie Caroline Jnjeyan Univeristy of Toronto (Toronta, Canada) and Kevin Scott Wayling for their instructive comments and suggestions concerning English proofs. All these succesful actions were supervised by Artur Łągwa, M.A. at Piotrków Trybunalski who also deserves the thanks.

Piotr Władysław Ostalczyk
Piotrków Tryb., 2015

Contents

List of Figures

List of Tables

Chapter 1

Discrete-variable real functions

In this chapter a notion of one discrete-variable real-valued functions is introduced. Fundamental operations are defined. A great emphasis is put on the discrete convolution of two functions [Ostalczyk (2001); O'Flynn and Moriarty (1987); Proakis (2007); Weeks (2011)]. Two special discrete-variable functions denoted as $a^{(\nu)(k)}$ and $c^{(\nu)(k)}$ are defined and discussed. The mentioned functions play an important role in fractional-order differences and sums. The link between two discrete-variable function and a digital image finishes this chapter.

1.1 Notation

Firstly one defines some sets:

$\mathbb{R}$ – the set of real numbers,
$\mathbb{R}_+ = [0 + \infty)$ – set of real non-negative numbers,
$\mathbb{R}_- = [0 + \infty)$ – set of real negative numbers,
$\mathbb{R}_{t_0} = [t_0, +\infty)$ – subset of real nonnegative numbers,
$\mathbb{R}_{t_0,t} = [t_0, t]$ – finite interval,
$\mathbb{C}$ – set of complex numbers,
$\mathbb{Z}$ – set of integers,
$\mathbb{Z}_{N_1,N_2} = [N_1, N_2]$ – finite set of integers,
$\mathbb{Z}_{k_0} = [k_0, \infty)$ – subset of integers greater or equal to $k_0 > 0$,
$\mathbb{Z}_+ = \mathbb{Z}_0$ – set of non-negative integers,
$\mathbb{Z}_{k_0,k} = [k_0, k]$ – finite set of integers greater or equal to $k_0 > 0$ and less or equal to k,
$\mathbb{T}_{k_0,k,h} = [k_0 h, (k_0 + 1)h, (k_0 + 2) \ldots, kh]$ – finite set of equally spaced real numbers $ih, i = k_0, k_0 + 1, \ldots, k_0 < k$. ($\mathbb{T}_{0,k,1} = \mathbb{Z}_+$),
$\mathbb{Q}_+$ – set of rational numbers,

$Re\{.\}$ – real part of a complex number,
$Im\{.\}$ – imaginary part of a complex number,
$\bar{z}$ – complex conjugate to z,

$h > 0$ – constant (sampling period),
t – continuous time,
kh – discrete time (k for $h = 1$),

$\mathbf{A}_k$ – real $(k+1) \times (k+1)$ matrix,
$\mathbf{1}_k$ – real $(k+1) \times (k+1)$ unit matrix,
$\mathbf{0}_k$ – real $(k+1) \times (k+1)$ zero matrix,
$\mathbf{a}_k, \mathbf{b}_k, \mathbf{c}_k$ – real $(k+1) \times 1$ column vectors,

T – matrix (vector) transposition,
* – discrete convolution operation,
tr – trase of the matrix,
$|.|$ – absolute value,

$[rad]$ – radian,
$[dB]$ – decibel,
$[dec]$ – decade,
$[V]$ – volt,
$[s]$ – second.

1.2   Class of one discrete-variable functions

Different signals encountered in science and engineering are described by real-valued functions of a single independent discrete real variable (discrete time instants, discrete distance), $k \in \mathbb{T}_{k_0,k,h}$ is denoted as

$$f : \mathbb{T}_{k_0,k,h} \to \mathbb{R} \tag{1.1}$$

where $\mathbb{T}_{k_0,k,h}$ is the function domain and $\mathbb{R}$ its image (codomain). There are several alternative representations of the discrete-variable functions. The most popular is the so-called functional representation. Below an example is given.

$$f(kh) = e^{ah(k-k_0)} \sin[bh(k - k_0)] \tag{1.2}$$

where $h > 0$ is the so-called sampling period (the time, or space between successive samples) and $a, b \in \mathbb{R}$ are function constants. The graphical representation of the function (1.2) for $h = 0.1$, $k_0 = 7$ and $a = -0.03, b = 0.125\pi$ is presented in Fig. 1.1. In many practical applications the discrete-variable functions are represented by an ordered set of real numbers. For every number there is a uniquely assigned corresponding index. Such a tabular representation collects usually a set of measured data. In Table 1.1 the function is tabularized. Very often the initial independent variable is set $k_0 = 0$. Such an assumption does not reduce generality of the considerations. Moreover, for known and constant sampling time h one can omit it considering function (1.2)

$$f(k) = e^{a'(k-k_0)} \sin[b'(k - k_0)] \tag{1.3}$$

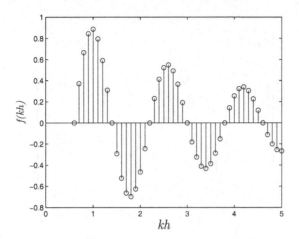

Fig. 1.1   Graphical representation of a function $f(k)$.

Table 1.1   Tabular representation of the discrete-variable function.

k	$f(k)$
0	0
1	3.713734277017140e-001
2	6.659280879784083e-001
3	8.443623161958684e-001
4	8.869204367171575e-001
5	7.951904828883181e-001
6	5.906252306120317e-001
7	3.101971614689395e-001
8	9.633412921752834e-017
9	-2.921326850899535e-001
10	-5.238375874705950e-001

with $a' = ah$ and $b' = bh$. The plot of (1.3) evaluated over interval $[0, k]$ $(k_0 = 0)$ is presented in Fig. 1.2. Notice that this time a horizontal scale is changed.

In many technical applications the function range is also quantized. Here one can mention a 2D digital signal (image) where $\mathbb{T}_{g_{\min}, g_{\max}, h_g}$, i.e.

$$g : \mathbb{T}_{k_0, k, h} \to \mathbb{T}_{g_{\min}, g_{\max}, h_g}. \qquad (1.4)$$

The function (1.4) defined by formula (1.3) is plotted in Fig. 1.3 for $h_g = 0.2$. Here one should mention that the immense progress in electronics causes that h in modern devices rapidly decreases achieving in 16bit A/C and C/A converters values $h_g = 2^{-16} = 1.52587890625e - 005[V]$. Applying this value for h one gets

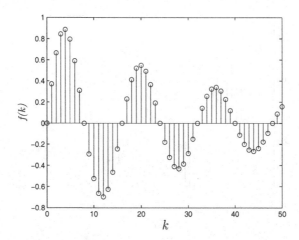

Fig. 1.2   Graphical representation of a function (1.3).

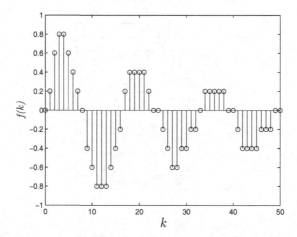

Fig. 1.3   Graphical representation of function (1.4).

the plot presented in Fig. 1.4. The difference between the plots 1.2 and 1.4 is almost negligible. The difference values are really very small. These are indicated in Fig. 1.5.

The discrete-variable function may be also represented by a discrete image characterized by the function value $f(k) \in \mathbb{R}_{y_{\min}, y_{\max}}$, whose codomain is represented by an appropriate element $g(k_1) \in \mathbb{R}_{g_{\min}, g_{\max}}$ where $\mathbb{R}_{g_{\min}, g_{\max}}$ is a set collecting the image gray levels. Hence, a one dimensional image (vector of pixels) described

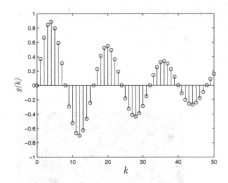

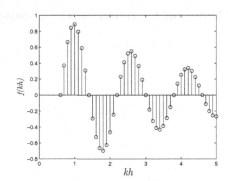

Fig. 1.4   Plot of the function (1.4).

Fig. 1.5   Plot of a difference between functions $f(k)$ (1.3) and $g(k)$ (1.4).

shortly as 1D image is described by a concatenation of two functions

$$f : \mathbb{T}_{k_0,k,h} \to \mathbb{T}_{f_{\min},f_{\max}}$$
$$g : \mathbb{T}_{f_{\min},f_{\max}} \to \mathbb{T}_{g_{\min},g_{\max}}. \tag{1.5}$$

Images related to the discrete-variable functions (1.3) and (1.4) are given in Figs. 1.6 and 1.7, respectively. In fact, for a better visibility of the vectors of pixels every vector pixel is magnified to a square box containing $l \times l$ identical pixels.

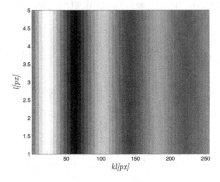

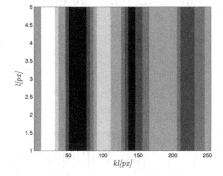

Fig. 1.6   Image of the function (1.3).

Fig. 1.7   1D image of the function (1.4).

1.3   Selected discrete-variable functions

1.3.1   *Fundamental discrete-variable functions*

The study of discrete-variable functions starts with the presentation of a number of basic functions that appear very often in analysis. These functions are defined below.

Definition 1.1 (Discrete Dirac pulse). The discrete Dirac pulse is defined as

$$\delta(k) = \begin{cases} 1 & \text{for } k = 0 \\ 0 & \text{for } k \neq 0 \end{cases}. \tag{1.6}$$

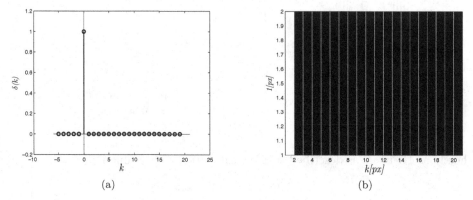

Fig. 1.8   Plot (a) and 1D image (b) of the discrete Dirac pulse (1.6).

The next function of a great importance in the discrete fractional calculus is the so-called unit step function. Its definition is given below, and its plot and image are presented in Figs. 1.9(a) and 1.9(b), respectively.

Definition 1.2 (Discrete unit step function). The discrete unit step function is defined as

$$\mathbf{1}(k) = \begin{cases} 0 & \text{for } k < 0 \\ 1 & \text{for } k \geqslant 0 \end{cases}. \tag{1.7}$$

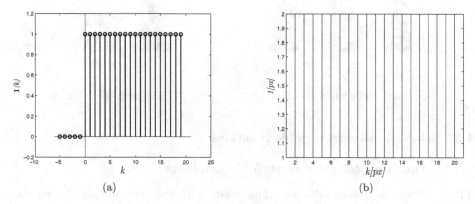

Fig. 1.9   Plot (a) and 1D image (b) of the discrete unit step function (1.7).

1.3.2 *Elementary discrete-variable functions*

Definition 1.3. *The general exponential function*

$$f(k) = a^{bk}\mathbf{1}(k) \text{ where } a, b, \in \mathbb{R}. \tag{1.8}$$

In linear discrete-time systems, an important role is played by the exponential function, which is a special case of the general exponential function (1.8) where $a = e$ ($e = 2.718281828459046$ is the Euler number). The plot of the exponential function is presented in Fig. 1.10.

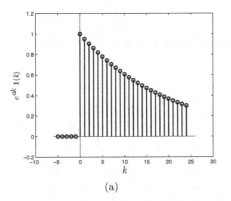
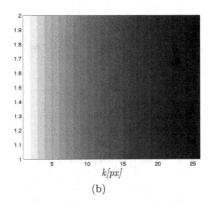

(a)   (b)

Fig. 1.10   Plot (a) and 1D image (b) of the exponential function evaluated for $b = -0.05$.

On the basis of the exponential function one defines discrete-variable trigonometric functions: sin and cos. Below their definitions are provided.

Definition 1.4. *The discrete sin function*

$$f(k) = \sin(ak + b)\mathbf{1}(k) \text{ where } a, b \in \mathbb{R}. \tag{1.9}$$

A typical shape of the discrete sin function is given in Fig. 1.11(a). Its image is given beside in Fig. 1.11(b). Here it is worth noting that in a discrete-variable function appropriate choice of its parameters may essentially change its plot. Such a situation occurs when in the sin function one puts $a = \frac{\pi}{2}$ or $a = \frac{\pi}{2}$. Appropriate plots and related 1D images are presented in Figs. 1.12(a) and Fig. 1.12(b) as well as Figs. 1.13(a) and Fig. 1.13(b), respectively. It must be admitted that the function graphs do not resemble the graphs of the sin function.

Even more surprising is the assumption that $a = 2\pi, b = \frac{\pi}{2}$ or $a = 2\pi, b = -\frac{\pi}{2}$. For such parameters one gets $\sin(2\pi k + 0.5\pi) = \mathbf{1}(k)$ and $\sin(2\pi k - 0.5\pi) = -\mathbf{1}(k)$.

The discrete cos function may be derived from formula (1.9) by the assumption that $b = \frac{\pi}{2}$. The discrete ramp function is defined as follows.

Definition 1.5. *The discrete ramp function defines a formula*

$$f(k) = k\mathbf{1}(k). \tag{1.10}$$

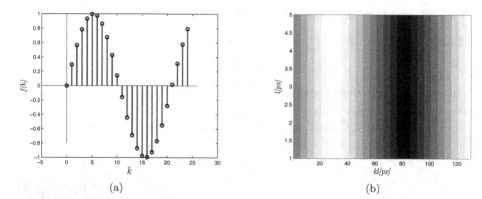

Fig. 1.11   Plot (a) and 1D image (b) of the sin function evaluated for $a = \frac{\pi}{2}, b = 0$.

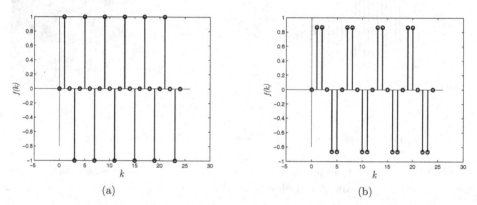

Fig. 1.12   Plot of the discrete sin function evaluated for special parameters. (a) $a = \frac{\pi}{2}, b = 0$. (b) $a = \frac{\pi}{3}, b = 0$.

This function is a special case of so-called discrete-variable power functions $f(k) = k^s \mathbf{1}(k)$ where $s \in \mathbb{Z}_+$. Realize that for $s = 0$ it is the unit step function. The plot and image of the ramp function are given in Figs. 1.14(a) and 1.14(b).

1.3.3   *Function* $a^{(\nu)}(k)$

Now one defines a discrete-variable function, which plays a crucial role in discrete fractional calculus. This discrete-variable function is characterized by one parameter denoted by ν. The function may be expressed by several equivalent formulas.

Definition 1.6. *The discrete function* $a^{(\nu)}(k)$ *is defined as*

$$a^{(\nu)}(k) = \begin{cases} 0 & \text{for } k < 0 \\ 1 & \text{for } k = 0 \\ (-1)^k \frac{\nu(\nu-1)\cdots(\nu-k+1)}{k!} & \text{for } k = 1, 2, 3, \ldots \end{cases} \qquad (1.11)$$

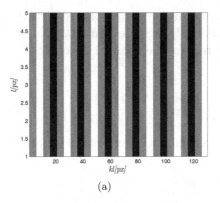

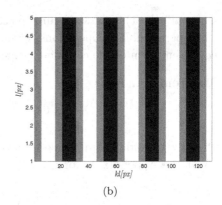

(a) (b)

Fig. 1.13 1D image of the discrete sin function evaluated for special parameters. (a) $a = \frac{\pi}{2}, b = 0$. (b) $a = \frac{\pi}{3}, b = 0$.

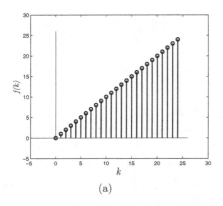

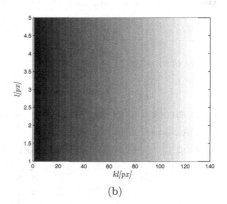

(a) (b)

Fig. 1.14 Plot (a) and 1D image (b) of the discrete ramp function.

By direct calculations one may verify that the function $a^{(\nu)}(k)$ may be evaluated in a recursive way

$$a^{(\nu)}(k) = \begin{cases} 1 & \text{for } k = 0 \\ (1 - \frac{\nu+1}{k})a^{(\nu)}(k-1) & \text{for } k = 1, 2, 3, \dots \end{cases}. \qquad (1.12)$$

This can be verified by direct calculation

$$a^{(\nu)}(k) = (-1)^k \frac{\nu(\nu-1)\cdots(\nu-k+1)}{k!}$$

$$\times (-1)^{k-1} \frac{\nu(\nu-1)\cdots(\nu-k+2)}{(k-1)!}(-1)\frac{(\nu-k+1)}{k}$$

$$= a^{(\nu)}(k-1)(-1)\frac{(\nu-k+1)}{k} = a^{(\nu)}(k-1)\left(1 - \frac{\nu+1}{k}\right).$$

where $a^{(\nu)}(k-1)$ is the same function as (1.11) but subjected to a forward horizontal shift operation. This operation is examined more fully in Subsection 1.4.2. The next equivalent form of the considered function introduces the following theorem.

Theorem 1.1. *For* $\nu \in \mathbb{R}$,

$$a^{(\nu)}(k) = \frac{\Gamma(k - \nu)}{\Gamma(-\nu)\Gamma(k + 1)}. \tag{1.13}$$

Proof. By fundamental property of the Euler Gamma function defined by formula (A.3) (see Appendix A) for $z = -\nu$ and $n = k$

$$\Gamma(-\nu) = \frac{\Gamma(-\nu + k)}{(-\nu)(-\nu + 1)\cdots(-\nu + k - 1)}. \tag{1.14}$$

$\square$

After a simple rearrangements one gets

$$(-1)^k(\nu)(\nu - 1)\cdots(\nu - k + 1) = \frac{\Gamma(-\nu + k)}{\Gamma(-\nu)}. \tag{1.15}$$

Moreover, by (A.3)

$$\Gamma(k + 1) = k!. \tag{1.16}$$

A substitution of (1.15) and (1.16) into (1.11) yields (1.12).

One should also note that function $a^{(\nu)}(k)$ values appear in the expansion of the function $(1 - x)^\nu$ into Maclaurin infinite series

$$(1 - x)^\nu = \sum_{i=0}^{+\infty} a^\nu(i)x^i \quad \text{for all } |x| \leqslant 1. \tag{1.17}$$

The plot and 1D image of the function $a^{(\nu)}(k)$ are given in Figs. 1.15(a) and 1.15(b). In order to fully present the nature of the analyzed function $a^{(\nu)}(k)$ in Figs. 1.16(a) and 1.16(b) plots of $a^{(\nu)}(k)$ related to different parameter ν values from a set $\{0.10.2\cdots0.9\}$ are given.

In Figs. 1.15(a) and 1.15(b) the plots of function $a^{(\nu)}(k)$ evaluated for coefficients ν from a set $\{0.1, 0.2, \ldots, 1.0\}$ and two different ranges $k = 0, 1, 2, \ldots, 9, 10$ and $k = 2, \ldots, 9, 10$ are presented, respectively. Every plot is uniquely identified by its value related to $k = 1$, where $a^{(\nu)}(1) = -\nu$. For better clarity of plots, its distinct points are connected by straight line segments. The second plot is a fragment of the first one. It shows that the individual segments can be crossed. In the next four figures, the plots of the $a^{(\nu)}(k)$ function, for further parameter ν ranges are shown. The plots reveal that function $a^{(\nu)}(k)$ sharply tends to zero for all $k > \nu+1$. This property will be discussed in details in the next subsection. The plots 1.16 and 1.17 also contain integer values of parameter ν. The following Figs. 1.18(a) and 1.18(b) present the function $a^{(\nu)}(k)$ plots evaluated only for integer parameters $\nu \in \{1, 2, \ldots, 9\}$ (a) and $\{10, 11, \ldots, 19\}$ (b), respectively.

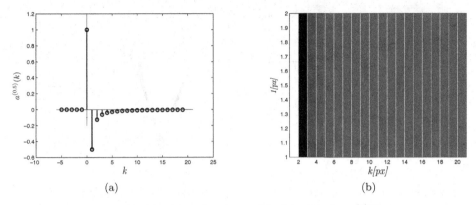

Fig. 1.15   Plot (a) and an image (b) of the function $a^{(\nu)}(k)$.

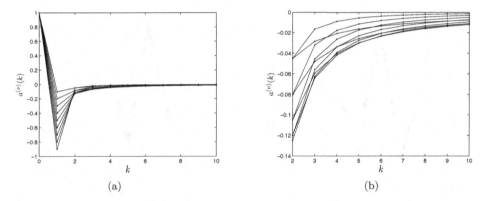

Fig. 1.16   Plots of the functions $a^{(\nu)}(k)$ for $\nu \in \{0.1, 0.2, \ldots, 1.0\}$ and two ranges: $k \in \{0, 1, \ldots, 10\}$ and $k \in \{2, 3, \ldots, 10\}$.

Now, the plots of function $a^{(\nu)}(k)$ related to negative parameter ν values are given. Similarly to the positive values of parameter ν four figures related to four negative parameter ranges are presented.

In contrast to the previous plots, which tend to zero for growing k all plots related to negative parameter ν values increase due to lowering ν. Properties of this important function are discussed in the next sub-section. In the following two examples one compares the results of the function $a^{(\nu)}(k)$ evaluation due to three equivalent formulas (1.11)–(1.13).

Example 1.1. For $\nu = 0.5$ one calculates the function $a^{(\nu)}(k)$ values according to formulas (1.11)–(1.13). In order to distinguish between the results calculated according to three mathematically equivalent formulas, the function $a^{(\nu)}(k)$ will be temporarily denoted as $a_{11}^{(\nu)}(k)$, $a_{12}^{(\nu)}(k)$ and $a_{13}^{(\nu)}(k)$, respectively. Error functions

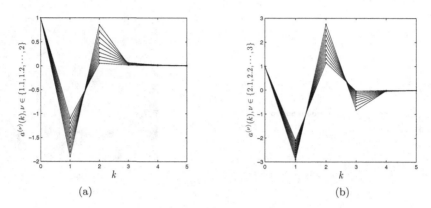

Fig. 1.17   Plots of the function (a) $a^{(\nu)}(k)$ for $\nu \in \{1.1, 1.2, \ldots, 2.0\}$ and (b) $\{2.1, 2.2, \ldots, 3.0\}$.

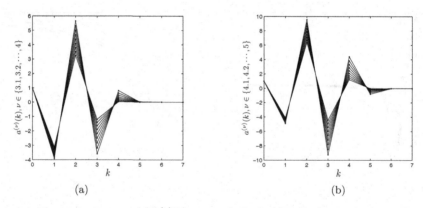

Fig. 1.18   Plots of the function (a) $a^{(\nu)}(k)$ for $\nu \in \{3.1, 3.2, \ldots, 4.0\}$ and (b) $\{4.1, 4.2, \ldots, 5.0\}$.

defined below will serve as tools to three considered functions shapes comparison

$$e_{11,12}^{(\nu)}(k) = |a_{11}^{(\nu)}(k) - a_{12}^{(\nu)}(k)|$$

$$e_{11,13}^{(\nu)}(k) = |a_{11}^{(\nu)}(k) - a_{13}^{(\nu)}(k)| \tag{1.18}$$

$$e_{12,13}^{(\nu)}(k) = |a_{12}^{(\nu)}(k) - a_{13}^{(\nu)}(k)|.$$

In Figs. 1.22–1.24 three pairs of the error functions evaluated for $\nu = 0.5$ and $\nu = -0.5$ are given, respectively.

The error plots reveal very similar results of $e_{11,13}^{(\nu)}(k)$ and $e_{12,13}^{(\nu)}(k)$ for positive and negative parameter ν. For negative value of ν, errors are two orders of magnitude greater then for positive values. The evaluation of $a^{(\nu)}(k)$ by formula

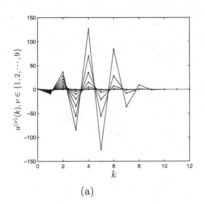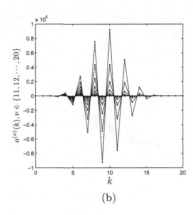

(a)   (b)

Fig. 1.19 Plots of the functions (a) $a^{(\nu)}(k)$ for $\nu \in \{1, 2, \ldots, 9\}$ and (b) $\{10, 11, \ldots, 20\}$.

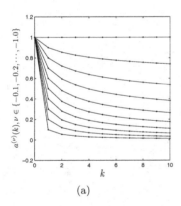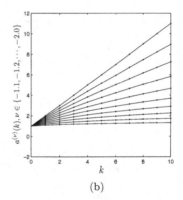

(a)   (b)

Fig. 1.20 Plots of the functions (a) $a^{(\nu)}(k)$ for $\nu \in \{-0.1, -0.2, \ldots, -1.0\}$ and (b) $\{-1.1, -1.2, \ldots, -2.0\}$.

1.13 disqualifies an explosion' of factors in this formula. This is caused by rapidly increasing values of gamma functions. (In the Matlab® program one can evaluate $a_{13}^{(\nu)}(k)$ only for $k < 170$).

1.3.3.1 *Function $a^{(\nu)}(k)$ properties*

Next some properties of the discrete-variable function $a^{(\nu)}(k)$indexfunction $a^{(\nu)}(k)$ with one parameter ν will be analyzed. They are formulated in the form of a few theorems. The most important range of parameter ν are $(0, 1)$ and $(-1, 0)$, thus the analysis will mainly concentrate on the mentioned ranges. Here one assumes that

$$0 < \nu < 1. \tag{1.19}$$

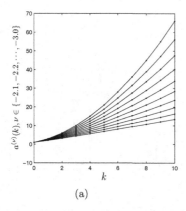 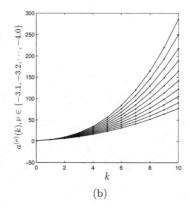

(a) (b)

Fig. 1.21 Plots of the functions (a) $a^{(\nu)}(k)$ for $\nu \in \{-2.1, -2.2, \ldots, -3.0\}$ and (b) $\{-3.1, -3.2, \ldots, -4.0\}$.

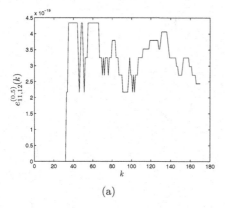 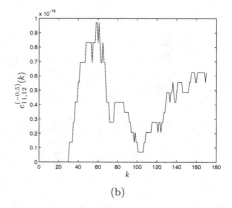

(a) (b)

Fig. 1.22 Plots of the error functions (a) $e^{(\nu)}_{11,12}(k)$ for $\nu = 0.5$ and (b) $\nu = -0.5$.

Theorem 1.2. *For $i = 1, 2, \ldots$ and $0 < \nu < 1$, the following inequalities hold*

$$a^{(\nu)}(k) < 0$$
$$a^{(-\nu)}(k) > 0. \tag{1.20}$$

Proof. To prove inequalities 1.20 one uses a recurrent formula (1.12)

$$a^{(\pm\nu)}(k) = \left(1 - \frac{\pm\nu + 1}{k}\right) a^{(\pm\nu)}(k) \text{ for } k = 1, 2, 3, \ldots \tag{1.21}$$

with

$$a^{(\pm\nu)}(1) = \mp\nu. \tag{1.22}$$

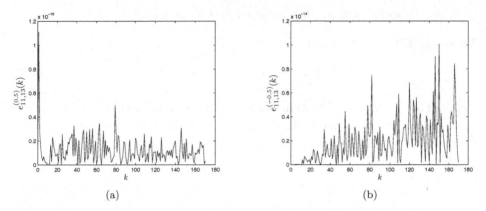

(a)   (b)

Fig. 1.23   Plots of the error functions (a) $e^{(\nu)}_{11,13}(k)$ for $\nu = 0.5$ and (b) $\nu = -0.5$.

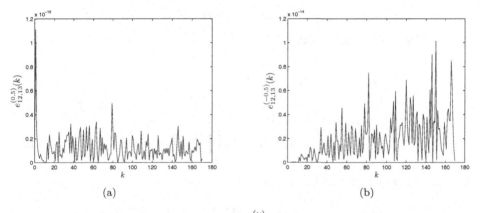

(a)   (b)

Fig. 1.24   Plots of the error functions (a) $e^{(\nu)}_{12,13}(k)$ for $\nu = 0.5$ and (b) $\nu = -0.5$.

Now one assumes that $a^{(\nu)}(k-1) < 0$ and $a^{(-\nu)}(k-1) > 0$ for $k-1 \geqslant 0$. One may then verify that

$$0 < \left(1 - \frac{\nu+1}{k}\right) = \frac{a^{(\nu)}(k)}{a^{(\nu)}(k-1)} \quad \text{for } k = 2, 3, \ldots \tag{1.23}$$

$$0 > \left(1 - \frac{-\nu+1}{k}\right) = \frac{a^{(-\nu)}(k)}{a^{(-\nu)}(k-1)} \quad \text{for } k = 2, 3, \ldots \tag{1.24}$$

because

$$\left(1 - \frac{-\nu+1}{k}\right) > \left(1 - \frac{\nu+1}{k}\right) > 0 \quad \text{for } k = 2, 3, \ldots. \tag{1.25}$$

From (1.23) and (1.24) one deduces (1.20). $\qquad\qquad\qquad\qquad\qquad\qquad\qquad\qquad\square$

A slightly different proof may be found in [Kaczorek (2011)].

Theorem 1.3. *For* $0 < \nu < 2$

$$\lim_{k \to +\infty} a^{(\nu)}(k) = a^{(\nu)}(+\infty) = 0. \tag{1.26}$$

Proof. For the discrete-variable function $a^{(\nu)}(k)$, consecutive values may be treated as a sequence of real numbers whose limit can be evaluated by the Cauchy formula theorem. A finite limit exists if and only if for every $\epsilon > 0$ there is an L such that $|a^{(\nu)}(k_1) - a^{(\nu)}(k_2)| < \epsilon$ for every $k_1, k_2 > L$. In fact for any $\epsilon > 0$ one can find L such that

$$|a^{(\nu)}(L)| < \epsilon. \tag{1.27}$$

Subsequently, any $k_1, k_2 > L$ can be expressed as $k_1 = L + L_1, k_2 = L + L_1 + L_2$ where $L, L_1, L - 2 \in \mathbb{Z}$ and

$$|a^{(\nu)}(k_1) - a^{(\nu)}(k_2)| = |a^{(\nu)}(L + L_1) - a^{(\nu)}(L + L_1 + L_2)|. \tag{1.28}$$

By definition from formula 1.12

$$|a^{(\nu)}(L + L_1 + L_2)| = \left| a^{(\nu)}(L + L_1) \prod_{i=1}^{L_2} \frac{\nu - L - L_1 - i + 1}{L + L_1 + i} \right| \tag{1.29}$$

$$|a^{(\nu)}(L + L_1)| = \left| a^{(\nu)}(L) \prod_{i=1}^{L_1} \frac{\nu - L - i + 1}{L + i} \right|. \tag{1.30}$$

A substitution of (1.29) and (1.30) into (1.28) yields

$$|a^{(\nu)}(L + L_1) - a^{(\nu)}(L + L_1 + L_2)|$$

$$= \left| a^{(\nu)}(L + L_1) - a^{(\nu)}(L + L_1) \prod_{i=1}^{L_2} \frac{\nu - L - L_1 - i + 1}{L + L_1 + i} \right|$$

$$= \left| a^{(\nu)}(L) \left(1 - \prod_{i=1}^{L_1} \frac{\nu - L - i + 1}{L + i} \prod_{i=1}^{L_2} \frac{\nu - L - L_1 - i + 1}{L + L_1 + i} \right) \right|$$

$$= |a^{(\nu)}(L)| \left| 1 - \prod_{i=1}^{L_1} \frac{\nu - L - i + 1}{L + i} \prod_{i=1}^{L_2} \frac{\nu - L - L_1 - i + 1}{L + L_1 + i} \right|. \tag{1.31}$$

Now one examines a second factor in the above expression. One can see that for $k = 2, 3, \ldots$

$$\left| \frac{\nu - k + 1}{k} \right| = 1 - \frac{\nu + 1}{k} < 1. \tag{1.32}$$

Hence

$$\left| a^{(\nu)}(L) \right| \left| 1 - \prod_{i=1}^{L_1} \frac{\nu - L - i + 1}{L + i} \prod_{i=1}^{L_2} \frac{\nu - L - L_1 - i + 1}{L + L_1 + i} \right| < |a^{(\nu)}(L)| < \epsilon \quad (1.33)$$

and

$$|a^{(\nu)}(k_1) - a^{(\nu)}(k_2)| < \epsilon. \quad (1.34)$$

The last inequality is also valid for $a^{(\nu)}(k_2) = a^{(\nu)}(+\infty) = 0$. This ends the proof. □

The next function $a^{(\nu)}(k)$ property is stated by the following theorem.

Theorem 1.4. *For the coefficient ν satisfying (1.19) a following equality holds*

$$\sum_{i=0}^{+\infty} a^{(\nu)}(i) = 0. \quad (1.35)$$

Proof. To prove equality (1.35) one refers to equality (1.17). For $x = 1$, one immediately obtains the result

$$(1 - x)^\nu \big|_{x=1} = (1 - 1)^\nu = 0 = \sum_{i=0}^{+\infty} a^\nu(i) 1^i = \sum_{i=0}^{+\infty} a^\nu(i). \quad (1.36)$$

□

As a consequence of Theorem 1.4 and the function $a^{(\nu)}(k)$ definition formula (1.11) one has

$$1 = -\sum_{i=1}^{+\infty} a^{(\nu)}(i). \quad (1.37)$$

Moreover, for (1.19) by (1.12) and (1.22)

$$\frac{a^{(\nu)}(i)}{a^{(\nu)}(i - 1)} = 1 - \frac{\nu + 1}{i} > 0 \quad (1.38)$$

which means that for $k = 2, 3, \ldots$

$$a^{(\nu)}(i - 1) < a^{(\nu)}(i) < 0. \quad (1.39)$$

Hence, (1.37) can be rewritten to the form

$$1 = \sum_{i=1}^{+\infty} |a^{(\nu)}(i)|. \quad (1.40)$$

Inspection of formula (1.38) reveals that $a^{(\nu)}(k) \to 0$ for $k \to +\infty$. This result is of practical use in real-time microprocessor calculation of the function $a^{(\nu)}(k)$ values. For any $L > 1$ formula (1.40) can be rewritten in the form

$$1 = \sum_{i=1}^{L} |a^{(\nu)}(i)| + \sum_{i=L+1}^{+\infty} |a^{(\nu)}(i)| \tag{1.41}$$

with

$$\lim_{L \to +\infty} \sum_{i=L+1}^{+\infty} |a^{(\nu)}(i)| \to 0. \tag{1.42}$$

Hence in practical applications one may define two parameter function $a_L^{(\nu)}(k)$.

Definition 1.7. *The discrete function $a_L^{(\nu)}(k)$ is defined as*

$$a_L^{(\nu)}(k) = \begin{cases} 0 & \text{for } k < 0 \\ 1 & \text{for } k = 0 \\ (-1)^k \frac{\nu(\nu-1)\cdots(\nu-k+1)}{k!} & \text{for } k = 1, 2, \ldots, L \\ 0 & \text{for } k = L+1, L+2, \ldots \end{cases} \tag{1.43}$$

Evidently for this simplified form of the function $a_L^{(\nu)}(k)$

$$\sum_{i=1}^{+\infty} |a_L^{(\nu)}(i)| = \sum_{i=1}^{L} |a_L^{(\nu)}(i)| + \sum_{i=L+1}^{+\infty} |a_L^{(\nu)}(i)| = \sum_{i=1}^{L} |a_L^{(\nu)}(i)| \approx 1. \tag{1.44}$$

To evaluate the impact of the neglected term one may use a two variables error function which may help in an approximation of the function $a^{(\nu)}(k)$ by function (1.43)

$$e(\nu, L) = \sum_{i=L+1}^{+\infty} |a_L^{(\nu)}(i)| = 1 - \sum_{i=1}^{L} |a_L^{(\nu)}(i)|. \tag{1.45}$$

Plot of the error function $e(\nu, L)$ for $\nu \in (01)$ and $L \in (0, 200)$ is given in Fig. 1.25(a).

The error function may be treated as a useful tool in establishing a relation between the coefficient ν and function (1.43) length L. Assuming that function (1.43) takes a constant value $e_{\nu, L}$ from a range $(0, 1]$ one may derive a relation between ν and L

$$\sum_{i=1}^{L} |a_L^{(\nu)}(i)| = 1 - e_{\nu, L} < 1. \tag{1.46}$$

Curves of a constant error function on a $L\nu$ plain are presented in Fig. 1.26. Establishing a horizontal line (dashed line in the above figure) one may find a relation between a coefficient ν and a length L for different constants.

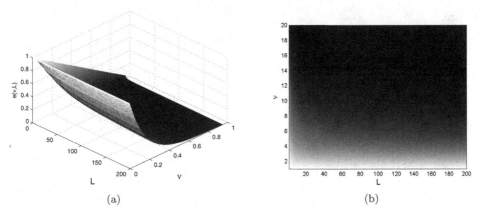

Fig. 1.25   Plot and 2D image of the error function $e(\nu, L)$.

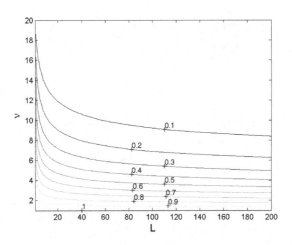

Fig. 1.26   Curves of constant values $e_{\nu, L}$ (1.45).

Theorem 1.5. *For $k = 1, 2, 3, \ldots$ and a coefficient $\nu \in (0, 1]$ the following inequality is valid*

$$|a^{(\nu)}(k)| \leqslant |a^{(-\nu)}(k)|. \tag{1.47}$$

Proof. For $k = 1$ evidently $|a^{(\nu)}(1)| = |-\nu| = \nu = |a^{(-\nu)(1)}| = \nu$. By induction, if (1.47) holds for some $k - 1 > 1$ then in view of the inequalities (1.20)

$$|a^{(-\nu)}(k-1)| - |a^{(\nu)}(k-1)| = a^{(-\nu)}(k-1) + a^{(\nu)}(k-1) \geqslant 0 \tag{1.48}$$

and evidently

$$a^{(-\nu)}(k-1) - a^{(\nu)}(k-1) > 0. \tag{1.49}$$

Hence,

$$\left|a^{(-\nu)}(k)\right| - \left|a^{(\nu)}(k)\right| = a^{(-\nu)}(k) + a^{(\nu)}(k)$$

$$= a^{(-\nu)}(k-1)\left(1 - \frac{-\nu+1}{k}\right) + a^{(\nu)}(k-1)\left(1 - \frac{\nu+1}{k}\right)$$

$$= \left(a^{(-\nu)}(k-1) + a^{(\nu)}(k-1)\right)\left(1 - \frac{1}{k}\right)$$

$$+ \left(a^{(-\nu)}(k-1) - a^{(\nu)}(k-1)\right)\frac{\nu}{k} \geqslant 0. \tag{1.50}$$

$$\square$$

For the sake of comparison, in Fig. 1.27 plots of functions $\left|a^{(\nu)}(k)\right|$ and $\left|a^{(-\nu)}(k)\right|$ for $\nu = 0.5$ are presented.

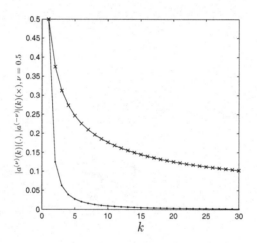

Fig. 1.27   Plots of functions $\left|a^{(\nu)}(k)\right|$ and $\left|a^{(-\nu)}(k)\right|$ for $\nu = 0.5$.

1.3.3.2   *Function $a^{(n)}(k)$ properties*

Now the function $a^{(\nu)}(k)$ for parameter $\nu \in \mathbb{Z}_0$ will be examined. To distinguish this case from the one considered before one introduces a notation $\nu = n \in \mathbb{Z}_0$. The function is defined as follows

Definition 1.8. *The discrete $a^{(n)}(k)$ is defined as*

$$a^{(n)}(k) = \begin{cases} 0 & \text{for } k < 0 \\ 1 & \text{for } k = 0 \\ (-1)^k \frac{n(n-1)\cdots(n-k+1)}{k!} & \text{for } k = 1, 2, \ldots, n-1, n \\ 0 & \text{for } k = n+1, n+2, \ldots \end{cases} \tag{1.51}$$

Table 1.2   Function $a^{(n)}(k)$ values.

n	$a^{(n)(k)}, k = 0, \ldots, n$
0	1
1	1 1
2	1 2 1
3	1 3 3 1
4	1 4 6 4 1
5	1 5 10 10 5 1
6	1 6 15 20 15 6 1
7	1 7 21 35 35 21 7 1
8	1 8 28 56 70 56 28 8 1
9	1 9 36 84 126 126 84 36 9 1

Table 1.3   Function $a^{(-n)}(k)$ values.

n	$a^{(-n)(k)}, k = 0, \ldots, n$
0	1 0 0 0 0 0 0 0 0 0
1	1 1 1 1 1 1 1 1 1 1
2	1 2 3 4 5 6 7 8 9 10
3	1 3 6 10 15 21 28 36 45 55
4	1 4 10 20 35 56 84 120 165 220
5	1 5 15 30 70 126 210 330 495 715
6	1 6 21 56 126 252 462 792 1287 2002
7	1 7 28 84 210 462 924 1716 3003 5005
8	1 8 36 120 330 792 1716 3432 6435 11440
9	1 9 45 165 495 1287 3003 6435 12870 24310

The function has only $n + 1$ non-zero values. Its absolute values for $n = 0, 1, \ldots, 9$ are given below. The integers create a so-called Pascal triangle.

From Table 1.1, one can see that the maximal value is reached for $\lfloor \frac{n}{2} \rfloor$, i.e.

$$\max_{k \in \mathbb{Z}_0} |a^{(n)}(k)| = \left| a^{(n)} \left(\left\lfloor \frac{n}{2} \right\rfloor \right) \right|. \tag{1.52}$$

For negative integers n the function $a^{(n)}(k)$ always takes non-zero values. To emphasize this case the condition $n > 0$ will be preserved but the function will be denoted as $a^{(-n)}(k)$. Values of $a^{(-n)}(k)$ for $n = 0, 1, \ldots, 9$ are collected in Table 1.3

1.3.4   *Function $c^{(\nu)}(k)$*

Another discrete-variable function of great importance is the function $c^{(\nu)}(k)$. It is defined as

$$c^{(\nu)}(k) = \begin{cases} 0 & \text{for } k < 0 \\ 1 & \text{for } k = 0 \\ 1 - \frac{\nu+1}{k} & \text{for } k = 1, 2, \ldots \end{cases} \tag{1.53}$$

Simple transformations show that

$$c^{(\nu)}(k) = \frac{a^{(\nu)}(k)}{a^{(\nu)}(k-1)} \qquad\qquad (1.54)$$

and

$$c^{(\nu)}(k) = c^{(\nu)}(k-1) + \frac{\nu+1}{k(k-1)}. \qquad\qquad (1.55)$$

Similarly to the analysis of function $a^{(\nu)}(k)$ plots of function $c^{(\nu)}(k)$ due to different parameter positive values from four sets are shown in Figs. 1.28(a), 1.28(b), 1.29(a) and 1.29(b).

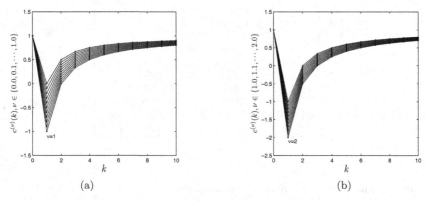

(a)   (b)

Fig. 1.28   Plots of function (a) $c^{(\nu)}(k)$ for $\nu \in \{0, 0.1, \dots, 1.0\}$ and (b) $\nu \in \{1.0, 1.1, \dots, 2.0\}$.

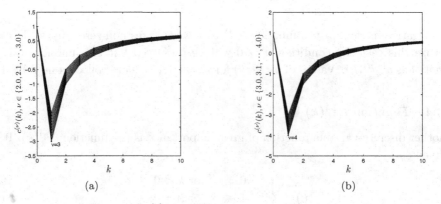

(a)   (b)

Fig. 1.29   Plots of function (a) $c^{(\nu)}(k)$ for $\nu \in \{2.0, 2.1, \dots, 3.0\}$ and (b) $\nu \in \{3.0, 3.1, \dots, 4.0\}$.

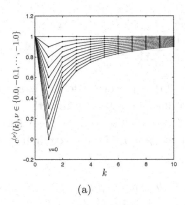 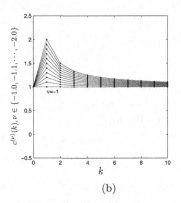

(a) (b)

Fig. 1.30 Plots of function (a) $c^{(\nu)}(k)$ for $\nu \in \{0, -0.1, \ldots, -1.0\}$ and (b) $\nu \in \{-1.0, -1.1, \ldots, -2.0\}$.

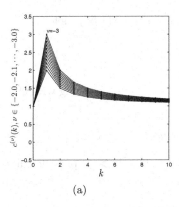 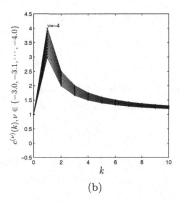

(a) (b)

Fig. 1.31 Plots of function (a) $c^{(\nu)}(k)$ for $\nu \in \{-2.0, -2.1, \ldots, -3.0\}$ and (b) $\nu \in \{-3.0, -3.1, \ldots, -4.0\}$.

1.3.4.1 *Function $c^{(\nu)}(k)$ properties*

In this section several properties of the function $c^{(\nu)}(k)$ will be proved. Immediate calculation reveals that

$$\lim_{k \to \infty} c^{(\nu)}(k) = \lim_{k \to \infty} \left(1 - \frac{\nu + 1}{k}\right) = 1. \tag{1.56}$$

By (1.54) from (1.56) one states that

$$\lim_{k \to \infty} \frac{a^{(\nu)}(k)}{a^{(\nu)}(k - 1)} = 1 \tag{1.57}$$

and

$$\lim_{k \to \infty} \frac{c^{(\nu)}(k)}{c^{(\nu)}(k-1)} = \lim_{k \to \infty} \frac{1 - \frac{\nu+1}{k}}{1 - \frac{\nu+1}{k-1}} = 1. \tag{1.58}$$

The function $c^{(\nu)}(k)$ is a spline function of degree 3 (i.e. piecewise defined over three intervals). The first interval is degenerated to one point $k = 0$, and the second for $k > 0$. By formulas (1.53) and (1.56), $c^{(\nu)}(0) = c^{(\nu)}(+\infty) = 1$. Assuming that $c^{(\nu)}(k)$ is non-constant one concludes that there must be an extremum. The following Theorems determine this property.

Theorem 1.6. *Function $c^{(\nu)}(k)$ is monotonically decreasing for $\nu \in (-\infty, -1)$ and $k = 1, 2, \ldots$ and monotonically increasing for $\nu \in (-1, +\infty)$ for $k = 1, 2, \ldots$.*

Proof. To prove this theorem one should check inequality for $k = 2, 3, \ldots$

$$c^{(\nu)}(k) - c^{(\nu)}(k-1) < 0 \quad \text{for } \nu \in (-\infty, -1). \tag{1.59}$$

A substitution of the second non-zero spline in (1.53) into (1.59) yields

$$1 - \frac{\nu+1}{k} - 1 + \frac{\nu+1}{k-1} = \frac{(\nu+1)}{k(k-1)} < 0 \quad \text{for } \nu \in (-\infty, -1). \tag{1.60}$$

Hence, for $\nu + 1 < 0$ the considered spline is monotonically decreasing with $c^{(\nu)}(1) = -\nu > 0$ as its maximal element. Next,

$$c^{(\nu)}(k) - c^{(\nu)}(k-1) > 0 \quad \text{for } \nu \in (-1, +\infty). \tag{1.61}$$

Similar substitution gives the opposite result for $\nu + 1 > 0$ proving that this time a spline is monotonically increasing with $c^{(\nu)}(1) = -\nu < 0$ as its minimal element. $\qquad\square$

Theorem 1.7. *Extreme values of the function $c^{(\nu)}(k)$ equal to $-\nu$*

$$\max_{k \in \mathbb{Z}_0} c^{(\nu)}(k) = c^{(\nu)}(1) = -\nu \quad \text{for } \nu \in (-\infty, -1) \tag{1.62}$$

$$c^{(\nu)}(k) = \text{const} = c^{(\nu)}(1) = 1, k \in \mathbb{Z}_0 \quad \text{for } \nu = -1 \tag{1.63}$$

$$\min_{k \in \mathbb{Z}_0} c^{(\nu)}(k) = c^{(\nu)}(1) = -\nu \quad \text{for } \nu \in (-1, +\infty). \tag{1.64}$$

Proof. For $\nu \in (-\infty, -1)$ the inequality $1 = c^{(\nu)}(0) < c^{(\nu)}(1) = -\nu$ holds. In view of the result stated by Theorem 1.6 $c^{(\nu)}(1) > c^{(\nu)}(k)$ for $k = 2, 3, \ldots$ and formula (1.62) is proved. By analogy, for $\nu \in (-1, -\infty)$ there is $1 = c^{(\nu)}(0) > c^{(\nu)}(1) = -\nu$ and the function $c^{(\nu)}(k)$ increases for $k = 2, 3, \ldots$. This shows the correctness of (1.64). Finally, for $\nu = -1$ the function $c^{(-1)}(k) = 1$ (is constant). This ends the proof. $\qquad\square$

The next theorem is of great practical importance especially in a real-time micro-processor functions $a^{(\nu)}(k)$ and $c^{(\nu)}(k)$ evaluations. Information about functions convergence provides, inter alia, differences and ration between its consecutive values.

Theorem 1.8. *For $0 < \nu < 1$ and $k = 1, 2, \ldots$ the following inequality holds*

$$a^{(\nu)}(k) - a^{(\nu)}(k - 1) \geqslant c^{(\nu)}(k) - c^{(\nu)}(k - 1). \qquad (1.65)$$

Proof. First inequality (1.65) will be proved for $k = 3, 4, \ldots$. By definition formula (1.11)

$$L = a^{(\nu)}(k) - a^{(\nu)}(k - 1) = a^{(\nu)}(k - 1)\left(1 - \frac{\nu + 1}{k}\right) - a^{(\nu)}(k - 1)$$

$$= -\frac{\nu + 1}{k} a^{(\nu)}(k - 1) > 0. \qquad (1.66)$$

Simple calculations reveal that

$$L = -\frac{\nu + 1}{k} a^{(\nu)}(k - 1) = -a^{(\nu+1)}(k) > 0. \qquad (1.67)$$

Inequality (1.66) results from the first formula (1.20) and is confirmed by inequality (1.67) which is valid on grounds of the second formula (1.20). Similarly, by (1.53)

$$R = c^{(\nu)}(k) - c^{(\nu)}(k - 1) = \left(1 - \frac{\nu + 1}{k}\right) - \left(1 - \frac{\nu + 1}{k - 1}\right)$$

$$= \frac{\nu + 1}{k(k - 1)} > 0. \qquad (1.68)$$

Up to now it was proved that $L, R > 0$. What remains to be determined is the relationship between L and R. The assumption that $L < R$ yields

$$(-1)^{k+1} \frac{(\nu + 1)\nu(\nu - 1) \cdots (\nu - k + 2)}{k!} < \frac{\nu + 1}{k(k - 1)}. \qquad (1.69)$$

Dividing both sides by positive factor $\frac{\nu+1}{k(k-1)}$ one obtains

$$-(-1)^{k-2} \frac{\nu(\nu - 1) \cdots (\nu - k + 2)}{(k - 2)!}$$

$$= -\nu\left(1 - \frac{\nu}{1}\right)\left(1 - \frac{\nu}{2}\right) \cdots \left(1 - \frac{\nu}{k - 2}\right) < 1. \qquad (1.70)$$

This is because by assumption, $\nu > 0$ and all terms in brackets are positive. For $k = 2$ there is

$$a^{(\nu)}(2) - a^{(\nu)}(1) = \frac{\nu(\nu + 1)}{2} < c^{(\nu)}(2) - c^{(\nu)}(1) = \frac{\nu + 1}{2} \qquad (1.71)$$

bearing in mind that $0 < \nu < 1$. Finally for $k = 1$

$$a^{(\nu)}(1) - a^{(\nu)}(0) = -1 + \nu = c^{(\nu)}(1) - c^{(\nu)}(0) = 1 - \nu \qquad (1.72)$$

which ends the proof. $\qquad \square$

The next Theorem estimates the functions $a^{(\nu)}(k)$ and $c^{(\nu)}(k)$ convergence.

Theorem 1.9. *For $0 < \nu < 1$ and $k = 3, 4, \ldots$ the following inequality holds*

$$\left| \frac{a^{(\nu)}(k)}{a^{(\nu)}(k-1)} \right| < \left| \frac{c^{(\nu)}(k)}{c^{(\nu)}(k-1)} \right|. \tag{1.73}$$

Proof. By definition formula (1.53) one has

$$\left| \frac{c^{(\nu)}(k)}{c^{(\nu)}(k-1)} \right| = \left| \frac{1 - \frac{\nu+1}{k}}{1 - \frac{\nu+1}{k-1}} \right| = \left| 1 - \frac{\nu+1}{k} \right| \left| \frac{1}{1 - \frac{\nu+1}{k-1}} \right|. \tag{1.74}$$

From (1.12) one obtains

$$\left| \frac{a^{(\nu)}(k)}{a^{(\nu)}(k-1)} \right| = \left| 1 - \frac{\nu+1}{k} \right|. \tag{1.75}$$

Hence

$$\left| \frac{c^{(\nu)}(k)}{c^{(\nu)}(k-1)} \right| = \left| \frac{a^{(\nu)}(k)}{a^{(\nu)}(k-1)} \right| \left| \frac{1}{1 - \frac{\nu+1}{k-1}} \right|. \tag{1.76}$$

Noting that

$$\left| \frac{1}{1 - \frac{\nu+1}{k-1}} \right| > 1 \text{ for } k = 3, 4, \ldots \tag{1.77}$$

one obtains the result (1.73).

$\square$

1.4 Fundamental operations on the discrete-variable functions

In this section fundamental operations that can be performed on discrete-variable functions will be introduced. On the basis of the numerous references one can state that the operations follow the well-known ones applied to the continuous-variable functions case. In both cases the fundamental assumption is the same function domain. Hence one assumes that these functions are defined over the same domain $\mathbb{T}_{k_0,k,h}$.

1.4.1 *Discrete-variable functions addition and multiplication*

Assuming that $f, g : \mathbb{T}_{k_0,k,h} \to \mathbb{R}$ one defines two operations: function addition and multiplication executed in a classical way and denoted by $f(k) + g(k)$ and $f(k) \cdot g(k)$, respectively.

1.4.2 *Discrete-variable function forward and backward shift*

In the discrete-variable function analysis it is very useful to define a function shift understood as a backward (left) and forward (right) shift performed along the horizontal axis. Using this operation one can easily define new non-typical discrete-variable functions. First one defines so-called shifted forward Dirac pulse. It is defined on the basis of function (1.6).

Definition 1.9. *The shifted forward discrete Dirac pulse is defined as*

$$\delta(k - k_0) = \begin{cases} 1 & \text{for } k = k_0 \\ 0 & \text{for } k \neq k_0 \end{cases}. \tag{1.78}$$

The plot and image of the shifted forward Dirac pulse are presented in Figs. 1.15(a) and 1.15(b), respectively.

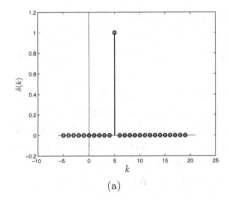
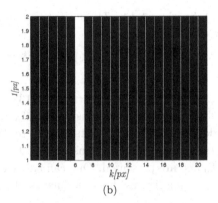

(a)   (b)

Fig. 1.32   Plot (a) and 1D image (b) of the shifted forward Dirac pulse for $k_0 = 5$.

Similarly, one defines the shifted forward unit step function. Its definition is given below.

Definition 1.10. *The shifted forward discrete unit step function defines the following formula*

$$\mathbf{1}(k - k_0) = \begin{cases} 0 & \text{for } k < k_0 \\ 1 & \text{for } k \geqslant k_0 \end{cases}. \tag{1.79}$$

The plot and image of the shifted forward discrete unit step function are presented in Figs. 1.33(a) and 1.33(b), respectively. Here it is worth pointing that the discrete unit step function can be equivalently defined by a sum of shifted forward discrete Dirac pulses.

$$\mathbf{1}(k - k_0) = \sum_{i=k_0}^{k} \delta(k - i). \tag{1.80}$$

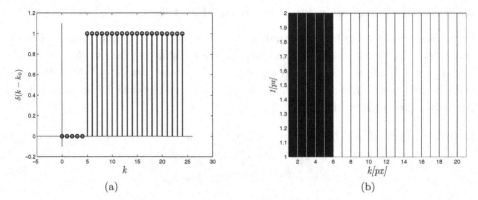

Fig. 1.33   Plot (a) and 1D image (b) of the shifted forward unit step function for $k_0 = 5$.

Definition 1.11. *The shifted forward general exponential function*

$$f(k) = a^{b(k-k_0)}\mathbf{1}(k - k_0) \text{ where } a, b, \in \mathbb{R}. \tag{1.81}$$

One should note that formula (1.13) is defined, in fact, by a product of two equally shifted forward functions: $a^{b(k-k_0)}$ and $\mathbf{1}(k - k_0)$. This is not obligatory. One can define two different shifts leading to different function shapes. In Figs. 1.16(a) and 1.16(b) plots of functions $e^{b(k-k_0)}\mathbf{1}(k - k_0)$ and $e^{bk}\mathbf{1}(k - k_0)$ with $b = -0.05$ and $k_0 = 5$ are presented. One can easily observe the difference between the plots. The following figures show the images of the considered functions

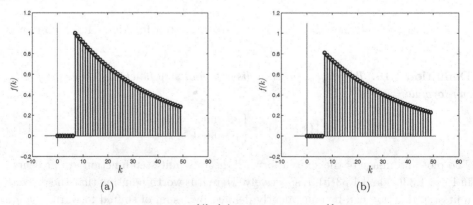

Fig. 1.34   Plots of functions (a) $e^{b(k-k_0)}\mathbf{1}(k - k_0)$ and (b) $e^{bk}\mathbf{1}(k - k_0)$ for $k_0 = 7$.

As was mentioned earlier the shift operations may be very useful in description of several discrete-variable signals. Next some examples are presented.

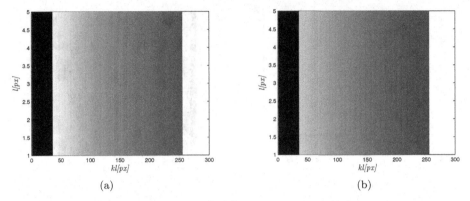

Fig. 1.35   Images of functions (a) $e^{b(k-k_0)}\mathbf{1}(k-k_0)$ and (b) $e^{bk}\mathbf{1}(k-k_0)$ for $k_0 = 7$.

Example 1.2. A function defined as

$$f(k) = \begin{cases} 0 & \text{for } k < 0 \\ 1 & \text{for } 0 \leqslant k \leqslant 5 \\ 0 & \text{for } k > 5 \end{cases} \qquad (1.82)$$

can be equivalently described as

$$f(k) = \mathbf{1}(k) - \mathbf{1}(k - 6). \qquad (1.83)$$

Example 1.3. Consider a function defined by this formula

$$f(k) = \begin{cases} 0 & \text{for } k < 0 \\ k & \text{for } 0 \leqslant k \leqslant 5 \\ 5 & \text{for } 6 \leqslant k \leqslant 10 \\ -k + 15 & \text{for } 11 \leqslant k \leqslant 14 \\ 0 & \text{for } k \geqslant 15 \end{cases} . \qquad (1.84)$$

It may be also described as a sum of shifted functions

$$f(k) = k\mathbf{1}(k) - (k-5)\mathbf{1}(k-5) - (k-10)\mathbf{1}(k-10) + (k-15)\mathbf{1}(k-15). \qquad (1.85)$$

A plot and a related image are given in Figs. 1.36(a) and 1.36(b).

1.4.3   *Convolution of two discrete-variable functions*

The most important operation one can perform on two discrete-variable functions is called a real convolution. It plays an important role in transient analysis of linear time-invariant dynamical systems.

Definition 1.12. For two scalar functions $f_1(k)$, $f_2(k)$ of a discrete-variable k satisfying the condition $f_1(k)$, $f_2(k) = 0$ for $k < 0$ one defines the third one $f(k)$ in the

Discrete Fractional Calculus

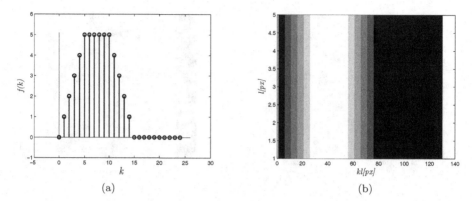

(a)   (b)

Fig. 1.36   Plot and an image of function (1.85).

following way

$$f(k) = \sum_{i=0}^{k} f_1(i) f_2(k-i). \tag{1.86}$$

The asterisk($*$) denotes the discrete convolution operation. Hence,

$$f(k) = f_1(k) * f_2(k). \tag{1.87}$$

The real discrete convolution satisfies three fundamental laws:

(I) Commutative law

$$f_1(k) * f_2(k) = f_2(k) * f_1(k). \tag{1.88}$$

(II) Distributive law

$$f_1(k) * [f_2(k) \pm f_3(k)] = f_1(k) * f_2(k) \pm *f_1(k) * f_3(k). \tag{1.89}$$

(III) Associative law

$$f_1(k) * [f_2(k) * f_3(k)] = [f_1(k) * f_2(k)] * f_3(k). \tag{1.90}$$

The proofs of the laws mentioned above can be found in numerous positions. Formula (1.87) can be also expressed in a vector-matrix product form. Using vectors
and $(k+1) \times (k+1)$ anti-diagonal matrix $\mathbf{N}_k$ defined in (A.19)

$$\mathbf{f}_1(k) = \begin{bmatrix} f_1(0) \\ f_1(1) \\ \vdots \\ f_1(k) \end{bmatrix}, \quad \mathbf{f}_2(k) = \begin{bmatrix} f_2(0) \\ f_2(1) \\ \vdots \\ f_2(k) \end{bmatrix} \tag{1.91}$$

the discrete convolution is expressed as a product

$$f(k) = f_1(k) * f_2(k) = [\mathbf{f_1}(k)]^{\mathrm{T}} \mathbf{N}_k \mathbf{f_2}(k). \tag{1.92}$$

The convolution sum is valid for all non-negative values of k. In evaluating all such sums for $k, k-1, \ldots, 1, 0$ one obtains a set of equalities. Collecting all of them in one matrix-vector formula yields:

$$
\mathbf{f}(k) = \begin{bmatrix} f(k) \\ f(k-1) \\ \vdots \\ f(1) \\ f(0) \end{bmatrix} = \begin{bmatrix} f_1(k) * f_2(k) \\ f_1(k-1) * f_2(k-1) \\ \vdots \\ f_1(1) * f_2(1) \\ f_1(0) * f_2(0) \end{bmatrix}
$$

$$
= \begin{bmatrix} f_1(0) & f_1(1) & \cdots & f_1(k-1) & f_1(k) \\ 0 & f_1(0) & \cdots & f_1(k-2) & f_1(k-1) \\ \vdots & \vdots & & \vdots & \vdots \\ 0 & 0 & \cdots & f_1(0)) & f_1(2) \\ 0 & 0 & \cdots & 0 & f_1(0) \end{bmatrix} \mathbf{N}_k \mathbf{f_2}(k)
$$

$$
= \begin{bmatrix} f_1(0) & f_1(1) & \cdots & f_1(k-1) & f_1(k) \\ 0 & f_1(0) & \cdots & f_1(k-2) & f_1(k-1) \\ \vdots & \vdots & & \vdots & \vdots \\ 0 & 0 & \cdots & f_1(0)) & f_1(2) \\ 0 & 0 & \cdots & 0 & f_1(0) \end{bmatrix} \begin{bmatrix} f_2(k) \\ f_2(k-1) \\ \vdots \\ f_2(1) \\ f_2(0) \end{bmatrix}. \tag{1.93}
$$

Formula (1.93) is very convenient in computer calculations giving the discrete convolution values ordered in a vector form. The numerical examples presented above explain the calculation procedure.

Example 1.4. Consider two simple functions

$$f_1(k) = \mathbf{1}(k) - \mathbf{1}(k-3) \tag{1.94}$$

$$f_2(k) = \mathbf{1}(k-2) - \mathbf{1}(k-5). \tag{1.95}$$

Plots of functions $f_1(k)$ and $f_2(k)$ are presented in Figs. 1.20(a) and Fig. 1.20(b), respectively. One should evaluate the discrete convolution of these functions due to formula (1.93).

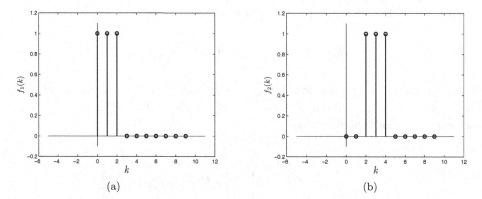

Fig. 1.37   Plot of functions (1.94) and (1.95).

$$\begin{bmatrix} \vdots & \vdots & \vdots & \vdots & \vdots & \vdots & \vdots & \vdots \\ \dots 1 & 1 & 1 & 0 & 0 & 0 & 0 & 0 \\ \dots 0 & 1 & 1 & 1 & 0 & 0 & 0 & 0 \\ \dots 0 & 0 & 1 & 1 & 1 & 0 & 0 & 0 \\ \dots 0 & 0 & 0 & 1 & 1 & 1 & 0 & 0 \\ \dots 0 & 0 & 0 & 0 & 1 & 1 & 1 & 0 \\ \dots 0 & 0 & 0 & 0 & 0 & 1 & 1 & 1 \\ \dots 0 & 0 & 0 & 0 & 0 & 0 & 1 & 1 \\ \dots 0 & 0 & 0 & 0 & 0 & 0 & 0 & 1 \end{bmatrix} \begin{bmatrix} \vdots \\ 0 \\ 0 \\ 0 \\ 1 \\ 1 \\ 1 \\ 0 \\ 0 \end{bmatrix} = \begin{bmatrix} \vdots \\ 0 \\ 1 \\ 2 \\ 3 \\ 2 \\ 1 \\ 0 \\ 0 \end{bmatrix}. \tag{1.96}$$

From (1.96) follows that

$$\begin{aligned} f(k) &= f_1(k) * f_2(k) \\ &= \mathbf{1}(k-2) + \mathbf{1}(k-3) + \mathbf{1}(k-4) - \mathbf{1}(k-5) - \mathbf{1}(k-6) - \mathbf{1}(k-7) \\ &= (k-1)\mathbf{1}(k-1) - 2(k-4)\mathbf{1}(k-4) + (k-7)\mathbf{1}(k-7). \end{aligned} \tag{1.97}$$

Plot of the convolution function $f(k)$ and its image are presented in Figs. 1.38(a) and Fig. 1.38(b), respectively.

In the next example one calculates a convolution sum of two trigonometric functions.

Example 1.5. For the two trigonomeric functions below, evaluate its convolution sum.

$$f_1(k) = \sin(ak)\mathbf{1}(k), \, f_2(k) = \cos(ak)\mathbf{1}(k). \tag{1.98}$$

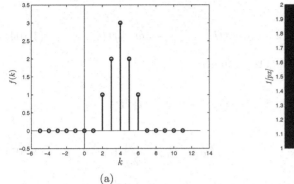 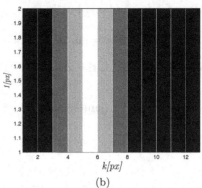

(a) (b)

Fig. 1.38 Plot (a) and an image (b) of function (1.97).

By definition (1.12)

$$f(k) = f_1(k) * f_2(k) = \sin(ak)\mathbf{1}(k) * \cos(ak)\mathbf{1}(k)$$
$$= \sum_{i=0}^{k} \sin(ai)\mathbf{1}(i)\cos[a(k-i)]\mathbf{1}(k-i). \tag{1.99}$$

Now one considers two cases where: $k -$ odd and $k -$ even. In the first case ($k -$ $odd, k > 0$), the right-hand terms in expression (1.99) can be re-grouped in the following way:

$$f(k) = \sum_{i=0}^{\frac{k-1}{2}} \{\sin(ai)\cos[a(k-i)] + \sin[a(k-i)]\cos(ai)\}\mathbf{1}(k-i). \tag{1.100}$$

Note that in the considered case $\frac{k-1}{2} -$ even and $\mathbf{1}(k)\mathbf{1}(k-i) = \mathbf{1}(k-i)$. From trigonometric identity for $i = 0, 1, \dots$ one obtains:

$$\sin(ai)\cos[a(k-i)] + \sin[a(k-i)]\cos(ai) = \sin(ak). \tag{1.101}$$

Substitution of (1.101) into (1.105) yields

$$f(k) = \sin(ak)\sum_{i=0}^{\frac{k-1}{2}} \mathbf{1}(k-i). \tag{1.102}$$

From Chapter 2, it will be clear that

$$\sum_{i=0}^{\frac{k-1}{2}} \mathbf{1}(k-i) = \frac{k+1}{2}\mathbf{1}(k). \tag{1.103}$$

Hence for $k - odd$ one finally gets

$$f(k) = \sin(ak)\mathbf{1}(k) * \cos(ak)\mathbf{1}(k) = \frac{k+1}{2}\sin(ak)\mathbf{1}(k). \qquad (1.104)$$

For $k - even, k > 0$ the formula slightly changes.

$$f(k) = \sum_{i=0}^{\frac{k}{2}-1}\{\sin(ai)\cos[a(k-i)] + \sin[a(k-i)]\cos(ai)\}\mathbf{1}(k-i)\}$$

$$+ \sin\left(a\frac{k}{2}\right)\cos\left(a\frac{k}{2}\right)\mathbf{1}\left(\frac{k}{2}\right). \qquad (1.105)$$

Analogous substitution of (1.101) into (1.105) leads to the final form:

$$f(k) = \sin(ak)\sum_{i=0}^{\frac{k}{2}-1}\mathbf{1}(k-i) + \sin\left(a\frac{k}{2}\right)\cos\left(a\frac{k}{2}\right)\mathbf{1}\left(\frac{k}{2}\right). \qquad (1.106)$$

By formula (1.103) evaluated over interval $[0, \frac{k}{2}]$ one finally obtains for $k - even$, $k > 0$:

$$f(k) = \sin(ak)\mathbf{1}(k) * \cos(ak)\mathbf{1}(k) = \frac{k}{2}\sin(ak)\mathbf{1}(k) + \frac{1}{2}\sin(ak)\mathbf{1}(k)$$

$$= \frac{k+1}{2}\sin(ak)\mathbf{1}(k). \qquad (1.107)$$

Finally, by direct calculation of the convolution sum for $k = 0$ one obtains:

$$f(k)|_{k=0} = [\sin(ak)\mathbf{1}(k) * \cos(ak)\mathbf{1}(k)]|_{k=0} = 0. \qquad (1.108)$$

Summing up the results shown above, the final form of evaluated convolution sum equals

$$f(k) = \frac{k+1}{2}\sin(ak)\mathbf{1}(k). \qquad (1.109)$$

The plot of the convolution function $f(k)$ and its 1D image are presented in Figs. 1.39(a) and 1.39(b), respectively.

Here it is worth commenting upon result (1.109). Though both functions $f_1(k)$ and $f_2(k)$ defined in (1.104) are bounded, i.e.: $-1 < \sin(ak), \cos(ak) < 1$ for all k its convolution sum values tend to $-\infty$ and $+\infty$. Such a transient behaviour can be explained on the basis of the discrete linear systems stability analysis.

1.5   Two discrete-variables functions

Now one considers two independent discrete-variable real function. The variables will be denoted as k_1 and k_2. Hence,

$$f : \mathbb{Z} \times \mathbb{Z} \to \mathbb{R} \qquad (1.110)$$

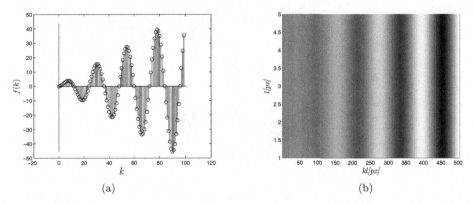

Fig. 1.39   Plot (a) and an image (b) of function (1.109).

In practice, considerations will be limited to

$$f : \mathbb{Z}_{N_{1,\min},N_{1,\max},h} \times \mathbb{Z}_{N_{2,\min},N_{2,\max},h} \to \mathbb{T}_{t_{\min},t_{\max}}. \tag{1.111}$$

Similarly to the one discrete-variable real function there are several representations of two discrete-variable functions. Here will be mentioned only two which are the most popular. The first one is a functional representation. The following numerical example gives such a representation.

Example 1.6. Consider a slightly modified "sombrero" function described by the formula

$$f(k_1, k_2) = \frac{\sin\left(\sqrt{0.5k_1^2 + k_2^2}\right)}{\sqrt{0.5k_1^2 + k_2^2}}. \tag{1.112}$$

Its 3D plot for $k_1 \in [-N_{1,\min}, N_{1,\max}] = [-10, 10]$ and $k_2 \in [-N_{2,\min}, N_{2,\max}] = [-10, 10]$ is given in Fig. 1.40.

Another very useful definition of a two discrete-variable function is similar to a tabularization of the one discrete-variable function. In the following example instead of the two column table one has an array. Such an array defining the function $f(k_1, k_2)$ is presented in Fig. 1.41.

A classical matrix with real elements is an example of such an array. In matrix algebra it is generally accepted that all matrix element indices are positive integers starting at the left upper corner. Appropriate shift and rotation of the coordinate axes enables a representation of any function $f(k_1, k_2)$ as a matrix $\mathbf{F}_{k_{1,\max}, k_{2,\max}}$ with $k_{1,\max} = N_{1,\min} + N_{1,\max}$, $k_{2,\max} = N_{2,\min} + N_{2,\max}$.

The referenced operations are depicted in in Figs. 1.42(a) and Fig. 1.42(b), respectively.

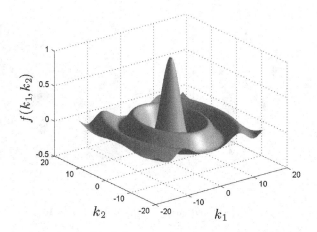

Fig. 1.40   3D plot of function (1.112).

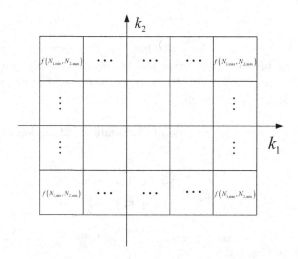

Fig. 1.41   Array containing data representing function $f(k_1, k_2)$ values.

1.5.1   *Fundamental two discrete-variable functions*

There is a two dimensional analogue of fundamental discrete-variable functions analyzed in Subsection 1.3.1 These are the two dimensional (2D) discrete Dirac pulse and 2D unit step function.

Definition 1.13 (2D discrete Dirac pulse). The 2D discrete Dirac pulse is defined as:

$$\delta(k_1, k_2) = \begin{cases} 1 & \text{for } k_1 = k_2 = 0 \\ 0 & \text{for } \mathbb{Z} \setminus (0,0) \end{cases}. \tag{1.113}$$

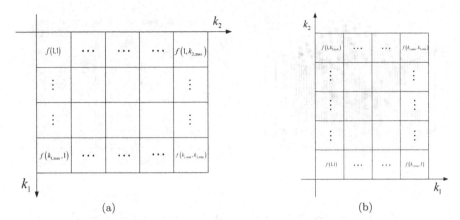

Fig. 1.42   Shift and rotation of an array.

In Figs. 1.43(a) and 1.43(b) the 3D plot and image of the discrete 2D Dirac pulse are presented.

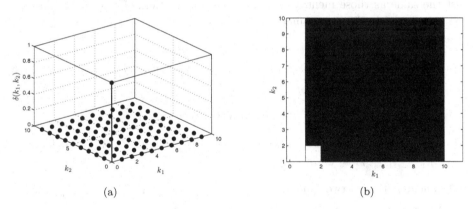

Fig. 1.43   Plot (a) and the image (b) of the 2D discrete Dirac pulse (1.113).

The function is the so-called 2D unit step function. Its definition is given below, and its plot and image are depicted in Figs. 1.44(a) and 1.44(b), respectively.

Definition 1.14 (2D discrete unit step function). The 2D discrete unit step function is defined as

$$1(k_1, k_2) = \begin{cases} 0 & \text{for } k_1, k_2 < 0 \\ 1 & \text{for } k_1, k_2 \geqslant 0 \end{cases}.$$ (1.114)

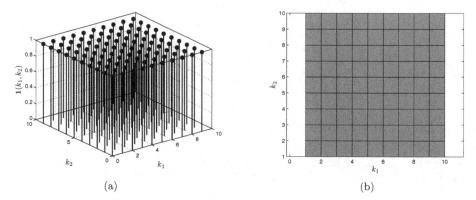

Fig. 1.44   Plot (a) and the image (b) of the discrete unit step function (1.7).

1.5.2   *Fundamental operations on the two discrete-variable functions*

Regarding the basic operations on the two discrete-variable functions they are almost the same as those mentioned in the previous section. The only condition is the same domain. Thus, for functions $f, g : \mathbb{Z}_{N_{1,\min},N_{1,\max},h} \times \mathbb{Z}_{N_{2,\min},N_{2,\max},h} \to \mathbb{R}$ one can evaluate:

(I) weighted sum of two (or more) functions $h(k_1, k_2) = af(k_1, k_2) + bg(k_1, k_2)$, where $a, b, \in \mathbb{R}$

(II) product of two functions $h(k_1, k_2) = f(k_1, k_2)g(k_1, k_2)$

(III) quotient of two functions $h(k_1, k_2) = \frac{f(k_1,k_2)}{g(k_1,k_2)}$ where $g(k_1, k_2) \neq 0$ for $k_1 \in \mathbb{Z}_{N_{1,\min},N_{1,\max},h}$, $k_2 \in \mathbb{Z}_{N_{2,\min},N_{2,\max},h}$

(IV) concatenation of two functions: $g(f) : \mathbb{Z}_{N_{1,\min},N_{1,\max},h} \times \mathbb{Z}_{N_{2,\min},N_{2,\max},h} \to \mathbb{R}$ where $f : \mathbb{Z}_{N_{1,\min},N_{1,\max},h} \times \mathbb{Z}_{N_{2,\min},N_{2,\max},h} \to \mathbb{Z}_{N_{3,\min},N_{3,\max},h} \times \mathbb{Z}_{N_{4,\min},N_{4,\max},h}$ and $g : \mathbb{Z}_{N_{3,\min},N_{3,\max},h} \times \mathbb{Z}_{N_{4,\min},N_{4,\max},h} \to \mathbb{R}$.

The next numerical example shows the result of a two functions' product.

Example 1.7. Consider slightly special two discrete-variables functions

$$f(k_1, k_2) = \sin\left(\frac{\pi k_1}{6}\right) \tag{1.115}$$

$$g(k_1, k_2) = \cos\left(\frac{\pi k_2}{6}\right). \tag{1.116}$$

Plot its product.

$$h(k_1, k_2) = f(k_1, k_2)g(k_1, k_2) = \sin\left(\frac{\pi k_1}{6}\right)\cos\left(\frac{\pi k_2}{6}\right). \tag{1.117}$$

Solution. First in Figs. 1.45(a) and 1.45(b) the plots of functions (1.113) and (1.114) are given, respectively. Next, its product is plotted in Fig. 1.46.

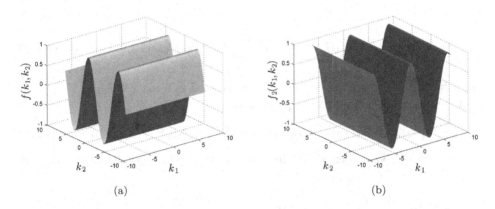

(a)　　　　　　　　　　　　　　(b)

Fig. 1.45　3D plot of function (a) (1.113) and (b) (1.114).

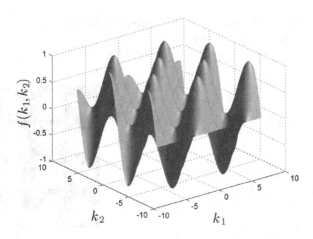

Fig. 1.46　3D plot of function (1.115).

1.5.3 *Two dimensional discrete convolution*

A discrete two dimensional convolution plays a very important role in the two dimensional signal and image processing. It is mainly defined as

$$g(k_1, k_2) = f_1(k_1, k_2) * f_1(k_1, k_2) = \sum_{i_1=-\infty}^{\infty} \sum_{i_2=-\infty}^{\infty} f_1(k_1, k_2) f_2(k_1 - i_1, k_2 - i_2).$$
(1.118)

However, for $f_1(k_1, k_2) = f_2(k_1, k_2) = 0$ for $k_1, k_2 < 0$ the formula simplifies to the form

$$g(k_1, k_2) = f_1(k_1, k_2) * f_1(k_1, k_2) = \sum_{i_1=0}^{k_1} \sum_{i_2=0}^{k_2} f_1(k_1, k_2) f_2(k_1 - i_1, k_2 - i_2). \quad (1.119)$$

Assuming that function values $f_i(k_1, k_2)$ are treated as a matrix $\mathbf{F}_{i, k_{1,\max}, k_{2,\max}}$ for $i = 1, 2$ formula (1.117) may be expressed as

$$g(k_1, k_2) = f_1(k_1, k_2) * f_1(k_1, k_2) = tr\left(\mathbf{F}_{1, k_1, k_2} \mathbf{N}_{k_2} \mathbf{F}_{2, k_2, k_1} \mathbf{N}_{k_1}\right) \qquad (1.120)$$

where a matrix $\mathbf{N}_k$ is defined by formula (11.19). The next example shows an application of the formula (1.118).

Example 1.8. Consider two functions

$$f_1(k_1, k_2) = f_2(k_1, k_2) = \sin\left(\frac{\pi k_1}{24}\right). \qquad (1.121)$$

Plot the discrete convolutions $f_1(k_1, k_2) * f_2(k_1, k_2), f_1(k_1, k_2) * \delta(k_1, k_2)$ and $f_1(k_1, k_2) * \mathbf{1}(k_1, k_2)$.

Solution. Appropriate plots and related images are given in Figs. 1.47(a)–1.47(b), 1.48(a)–1.48(b), 1.49(a) and 1.49(b).

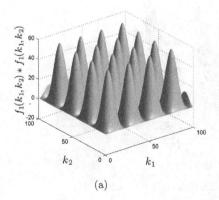

(a)

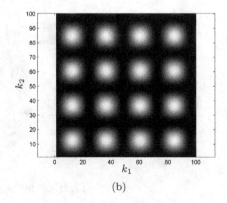

(b)

Fig. 1.47   3D plot and image of 2D convolution of $f_1(k_1, k_2) * f_1(k_1, k_2)$.

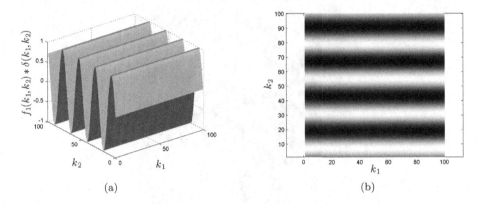

Fig. 1.48   3D plot and image of 2D convolution of $f_1(k_1, k_2) * \delta(k_1, k_2)$.

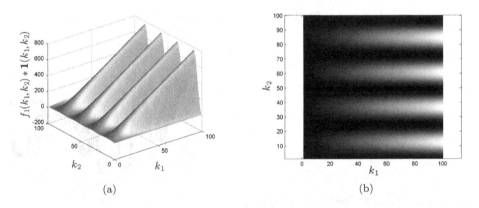

Fig. 1.49   3D plot and image of 2D convolution of $f_1(k_1, k_2) * \mathbf{1}(k_1, k_2)$.

1.5.4   *Discrete-image as the two discrete-variables function*

Up to now there have been considered explicit functions of one or two discrete-variables. Such functions can be represented as discrete images when their scaled values are represented by levels of a specified gray scale. Such an image is considered as a real two discrete-variable function $F(k_1, k_2)$ for $k_1 \in [1k_{1,\max}]$, $k_2 \in [1k_{2,\max}]$ with elements coding the gray scale (or in general color). An example of a real digital image is given in Fig. 1.50, [MF (2015)].

Chosen fragments of the function $F(k_1, k_2)$ can be plotted as 1D plots. So, for $k_2 = \text{const} \in [103, 113]$ one gets a sub-image and a set of plots.

The plots in Fig. 1.51(b) clearly show the "Small Dogs" legs. The "Small Dog" ears are more clearly indicated in Figs. 1.52(a)–1.52(b) which present a sub-image and related plots.

Fig. 1.50   Discrete image of the "Small Dog".

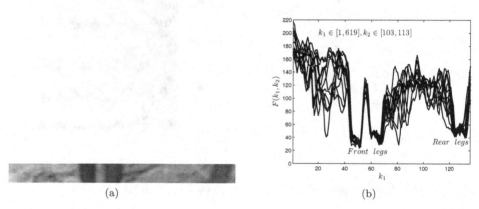

(a)   (b)

Fig. 1.51   Sub-image of Fig. 1.50 and related plot of $F(k_1, k_2)$ for $k_2 = \text{const} \in [103, 113]$.

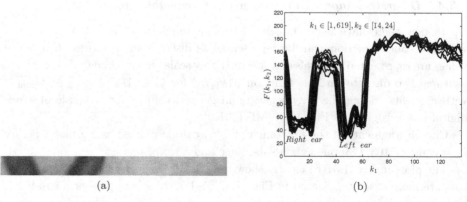

(a)   (b)

Fig. 1.52   Sub-image of Fig. 1.50 and related plots of $F(k_1, k_2)$ for $k_2 = \text{const} \in [14, 24]$.

Chapter 2

The n-th order backward difference/sum of the discrete-variable function

2.1 The n-th order backward difference of the discrete-variable function

Every continuous-time dynamical process is described by different types of difference equations [Bayin (2006)]. The left-hand side n-th order derivative of a continuous-variable real function has its discrete counterpart. It is commonly called the integer (n-th) order backward-difference (IOBD) of the discrete-variable function [Bose (2004)], [Butt (2009)], [Ralston (1965)], [Dahlquist (1974)], [Linz and Wang (2003)], [Ostalczyk (2001)]. This chapter recalls some basic properties of the IOBD as it is a very important mathematical tool in various areas of the technical sciences. Here one should mention discrete signal processing where the sampled signals are described by the discrete-time functions [Bose (2004)], [Ifeachor and Jervis (1993)], [Proakis (2007)], [Smith (2003)], [Weeks (2011)]. An important area of technical application of the IOBD is the control of closed-loop discrete-time systems [Kaczorek (1992)], [Kailath (1980)], [O'Flynn and Moriarty (1987)], [Ogata (1987)] and modeling of discrete systems [Fulford *et al.* (1997)], [Ljung (1999)].

Definition 2.1 (IOBD recursive form). The IOBD of a discrete-variable real function $f(k)$ for $k \in \mathbb{Z}_+$ is defined by the following recursive formula

$$\Delta^{(n)} f(k) = \begin{cases} f(k) & \text{for } n = 0 \\ \Delta^{(n-1)} f(k) - \Delta^{(n-1)} f(k-1) & \text{for } n = 1, 2, \dots \end{cases} \tag{2.1}$$

where:

$\quad$ $n \in \mathbb{Z}_+$ – integer-order of the BD

$\quad$ $f(k)$ – discrete-variable function.

One should note that following the notation used in the calculus where the differentiation order n is labeled in parenthesis $f^{(n)}(t)$, $t \in \mathbb{R}_+$ in the IOBD case there will be applied a similar notation. Below a few IOBDs of the discrete-variable function

$f(k)$ are evaluated

$$
\begin{aligned}
\Delta^{(1)} f(k) &= \Delta f(k) = f(k) - f(k-1), \\
\Delta^{(2)} f(k) &= f(k) - 2f(k-1) + f(k-2), \\
\Delta^{(3)} f(k) &= f(k) - 3f(k-1) + 3f(k-2) - f(k-3), \\
\Delta^{(4)} f(k) &= f(k) - 4f(k-1) + 6f(k-2) - 4f(k-3) + f(k-4).
\end{aligned}
\tag{2.2}
$$

One can easily verify that the formulas given above may be also expressed in a more compact form known as the Grünwald-Letnikov form of the (n-th order, $n \in \mathbb{Z}_+$) BD.

Definition 2.2 (Grünwald-Letnikov form of the IOBD). The IOBD of order n-th of a discrete-variable real function $f(k)$ for $k \in \mathbb{Z}_+$ is defined as

$$
\Delta^{(n)} f(k) = \sum_{i=0}^{k} a^{(n)}(i) f(k-i)
\tag{2.3}
$$

where the function $a^{(n)}(k)$ was discussed in the previous chapter. The formula given above may be equivalently expressed in a vector form

$$
\Delta^{(n)} f(k) = \left[a^{(n)}(0) \cdots a^{(n)}(n)\, a^{(n)}(n+1) \cdots a^{(n)}(k) \right]
\begin{bmatrix}
f(k) \\
\vdots \\
f(n) \\
f(n-1) \\
\vdots \\
f(0)
\end{bmatrix}
$$

$$
= \left[a^{(n)}(0)\, a^{(n)}(1) \cdots a^{(n)}(n) \right]
\begin{bmatrix}
f(k) \\
\vdots \\
f(k-n+1) \\
f(k-n)
\end{bmatrix}
$$

$$
+ \left[a^{(n)}(n+1)\, a^{(n)}(n+2) \cdots a^{(n)}(k) \right]
\begin{bmatrix}
f(k-n-1) \\
\vdots \\
f(1) \\
f(0)
\end{bmatrix}
$$

$$
= \left[a^{(n)}(n+1)\, a^{(n)}(n+2) \cdots a^{(n)}(k) \right]
\begin{bmatrix}
f(k) \\
\vdots \\
f(k-n+1) \\
f(k-n)
\end{bmatrix}.
\tag{2.4}
$$

The above transformations could be done noting that for $n \in \mathbb{Z}_+$ $a^{(n)}(k) = 0$ for $k > n$ (see formula (1.51)). Hence, the IOBD sum (2.3) simplifies to

$$\Delta^{(n)} f(k) = \sum_{i=0}^{n} a^{(n)}(i) f(k - i). \tag{2.5}$$

Simple manipulations with summation ranges lead to another equivalent form of the IOBD sum

$$\Delta^{(n)} f(k) = \sum_{i=k-n}^{k} a^{(n)}(i - k + n) f(2k - i - n). \tag{2.6}$$

The three forms of the IOBD ((2.3),(2.5) and (2.6)) suggest that it would be worthwhile to introduce in the IOBD a unified notation containing subscripts indicating a discrete differentiation range $[k_0, k]$ and superscripts pointing to the Grünwald-Letnikov form (GL) and the IO (n)

$$_{k_0}^{GL} \Delta_k^{(n)} f(k) = \sum_{i=k_0}^{k} a^{(n)}(i - k_0) f(k + k_0 - i). \tag{2.7}$$

Here one should also unambiguously determine the effect of summation in the IOBD defined by formula (2.7) with respect to (2.6)

$$_{k_0}^{GL} \Delta_k^{(n)} f(k) = \begin{cases} \sum_{i=k_0}^{k} a^{(n)}(i - k_0) f(k + k_0 - i) & \text{for } k_0 \leqslant k \leqslant n \\ \Delta^{(n)} f(k) & \text{for } n < k \end{cases}. \tag{2.8}$$

The last form can be simplified by noting the function $f(k) = 0$ for $k < 0$:

$$_{k_0}^{GL} \Delta_k^{(n)} \left[f(k - k_0) \mathbf{1}(k - k) \right] = \Delta^n \left[f(k - k_0) \mathbf{1}(k - k_0) \right]. \tag{2.9}$$

Now one assumes that $k_0 = 0$ and defines a new form of the IOBD.

Definition 2.3. The IOBD of order n of a discrete-variable function $f(k)$ over an interval $[0, k]$ is defined as a discrete convolution

$$_{0}^{GL} \Delta_k^{(n)} f(k) = a^{(n)}(k) * f(k). \tag{2.10}$$

The equivalence between forms (2.10) and (2.7) is immediate, but it would be worth applying a discrete convolution commutative law (see formula (1.88)) to show that

$$a^{(n)}(k) * f(k) = \left[a^{(n)}(0) \, a^{(n)}(1) \, \cdots \, a^{(n)}(k)\right] \begin{bmatrix} f(k) \\ f(k-1) \\ \vdots \\ f(0) \end{bmatrix}$$

$$= \left[\left[a^{(n)}(0) \, a^{(n)}(1) \, \cdots \, a^{(n)}(k)\right] \begin{bmatrix} f(k) \\ f(k-1) \\ \vdots \\ f(0) \end{bmatrix}\right]^T$$

$$= \left[f(k) \, f(k-1) \, \cdots \, f(0)\right] \begin{bmatrix} a^{(n)}(0) \\ a^{(n)}(1) \\ \vdots \\ a^{(n)}(k) \end{bmatrix}$$

$$= \left[f(k) \, f(k-1) \, \cdots \, f(0)\right] \mathbf{N}_k \mathbf{N}_k \begin{bmatrix} a^{(n)}(0) \\ a^{(n)}(1) \\ \vdots \\ a^{(n)}(k) \end{bmatrix}$$

$$= \left[f(0) \, f(1) \, \cdots \, f(k)\right] \begin{bmatrix} a^{(n)}(k) \\ a^{(n)}(k-1) \\ \vdots \\ a^{(n)}(0) \end{bmatrix}$$

$$= f(k) * a^{(n)}(k) = \sum_{i=0}^{k} f(i) a^{(n)}(k-i) \qquad (2.11)$$

where an anti-diagonal unit matrix $\mathbf{N}_k$ is defined in Appendix A.

Formula (2.7) can be also expressed in a vector form

$$\genfrac{}{}{0pt}{}{GL}{k_0} \Delta_k^{(n)} f(k) = \left[a^{(n)}(0) \, a^{(n)}(1) \, \cdots \, a^{(n)}(k-k_0)\right] \begin{bmatrix} f(k) \\ f(k-1) \\ \vdots \\ f(k_0) \end{bmatrix}. \qquad (2.12)$$

It is also valid for consecutive time instants $k-1, k-2, \ldots, k_0$. This means that

$$\genfrac{}{}{0pt}{}{GL}{k_0} \Delta_{k-1}^{(n)} f(k-1) = \left[a^{(n)}(0) \, a^{(n)}(1) \, \cdots \, a^{(n)}(k-k_0-1)\right] \begin{bmatrix} f(k-1) \\ f(k-2) \\ \vdots \\ f(k_0) \end{bmatrix}. \qquad (2.13)$$

Formulas (2.13) and (2.14) are similarly defined for $k-2, k-3, \ldots, k_0$ an may be written compactly as one vector-matrix equality

$$
{}^{GL}_{k_0}\Delta_k^{(n)}\mathbf{f}(k) = \mathbf{A}^{(n)}_{k-k_0}\mathbf{f}(k) \tag{2.14}
$$

where

$$
\mathbf{f}(k) = \begin{bmatrix} f(k) \\ f(k-1) \\ \vdots \\ f(k_0) \end{bmatrix}, \quad {}^{GL}_{k_0}\Delta_k^{(n)}\mathbf{f}(k) = \begin{bmatrix} {}^{GL}_{k_0}\Delta_k^{(n)}f(k) \\ {}^{GL}_{k_0}\Delta_{k-1}^{(n)}f(k-1) \\ \vdots \\ {}^{GL}_{k_0}\Delta_{k_0}^{(n)}f(k_0) \end{bmatrix} \tag{2.15}
$$

$$
\mathbf{A}^{(n)}_{k-k_0} = \begin{bmatrix} a^{(n)}(0) & a^{(n)}(1) & \cdots & a^{(n)}(k-1-k_0) & a^{(n)}(k-k_0) \\ 0 & a^{(n)}(0) & \cdots & a^{(n)}(k-2-k_0) & a^{(n)}(k-1-k_0) \\ \vdots & \vdots & & \vdots & \vdots \\ 0 & 0 & \cdots & a^{(n)}(0) & a^{(n)}(1) \\ 0 & 0 & \cdots & 0 & a^{(n)}(0) \end{bmatrix}. \tag{2.16}
$$

Matrix (2.16) is a so-called Toeplitz matrix (A.20). One should note that this $(k-k_0+1) \times (k-k_0+1)$ matrix is not only an upper triangular matrix but also a band one. Recalling properties of function $a^{(n)}(k)$ (see Table 1.2) it is always non-singular. It can be simplified to the form

$$
\mathbf{A}^{(n)}_{k-k_0} = \begin{bmatrix} 1 & a^{(n)}(1) & \cdots & a^{(n)}(n) & 0 & 0 & 0\cdots0 & 0 \\ 0 & 1 & \cdots & a^{(n)}(n-1) & a^{(n)}(n) & 0 & 0\cdots0 & 0 \\ 0 & 0 & \cdots & a^{(n)}(n-2) & a^{(n)}(n-1) & a^{(n)}(n) & 0\cdots0 & 0 \\ \vdots & \vdots & & \vdots & \vdots & \vdots & \vdots \cdots & \cdots \\ 0 & 0 & \cdots & 0 & 0 & 0 & 0\cdots1 & a^{(n)}(1) \\ 0 & 0 & \cdots & 0 & 0 & 0 & 0\cdots0 & 1 \end{bmatrix}. \tag{2.17}
$$

Now one assumes that $n_1, n_2 \in \mathbb{Z}_+$. Formula (2.15) allows to write two equalities

$$
{}^{GL}_{k_0}\Delta_k^{(n_1)}\mathbf{f}(k) = \mathbf{A}^{(n_1)}_{k-k_0}\mathbf{f}(k) \tag{2.18}
$$

$$
{}^{GL}_{k_0}\Delta_k^{(n_2)}\mathbf{f}(k) = \mathbf{A}^{(n_2)}_{k-k_0}\mathbf{f}(k). \tag{2.19}
$$

Pre-multiplication of equality (2.18) by a matrix $\mathbf{A}^{(n_2)}_{k-k_0}$ yields:

$$
\mathbf{A}^{(n_2)}_{k-k_0}{}^{GL}_{k_0}\Delta_k^{(n_1)}\mathbf{f}(k) = \mathbf{A}^{(n_2)}_{k-k_0}\mathbf{A}^{(n_1)}_{k-k_0}\mathbf{f}(k). \tag{2.20}
$$

Discrete Fractional Calculus

Taking the left-hand side of (2.20) one obtains

$$
\mathbf{A}_{k-k_0}^{(n_2)}{}_{k_0}^{GL}\Delta_k^{(n_1)}\mathbf{f}(k) =
\begin{bmatrix}
{}_{k_0}^{GL}\Delta_k^{(n_2)}\left[{}_{k_0}^{GL}\Delta_k^{(n_1)}f(k)\right] \\
{}_{k_0}^{GL}\Delta_{k-1}^{(n_2)}\left[{}_{k_0}^{GL}\Delta_{k-1}^{(n_1)}f(k-1)\right] \\
\vdots \\
{}_{k_0}^{GL}\Delta_1^{(n_2)}\left[{}_{k_0}^{GL}\Delta_1^{(n_1)}f(1)\right] \\
{}_{k_0}^{GL}\Delta_0^{(n_2)}\left[{}_{k_0}^{GL}\Delta_0^{(n_1)}f(0)\right]
\end{bmatrix}.
\tag{2.21}
$$

Elementary calculations reveal

$$
\mathbf{A}_{k-k_0}^{(n_1+n_2)} = \mathbf{A}_{k-k_0}^{(n_2)}\mathbf{A}_{k-k_0}^{(n_1)} = \mathbf{A}_{k-k_0}^{(n_1)}\mathbf{A}_{k-k_0}^{(n_2)}
\tag{2.22}
$$

and

$$
{}_{k_0}^{GL}\Delta_k^{(n_2)}\left[{}_{k_0}^{GL}\Delta_k^{(n_1)}f(k)\right] = {}_{k_0}^{GL}\Delta_k^{(n_1)}\left[{}_{k_0}^{GL}\Delta_k^{(n_2)}f(k)\right] = {}_{k_0}^{GL}\Delta_k^{(n_1+n_2)}f(k).
\tag{2.23}
$$

Now, the IOBD of discrete fundamental functions is evaluated. First, one considers the IOBD of a discrete Dirac function. By definition (2.7), for $k_0 = 0$:

$$
{}_0^{GL}\Delta_k^{(n)}\delta(k) = \sum_{i=0}^{k} a^{(n)}(i)\delta(k-i) = \sum_{i=0}^{n} a^{(n)}(i)\delta(k-i).
\tag{2.24}
$$

Taking into account the properties of the discrete Dirac pulse one can write:

$$
{}_0^{GL}\Delta_k^{(n)}\delta(k) = a^{(n)}(k) * \delta(k) = a^{(n)}(k).
\tag{2.25}
$$

For $n = 1$ one immediately gets:

$$
{}_0^{GL}\Delta_k^{(1)}\delta(k) = \sum_{i=0}^{1} a^{(1)}(i)\delta(k-i) = a^{(1)}(0)\delta(k) + a^{(1)}(1)\delta(k-1)
$$

$$
= \delta(k) - \delta(k-1) = a^{(1)}(k).
\tag{2.26}
$$

This result is consistent with that obtained by formula (2.2). The first order BD of the unit step function

$$
{}_0^{GL}\Delta_k^{(1)}\mathbf{1}(k) = \sum_{i=0}^{k} a^{(1)}(i)\mathbf{1}(k-i) = \mathbf{1}(k) - \mathbf{1}(k-1) = \delta(k).
\tag{2.27}
$$

The results stated by equality (2.28) are similar to that obtained for the Dirac function first order derivative ([O'Flynn and Moriarty (1987)]). A similar result may be obtained using the one-sided $\mathcal{Z}$-transform [Jury (1964)], [Ostalczyk (2002)] of the function $a^{(n)}(k)$

$$\mathcal{Z}\left\{a^{(n)}(k)\right\} = \sum_{i=0}^{+\infty} a^{(n)}(i)z^{-i} = \left(1 - z^{-1}\right)^n. \tag{2.28}$$

Note that the result given above coincides with formula (1.17). Hence,

$$\mathcal{Z}\left\{{}_0^{GL}\Delta_k^{(n)}\delta(k)\right\} = \mathcal{Z}\left\{a^{(n)}(k) * \delta(k)\right\} = \mathcal{Z}\left\{a^{(n)}(k)\right\}\mathcal{Z}\left\{\delta(k)\right\}$$

$$= \mathcal{Z}\left\{a^{(n)}(k)\right\} \cdot 1 = \left(1 - z^{-1}\right)^n. \tag{2.29}$$

This result can be immediately generalized to any transformable discrete-variable function $f(k)$

$$\mathcal{Z}\left\{{}_0^{GL}\Delta_k^{(n)}f(k)\right\} = \mathcal{Z}\left\{a^{(n)}(k) * f(k)\right\} = \mathcal{Z}\left\{a^{(n)}(k)\right\}\mathcal{Z}\left\{f(k)\right\}$$

$$= \mathcal{Z}\left\{a^{(n)}(k)\right\} \cdot F(z) = \left(1 - z^{-1}\right)^n F(z). \tag{2.30}$$

where $F(z) = \mathcal{Z}\left\{f(k)\right\}$.

Further analysis will focus on the so-called discrete power function $f(k) = k^m$ where $m \in \mathbb{Z}_+$. In order to avoid ambiguity, actually the function $f(k) = k^m \mathbf{1}(k)$ will be considered. By definition of the IOBD

$$
{}_0^{GL}\Delta_k^{(n)}f(k) = {}_0^{GL}\Delta_k^{(n)}k^m\mathbf{1}(k) = \sum_{i=0}^{k} a^{(n)}(i)[(k - i)^n \mathbf{1}(k - i)]^m
$$

$$
= \sum_{i=0}^{k} a^{(n)}(i)(k - i)^m \mathbf{1}(k - i) = \sum_{i=0}^{n} a^{(n)}(i)(k - i)^m \mathbf{1}(k - i). \tag{2.31}
$$

The function $(k - i)^m$ is a polynomial in a variable k. It can be expressed as a sum as follows

$$(k - i)^m = \sum_{j=0}^{m} a^{(m)}(j)k^{m-j}i^j. \tag{2.32}$$

Substituting this form to (2.32) one obtains

$$
{}_0^{GL}\Delta_k^{(n)}f(k) = \sum_{i=0}^{n} a^{(n)}(i)\left[\sum_{j=0}^{m} a^{(m)}(j)k^{m-j}i^j\mathbf{1}(k - i)\right]. \tag{2.33}
$$

Straightforward calculations lead to an equivalent vector-matrix form:

$$
{}_0^{GL}\Delta_k^{(n)}f(k)
$$

$$
= \left[a^{(n)}(0)\ a^{(n)}(0)\ \cdots\ a^{(n)}(n)\right]
\begin{bmatrix}
k^m0^0\mathbf{1}(k) & k^{m-1}0^1\mathbf{1}(k) & \cdots & k^00^m\mathbf{1}(k) \\
k^m1^0\mathbf{1}(k) & k^{m-1}1^1\mathbf{1}(k) & \cdots & k^01^m\mathbf{1}(k) \\
\vdots & \vdots & & \vdots \\
k^mn^0\mathbf{1}(k) & k^{m-1}n^1\mathbf{1}(k) & \cdots & k^0n^m\mathbf{1}(k)
\end{bmatrix}
\begin{bmatrix}
a^{(m)}(0) \\
a^{(m)}(1) \\
\vdots \\
a^{(m)}(m)
\end{bmatrix}.
$$

$$\tag{2.34}$$

Formulas (2.33) and (2.34) contain terms equal to zero. For $n = 1$ and $m \in \mathbb{Z}_+$ the formula (2.34) simplifies to

$$
{}^{GL}_{0}\Delta^{(1)}_{k} f(k)
$$

$$
= \begin{bmatrix} 1 & -1 \end{bmatrix} \begin{bmatrix} k^m 0^0 \mathbf{1}(k) & k^{m-1} 0^1 \mathbf{1}(k) & \cdots & k^0 0^m \mathbf{1}(k) \\ k^m 1^0 \mathbf{1}(k) & k^{m-1} 1^1 \mathbf{1}(k) & \cdots & k^0 1^m \mathbf{1}(k) \end{bmatrix} \begin{bmatrix} a^{(m)}(0) \\ a^{(m)}(1) \\ \vdots \\ a^{(m)}(m) \end{bmatrix}
$$

$$
= \begin{bmatrix} 1 & -1 \end{bmatrix} \begin{bmatrix} k^m \mathbf{1}(k) & 0 & \cdots & 0 \\ k^m 1^0 \mathbf{1}(k) & k^{m-1} 1^1 \mathbf{1}(k) & \cdots & k^0 1^m \mathbf{1}(k) \end{bmatrix} \begin{bmatrix} a^{(m)}(0) \\ a^{(m)}(1) \\ \vdots \\ a^{(m)}(m) \end{bmatrix}
$$

$$
= \begin{bmatrix} k^m 1^0 \mathbf{1}(k) - k^m \mathbf{1}(k-1) & -k^{m-1} 1^1 \mathbf{1}(k-1) & \cdots & -k^0 1^m \mathbf{1}(k-1) \end{bmatrix} \begin{bmatrix} a^{(m)}(0) \\ a^{(m)}(1) \\ \vdots \\ a^{(m)}(m) \end{bmatrix}.
$$

$$(2.35)$$

Noting that

$$
k^m \left[\mathbf{1}(k) - \mathbf{1}(k-1) \right] = 0 \quad \text{for } k = 0, 1, 2, \ldots \tag{2.36}
$$

formula (2.35) further simplifies to

$$
{}^{GL}_{0}\Delta^{(1)}_{k} f(k) = - \begin{bmatrix} k^{m-1} \mathbf{1}(k-1) & k^{m-2} \mathbf{1}(k-1) & \cdots & k^0 \mathbf{1}(k-1) \end{bmatrix} \begin{bmatrix} a^{(m)}(1) \\ a^{(m)}(2) \\ \vdots \\ a^{(m)}(m) \end{bmatrix}
$$

$$
= mk^{m-1} \mathbf{1}(k-1) - \begin{bmatrix} k^{m-2} & k^{m-3} & \cdots & k^0 \end{bmatrix} \begin{bmatrix} a^{(m)}(2) \\ a^{(m)}(3) \\ \vdots \\ a^{(m)}(m) \end{bmatrix} \mathbf{1}(k-1). \tag{2.37}
$$

The correctness of the result given above will be confirmed in a numerical example.

Example 2.1. Evaluate first-order BD of the power function $f(k) = k^3 \mathbf{1}(k)$.

Solution. In this example, $m = 3$. Hence, by (2.37)

$$
{}^{GL}_0 \Delta^{(1)}_k f(k) = {}^{GL}_0 \Delta^{(1)}_k \left[k^3 \mathbf{1}(k) \right]
$$

$$
= 3k^{3-1} \mathbf{1}(k-1) - \left[k^{3-2} \ k^{3-3} \right] \begin{bmatrix} a^{(3)}(2) \\ a^{(3)}(3) \end{bmatrix} \mathbf{1}(k-1)
$$

$$
= 3k^2 \mathbf{1}(k-1) - \left[k \ 1 \right] \begin{bmatrix} a^{(3)}(2) \\ a^{(3)}(3) \end{bmatrix} \mathbf{1}(k-1)
$$

$$
= \left(3k^2 - 3k + 1 \right) \mathbf{1}(k-1). \tag{2.38}
$$

An analogous result is obtained by direct calculations of the first-order BD

$$
{}^{GL}_0 \Delta^{(1)}_k \left[k^3 \mathbf{1}(k) \right] = \sum_{i=0}^{k} a^{(1)}(k)(k-i)^3 \mathbf{1}(k-i)
$$

$$
= k^3 \mathbf{1}(k) - (k-1)^3 \mathbf{1}(k-1)
$$

$$
= k^3 \mathbf{1}(k) - \left(k^3 - 3k^2 + 3k - 1 \right) \mathbf{1}(k-1)
$$

$$
= k^3 \left[\mathbf{1}(k) - \mathbf{1}(k-1) \right] + 3k^2 \mathbf{1}(k-1) - (3k-1)\mathbf{1}(k-1)
$$

$$
= k^3 \delta(k) + 3k^2 \mathbf{1}(k-1) - (3k-1)\mathbf{1}(k-1)
$$

$$
3k^2 \mathbf{1}(k-1) - (3k-1)\mathbf{1}(k-1). \tag{2.39}
$$

Here it is very important to note that contrary to differentiation of continuous-variable power functions, in the case of discrete-variable functions, there are additional terms that are represented by the vector product in formula (2.37). Hence,

$$
\left. \frac{df(t)}{dt} \right|_{t=kh} = 3t^2 \big|_{t=kh} = 3(kh)^2 \mathbf{1}(k) \neq \frac{{}^{GL}_0 \Delta^{(1)}_k f(k)}{h} = \frac{{}^{GL}_0 \Delta^{(1)}_k \left[k^3 \mathbf{1}(k) \right]}{h} \tag{2.40}
$$

which means that results obtained for the continuous function cannot be simply discretized to the discrete-variable case. However, the nature of the solution (2.39) reveals that the relative error tends to zero

$$
\lim_{k \to +\infty} \left[1 - \frac{\left. \frac{d(t^3)}{dt} \right|_{t=kh}}{{}^{GL}_0 \Delta^{(1)}_k (kh)^3 \mathbf{1}(kh)} \right] = 0. \tag{2.41}
$$

In the next numerical example disparity between the first-order BD of a function $f(k) = k^3 \mathbf{1}(k)$ evaluated by formulas (2.38) and (2.40) with $h = 1$ are presented.

Example 2.2. Evaluate error between first-order BDs calculated according to formulas (2.40) and (2.38). Assume $h = 1$.

Solution. For $h = 1$,

$$
f_1(k) = {}^{GL}_0 \Delta^{(1)}_k f(k) = {}^{GL}_0 \Delta^{(1)}_k \left[k^3 \mathbf{1}(k) \right] \tag{2.42}
$$

$$f_2(k) = \left.\frac{df(t)}{dt}\right|_{t=kh} = 3t^2\big|_{t=kh} = 3(kh)^2 \mathbf{1}(k) \tag{2.43}$$

and

$$e(k) = \frac{f_1(k) - f_2(k)}{f_1(k)} = \frac{-3k+1}{3k^2 - 3k + 1}\mathbf{1}(k). \tag{2.44}$$

Plots of functions $f_1(k)$ and $f_2(k)$ are given in Fig. 2.1, whereas the error function is presented in Fig. 2.2.

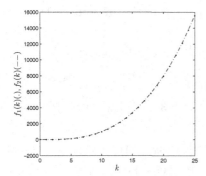
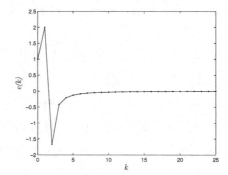

Fig. 2.1 Plot of functions (2.43) and (2.44).

Fig. 2.2 Plot of the error function $e(k)$ (2.45).

The plots reveal that the error function rapidly decreases to zero. In Fig. 2.1, the difference between plots of $f_1(k)$ and $f_2(k)$ is mostly negligible. Though the absolute error $|f_1(k) - f_2(k)|$ grows linearly, the relative error tends to zero.

2.2 The n-fold order backward sum of the discrete-variable function

As in the IOBD case, the definite n-fold integral of a continuous-variable real function also has its discrete counterpart. It is called the n-fold backward sum of the discrete-variable function. For reasons that will be explained later, it will be further called the n-th (integer) order backward sum (IOBS). First, a recursive definition of the IOBS is given.

Definition 2.4 (Grünwald-Letnikov form of the IOBS). The n-th order BS of a discrete-variable real and bounded function $f(k)$ defined over interval $[k_0, k]$, with $k_0, k \in \mathbb{Z}_+$ is defined by a following recursive formula:

$$\underset{k_0}{\overset{GL}{\sum}}{}_k^{(n)} f(k) = \begin{cases} f(k) & \text{for } n = 0 \\ \sum_{i=k_0}^{k} \underset{k_0}{\overset{GL}{\sum}}{}_i^{(n-1)} f(i) & \text{for } n = 1, 2, \dots \end{cases} \tag{2.45}$$

where ${}^{GL}_{k_0}\sum_{k}^{(n)} f(k)$ is a symbol of the IOBS. Here, similarly to the IOBD notation, left and right superscripts denote Grünwald-Letniokv (GL) and a summation order (n), respectively. Subscripts k_0 and k denote a summation range. The recursive formula (2.46) gives the following IOBSs

$$\begin{matrix} GL \\ k_0 \end{matrix} \sum_{k}^{(1)} f(k) = \sum_{i=k_0}^{k} f(i) \tag{2.46}$$

$$\begin{matrix} GL \\ k_0 \end{matrix} \sum_{k}^{(2)} f(k) = \sum_{j_1=k_0}^{k} \begin{matrix} GL \\ k_0 \end{matrix} \sum_{j_1}^{(1)} f(j_1) = \sum_{j_1=k_0}^{k} \sum_{j_2=k_0}^{j_1} f(j_2) \tag{2.47}$$

$$\begin{matrix} GL \\ k_0 \end{matrix} \sum_{k}^{(3)} f(k) = \sum_{j_1=k_0}^{k} \begin{matrix} GL \\ k_0 \end{matrix} \sum_{j_1}^{(2)} f(j_1) = \sum_{j_1=k_0}^{k} \sum_{j_2=k_0}^{j_1} \sum_{j_3=k_0}^{j_2} f(j_3). \tag{2.48}$$

The above forms lead to an equivalent definition of the IOBS.

Definition 2.5 (The n-fold BS). The n-fold BS of a discrete-variable real and bounded function $f(k)$ defined over interval $[k_0, k]$ with $k_0, k \in \mathbb{Z}_+$ is defined by the following recursive formula:

$$\begin{matrix} GL \\ k_0 \end{matrix} \sum_{k}^{(n)} f(k) = \begin{cases} f(k) & \text{for} \quad n = 0 \\ \sum_{j_1=k_0}^{k} \sum_{j_2=k_0}^{j_1} \cdots \sum_{j_n=k_0}^{j_{n-1}} f(j_n) & \text{for } n = 1, 2, \ldots \end{cases} \tag{2.49}$$

The IOBS may be also equivalently expressed as a sum.

Definition 2.6 (The Grünwald-Letnikow form of the IOBS). The n-fold BS of a discrete-variable real and bounded function $f(k)$ defined over interval $[k_0, k]$ with $k_0, k \in \mathbb{Z}_+$ is defined by the IOBD evaluated for negative order $-n$ $(n \in \mathbb{Z}_+)$

$$\begin{matrix} GL \\ k_0 \end{matrix} \sum_{k}^{(n)} f(k) = {}^{GL}_{k_0}\Delta_{k}^{(-n)} f(k) = \sum_{i=k_0}^{k} a^{(-n)}(i - k_0) f(k - i + k_0). \tag{2.50}$$

Remark 2.1. The function $a^{(-n)}(k)$ is the same as defined by formula (1.51) but for negative parameter $-n$ where $n \in \mathbb{Z}_+$. Such a notation is introduced to indicate that one considers the IOBS not IOBD.

Now one examines the correctness and equivalence of formulas (2.49) and (2.50). For $n = 1$ one obtains:

$$\begin{matrix} GL \\ k_0 \end{matrix} \sum_{k}^{(1)} f(k) = \sum_{i=k_0}^{k} a^{(-1)}(i - k_0) f(k - i + k_0)$$

$$= \sum_{i=k_0}^{k} 1 \cdot f(k - i + k_0) = \sum_{i=k_0}^{k} f(k - i + k_0) = \sum_{i=k_0}^{k} f(i). \tag{2.51}$$

For $n = 2$, one obtains

$$
{}^{GL}_{k_0}\sum\nolimits_{k}^{(2)} f(k) = \sum_{i=k_0}^{k} a^{(-2)}(i - k_0)f(k - i + k_0)
$$

$$
= \sum_{i=k_0}^{k} (i - k_0 + 1)f(k - i + k_0)
$$

$$
= f(k) + 2f(k-1) + \cdots + (k - k_0)f(k_0 + 1) + (k - k_0 + 1)f(k_0)
$$

$$
= \sum_{j_1=k_0}^{k} \sum_{j_2=k_0}^{j_1} f(j_2) \qquad (2.52)
$$

$$
{}^{GL}_{k_0}\sum\nolimits_{k}^{(3)} f(k) = \sum_{i=k_0}^{k} a^{(-3)}(i - k_0)f(k - i + k_0)
$$

$$
= f(k) + 3f(k-1) + 6f(k-2)\cdots + \frac{(k - k_0 + 1)(k - k_0 + 2)}{2}f(k_0)
$$

$$
= \sum_{j_1=k_0}^{k} \sum_{j_2=k_0}^{j_1} \sum_{j_3=k_0}^{j_2} f(j_3). \qquad (2.53)
$$

For $k_0 = 0$, the BS can be expressed as a convolution sum. This form is defined as follows.

Definition 2.7 (Convolution form of the n-fold BS). The n-fold BS of a discrete-variable real and bounded function $f(k)$ defined over interval $[k_0, k]$ with $k_0, k \in \mathbb{Z}_+$ is defined as a discrete convolution

$$
{}^{GL}_{k_0}\sum\nolimits_{k}^{(n)} f(k) = a^{(-n)}(k) * f(k). \qquad (2.54)
$$

As mentioned earlier, the function $a^{(-n)}(k)$ is defined by formula (1.51) which for negative parameter $-n$ may be transformed to the form

$$
a^{(-n)}(k) = \begin{cases} 1 & \text{for } k = 0 \\ \frac{n(n+1)\cdots(n+k-1)}{k!} & \text{for } k = 1, 2, \ldots \end{cases} \qquad (2.55)
$$

Analogous to the IOBD, the IOBS can be also expressed in a vector form

$$
{}^{GL}_{k_0}\Sigma_k^{(n)} f(k) = {}^{GL}_{k_0}\Delta_k^{(-n)} f(k) = \begin{bmatrix} a^{(-n)}(0) & a^{(-n)}(1) & \cdots & a^{(-n)}(k - k_0) \end{bmatrix} \begin{bmatrix} f(k) \\ f(k-1) \\ \vdots \\ f(k_0) \end{bmatrix}
$$

$$
(2.56)
$$

which can be also evaluated for $k - 1, k - 2, \ldots, k_0$. All such formulas collected in one matrix form yield

$$\substack{GL\\k_0}\Sigma_k^{(n)} \mathbf{f}(k) = \mathbf{A}_{k-k_0}^{(-n)} \mathbf{f}(k) \tag{2.57}$$

where

$$\mathbf{f}(k) = \begin{bmatrix} f(k) \\ f(k-1) \\ \vdots \\ f(k_0) \end{bmatrix}, \quad \substack{GL\\k_0}\Sigma_k^{(n)} \mathbf{f}(k) = \begin{bmatrix} \substack{GL\\k_0}\Sigma_k^{(n)} f(k) \\ \substack{GL\\k_0}\Sigma_{k-1}^{(n)} f(k-1) \\ \vdots \\ \substack{GL\\k_0}\Sigma_{k_0}^{(n)} f(k_0) \end{bmatrix}, \tag{2.58}$$

$$\mathbf{A}_{k-k_0}^{(-n)} = \begin{bmatrix} a^{(-n)}(0) & a^{(-n)}(1) & \cdots & a^{(-n)}(k-1-k_0) & a^{(-n)}(k-k_0) \\ 0 & a^{(-n)}(0) & \cdots & a^{(-n)}(k-2-k_0) & a^{(-n)}(k-1-k_0) \\ \vdots & \vdots & & \vdots & \vdots \\ 0 & 0 & \cdots & a^{(-n)}(0) & a^{(-n)}(1) \\ 0 & 0 & \cdots & 0 & a^{(-n)}(0) \end{bmatrix}. \tag{2.59}$$

Contrary to matrices (2.17) and (2.18) all elements $a^{(-n)}(i) \neq 0$ for $i = 0, 1, \ldots . k - k_0$. The matrix is in the upper triangular form with all elements on a main diagonal equal to 1 because by (2.56) $a^{(-n)}(0) = 1$. Hence, the matrix is always non-singular. Now one assumes that $n_1, n_2 \in \mathbb{Z}_+$. Equality (2.57) is valid for orders n_1, n_2. Thus,

$$\substack{GL\\k_0}\Sigma_k^{(n_j)} \mathbf{f}(k) = \mathbf{A}_{k-k_0}^{(-n_j)} \mathbf{f}(k), \text{ for } j = 1, 2. \tag{2.60}$$

Pre-multiplication of equation (2.57) evaluated for $i = 1$ by a matrix $\mathbf{A}_{k-k_0}^{(-n_2)}$ yields

$$\mathbf{A}_{k-k_0}^{(-n_2)} \substack{GL\\k_0}\Sigma_k^{(n_1)} \mathbf{f}(k) = \mathbf{A}_{k-k_0}^{(-n_2)} \mathbf{A}_{k-k_0}^{(-n_1)} \mathbf{f}(k). \tag{2.61}$$

The left-hand side of the above equality means

$$\mathbf{A}_{k-k_0}^{(-n_2)} \substack{GL\\k_0}\Sigma_k^{(n_1)} \mathbf{f}(k) = \mathbf{A}_{k-k_0}^{(-n_2)} \begin{bmatrix} \substack{GL\\k_0}\Sigma_k^{(n_1)} f(k) \\ \substack{GL\\k_0}\Sigma_{k-1}^{(n_1)} f(k-1) \\ \vdots \\ \substack{GL\\k_0}\Sigma_{k_0}^{(n_1)} f(k_0) \end{bmatrix} = \begin{bmatrix} \substack{GL\\k_0}\Sigma_k^{(n_2)} \left[\substack{GL\\k_0}\Sigma_k^{(n_1)} f(k) \right] \\ \substack{GL\\k_0}\Sigma_{k-1}^{(n_2)} \left[\substack{GL\\k_0}\Sigma_{k-1}^{(n_1)} f(k-1) \right] \\ \vdots \\ \substack{GL\\k_0}\Sigma_{k_0}^{(n_2)} \left[\substack{GL\\k_0}\Sigma_{k_0}^{(n_1)} f(k_0) \right] \end{bmatrix}. \tag{2.62}$$

Simple matrix multiplication reveals that

$$\mathbf{A}_{k-k_0}^{(-n_2)} \mathbf{A}_{k-k_0}^{(-n_1)} = \mathbf{A}_{k-k_0}^{(-n_2-n_1)}. \tag{2.63}$$

Hence, the right-hand side of (2.61) may be expressed as

$$\mathbf{A}_{k-k_0}^{(-n_2-n_1)}\mathbf{f}(k) = \begin{bmatrix} {}^{GL}_{k_0}\Sigma_k^{(n_1+n_2)}f(k) \\ {}^{GL}_{k_0}\Sigma_{k-1}^{(n_1+n_2)}f(k-1) \\ \vdots \\ {}^{GL}_{k_0}\Sigma_{k_0}^{(n_1+n_2)}f(k_0) \end{bmatrix}. \tag{2.64}$$

Comparison of (2.62) with (2.64) leads to a conclusion that

$$ {}^{GL}_{k_0}\Sigma_k^{(n_2)} \left[{}^{GL}_{k_0}\Sigma_k^{(n_1)}f(k) \right] = {}^{GL}_{k_0}\Sigma_k^{(n_1+n_2)}f(k). \tag{2.65}$$

Finally, some n-folds BSs of chosen functions are calculated. One begins with the IOBS of the Kronecker discrete delta function

$$ {}^{GL}_{0}\Sigma_k^{(n)}\delta(k) = \sum_{i=0}^{k} a^{(-n)}(i)\delta(k-i) = a^{(-n)}(k) = \frac{n(n+1)\cdots(k+n_1)}{k!}. \tag{2.66}$$

It can be further transformed to an elegant equivalent form

$$\begin{aligned} {}^{GL}_{0}\Sigma_k^{(n)}\delta(k) &= \frac{n(n+1)\cdots(k-1)k(k+1)\cdots(k+n-1)}{1\cdot2\cdot3\cdots\cdots(n-1)n(n+1)\cdots(k-1)k} \\ &= \frac{1n(n+1)\cdots(k-1)k}{n(n+1)\cdots(k-1)k}\cdot\frac{(k+1)\cdots(k+n-1)}{1\cdot2\cdot3\cdots\cdots(n-1)} \\ &= \frac{k+1}{1}\cdot\frac{k+2}{2}\cdots\frac{k+n-1}{n-1} = \prod_{i=1}^{n-1}\left(\frac{k}{i}+1\right) = \frac{\Gamma(k+n)}{\Gamma(k+1)\Gamma(n)}. \end{aligned} \tag{2.67}$$

From the above form one can derive the well-known relation between the discrete Kronecker delta function and the discrete Heaviside one. From (2.68), for $n = 1$ one immediately obtains

$$ {}^{GL}_{0}\Sigma_k^{(1)}\delta(k) = \left.\frac{\Gamma(k+n)}{\Gamma(k+1)\Gamma(n)}\right|_{n=1} = \frac{\Gamma(k+1)}{\Gamma(k+1)\Gamma(1)} = 1(k). \tag{2.68}$$

By formula (2.65) and the last result one can easily get the IOBS of the first-order of the discrete unit step function.

$$ {}^{GL}_{k_0}\Sigma_k^{(1)}1(k) = {}^{GL}_{k_0}\Sigma_k^{(1)}\left[{}^{GL}_{k_0}\Sigma_k^{(1)}\delta(k) \right] = \prod_{i=1}^{1}\left(\frac{k}{i}+1\right) = (1+k)1(k). \tag{2.69}$$

Continuing with such a substitution, one obtains a set of formulas being n-degree polynomials in a variable k. Collecting them in one matrix vector form, one obtains

$$ {}^{GL}_{k_0}\mathbf{\Delta}_k^{\mathbf{m}}1(k) = \mathbf{C}_m\mathbf{k}_m \tag{2.70}$$

where

$$
{}^{GL}_{k_0}\boldsymbol{\Delta}^{\mathbf{m}}_k\mathbf{1}(k) =
\begin{bmatrix}
{}^{GL}_{k_0}\Delta^{(0)}_k\mathbf{1}(k) \\
{}^{GL}_{k_0}\Delta^{(1)}_k\mathbf{1}(k) \\
\vdots \\
{}^{GL}_{k_0}\Delta^{(m-1)}_k\mathbf{1}(k) \\
{}^{GL}_{k_0}\Delta^{(m)}_k\mathbf{1}(k)
\end{bmatrix}
, \quad
\mathbf{k}_m =
\begin{bmatrix}
k^0 \\
k^1 \\
\vdots \\
k^{m-1} \\
k^m
\end{bmatrix}
\mathbf{1}(k), \quad
\mathbf{m} =
\begin{bmatrix}
0 \\
1 \\
\vdots \\
m-1 \\
m
\end{bmatrix}
\tag{2.71}
$$

$$
\mathbf{C}_m =
\begin{bmatrix}
1 & 0 & \cdots & 0 & 0 \\
c_{1,2} & \frac{1}{1!} & \cdots & 0 & 0 \\
\vdots & \vdots & & \vdots & \vdots \\
c_{1,m-1} & c_{2,m-1} & \cdots & \frac{1}{(m-1)!} & 0 \\
c_{1,m} & c_{2,m} & \cdots & c_{m-1,m} & \frac{1}{m!}
\end{bmatrix}
\tag{2.72}
$$

and elements $c_{i,j}$ for $j = 1, 2, \ldots, m$, $i < j$ are coefficients resulting from products in (2.69). For example, for $m = 8$ matrix (2.72) takes the form

$$
\mathbf{C}_8 =
\begin{bmatrix}
1 & 0 & 0 & 0 & 0 & 0 & 0 & 0 & 0 \\
1 & 1 & 0 & 0 & 0 & 0 & 0 & 0 & 0 \\
1 & \frac{1}{3} & \frac{1}{2} & 0 & 0 & 0 & 0 & 0 & 0 \\
1 & \frac{11}{6} & 1 & \frac{1}{6} & 0 & 0 & 0 & 0 & 0 \\
1 & \frac{25}{12} & \frac{35}{24} & \frac{5}{12} & \frac{1}{24} & 0 & 0 & 0 & 0 \\
1 & \frac{137}{60} & \frac{15}{8} & \frac{17}{24} & \frac{1}{8} & \frac{1}{120} & 0 & 0 & 0 \\
1 & \frac{49}{20} & \frac{203}{90} & \frac{49}{48} & \frac{35}{144} & \frac{7}{240} & \frac{1}{720} & 0 & 0 \\
1 & \frac{365}{140} & \frac{469}{180} & \frac{967}{720} & \frac{7}{18} & \frac{23}{360} & \frac{1}{180} & \frac{1}{5040} & 0 \\
1 & \frac{761}{280} & \frac{958}{327} & \frac{267}{160} & \frac{1069}{1920} & \frac{9}{80} & \frac{13}{960} & \frac{1}{1120} & \frac{1}{40320}
\end{bmatrix}.
\tag{2.73}
$$

Interesting is also an inverse relation

$$
\mathbf{k}_m = \left(\mathbf{C}_m^{-1}\right) {}^{GL}_{k_0}\boldsymbol{\Delta}^{\mathbf{m}}_k\mathbf{1}(k)
\tag{2.74}
$$

stating that one can express a power function as a linear combination of the IOBSs of consecutive orders. Below is an inverse matrix

$$
\mathbf{C}_8 =
\begin{bmatrix}
1 & 0 & 0 & 0 & 0 & 0 & 0 & 0 & 0 \\
-1 & 1 & 0 & 0 & 0 & 0 & 0 & 0 & 0 \\
1 & -3 & 2 & 0 & 0 & 0 & 0 & 0 & 0 \\
-1 & 7 & -12 & 6 & 0 & 0 & 0 & 0 & 0 \\
1 & -15 & 50 & -60 & 24 & 0 & 0 & 0 & 0 \\
-1 & 31 & -180 & 390 & -360 & 120 & 0 & 0 & 0 \\
1 & -63 & 602 & -2100 & 3360 & -2520 & 720 & 0 & 0 \\
-1 & 127 & -1932 & 10206 & -25200 & 31920 & -20160 & 5040 & 0 \\
1 & -255 & 6050 & -46620 & 166824 & -317520 & 332640 & -181440 & 40320
\end{bmatrix}.
\tag{2.75}
$$

Note that both equations (2.70) and (2.74) are rather difficult in numerical calculations due to their sensitivity to elements' changes. Firstly, in matrix (2.73), coefficients rapidly decrease in accordance to the growth of m, whereas in the second matrix elements increase in accordance to the factorial function growth.

2.3   The n-th order differ-sum of the discrete-variable function

In this section one formally summarizes the properties of the IOBD and IODS. Comparing formulas (2.7) and (2.50) leads to the conclusion that both notions are defined by the same formula with different parameters. Positive orders are assigned to the IOBD, while negative orders are assigned to the IOBS. Hence, one can write

$$
{}^{GL}_{k_0}\Delta_k^{(\pm n)}f(k) = {}^{GL}_{k_0}\Sigma_k^{(\mp n)}f(k) = {}^{GL}_{k_0}\Sigma_{i=k_0}^k a^{(\pm n)}(i)f(k-i) \tag{2.76}
$$

or

$$
{}^{GL}_0\Delta_k^{(\pm n)}f(k) = {}^{GL}_0\Sigma_k^{(\mp n)}f(k) = a^{(\pm n)}(k) * f(k). \tag{2.77}
$$

In this book the symbol Δ will be preferred even when a summation operation Σ is considered. The IOBD/S satisfy the linearity property. This formulates a theorem.

Theorem 2.1. *Consider two discrete-variable functions $f(k)$ and $g(k)$ which are defined over the same interval $[k_0, +\infty)$ and two real numbers $a, b \in \mathbb{R}$. Assuming that ${}^{GL}_0\Delta_k^{(\pm n)}f(k)$ and ${}^{GL}_0\Delta_k^{(\pm n)}g(k)$ exist, one can state that*

$$
a\left[{}^{GL}_0\Delta_k^{(\pm n)}f(k)\right] + b\left[{}^{GL}_0\Delta_k^{(\pm n)}g(k)\right] = {}^{GL}_0\Delta_k^{(\pm n)}\left[af(k)+bg(k)\right]. \tag{2.78}
$$

The additivity property follows immediately from the IOBD/S definition sum.

Now, the IOBD/S concatenation property is analyzed. Formulas (2.24) and (2.65) already show the action of this operation. Under similar assumptions, regarding orders $n_1, n_2 \in \mathbb{Z}_+$ one analyses the following

$$
{}^{GL}_{k_0}\Sigma_k^{(n_1)}\left[{}^{GL}_{k_0}\Delta_k^{(n_2)}f(k)\right] = {}^{GL}_{k_0}\Delta_k^{(-n_1)}\left[{}^{GL}_{k_0}\Delta_k^{(n_2)}f(k)\right]
$$

$$
= \sum_{i=k_0}^k a^{(-n_1)}(i-k_0)\left[\Sigma_{j=k_0}^{k-i} a^{(n_2)}(j-k_0)f(k-i-j+k_0)\right] \tag{2.79}
$$

$$
{}^{GL}_{k_0}\Sigma_k^{(n_1)}\left[{}^{GL}_{k_0}\Delta_k^{(n_2)}\mathbf{f}(k)\right] = {}^{GL}_{k_0}\Delta_k^{(-n_1)}\left[{}^{GL}_{k_0}\Delta_k^{(n_2)}\mathbf{f(k)}\right]. \tag{2.80}
$$

Recalling formulas (2.19) and (2.60) one can write

$$\substack{GL\\k_0}\Sigma_k^{(n_1)}\left[\substack{GL\\k_0}\Delta_k^{(n_2)}\mathbf{f}(k)\right] = \substack{GL\\k_0}\Sigma_k^{(n_1)}\left[\mathbf{A}_{k-k_0}^{(n_2)}\mathbf{f}(k))\right]$$

$$= \mathbf{A}_{k-k_0}^{(-n_1)}\left[\mathbf{A}_{k-k_0}^{(n_2)}\mathbf{f}(k))\right] = \mathbf{A}_{k-k_0}^{(-n_1)}\mathbf{A}_{k-k_0}^{(n_2)}\mathbf{f}(k) = \mathbf{A}_{k-k_0}^{(n_2-n_1)}\mathbf{f}(k). \qquad (2.81)$$

The result means that

$$\substack{GL\\k_0}\Sigma_k^{(n_1)}\left[\substack{GL\\k_0}\Delta_k^{(n_2)}\mathbf{f}(k)\right] = \begin{cases} \substack{GL\\k_0}\Sigma_k^{(n_1-n_2)}f(k) & \text{for } n_1 > n_2 \\ f(k) & \text{for } n_1 = n_2 \\ \substack{GL\\k_0}\Delta_k^{(n_2-n_1)}f(k) & \text{for } n_1 < n_2 \end{cases}. \qquad (2.82)$$

The result obtained above was achieved under the assumption that both summation operations are performed by definite sums with the same lower limits $i, j = k_0$ in (2.80). Now, one slightly changes the assumptions concerning the concatenation of summation and discrete-differentiation operations. This time one considers the following

$$\substack{GL\\k_0}\Sigma_k^{(n_1)}\left[\Delta^{n_2}f(k)\right]$$

$$= \substack{GL\\k_0}\Sigma_k^{(n_1)}\left[\substack{GL\\k-n_2}\Delta_k^{(n_2)}f(k)\right]$$

$$= \substack{GL\\k_0}\Delta_k^{(-n_1)}\left[\substack{GL\\k-n_2}\Delta_k^{(n_2)}f(k)\right]$$

$$= \sum_{i=k_0}^{k} a^{(-n_1)}(i-k_0)\left[\Sigma_{j=k-n_2}^{k-i}a^{(n_2)}(j-k+n_2)f(k-i-j+k-n_2)\right]. \qquad (2.83)$$

This relation may be expressed in a vector form

$$\substack{GL\\k_0}\Sigma_k^{(n_1)}\left[\Delta^{n_2}f(k)\right] = \left[a^{(-n_1)}(0)\ a^{(-n_1)}(0)\ \cdots\ a^{(-n_1)}(k-k_0-1)\ a^{(-n_1)}(k-k_0)\right]$$

$$\times\begin{bmatrix} \substack{GL\\k-n_2}\Delta_k^{(n_2)}f(k) \\ \substack{GL\\k-1-n_2}\Delta_{k-1}^{(n_2)}f(k-1) \\ \vdots \\ \substack{GL\\k_0+1-n_2}\Delta_{k_0+1}^{(n_2)}f(k_0+1) \\ \substack{GL\\k_0-n_2}\Delta_{k_0}^{(n_2)}f(k_0) \end{bmatrix}$$

$$= \left[a^{(-n_1)}(0)\ a^{(-n_1)}(0)\ \cdots\ a^{(-n_1)}(k-k_0-1)\ a^{(-n_1)}(k-k_0)\right]$$

$$\times\left[\mathbf{A}_{k-k_0}^{(n_2)}\ \mathbf{A}_{k-k_0,n_2-1}^{(n_2)}\right]\begin{bmatrix} f(k) \\ \vdots \\ f(k_0) \\ f(k_0-1) \\ \vdots \\ f(k_0-n_2) \end{bmatrix}.$$

$$(2.84)$$

where matrix $\mathbf{A}_{k-k_0+1}^{(n_2)}$ has been already defined by formula (2.59) and matrix $\mathbf{A}_{k-k_0,n_2}^{(n_2)}$ is of the form

$$\mathbf{A}_{k-k_0,n_2-1}^{(n_2)} = \begin{bmatrix} 0 & 0 & \cdots & 0 & 0 \\ \vdots & \vdots & & \vdots & \vdots \\ 0 & 0 & \cdots & 0 & 0 \\ a^{(n_2)}(n_2) & 0 & \cdots & 0 & 0 \\ a^{(n_2)}(n_2-1) & a^{(n_2)}(n_2) & \cdots & 0 & 0 \\ \vdots & \vdots & & \vdots & \vdots \\ a^{(n_2)}(2) & a^{(n_2)}(3) & \cdots & a^{(n_2)}(n_2) & 0 \\ a^{(n_2)}(1) & a^{(n_2)}(2) & \cdots & a^{(n_2)}(n_2-1) & a^{(n_2)}(n_2) \end{bmatrix}$$

$$= \begin{bmatrix} \mathbf{0}_{k-k_0-n_2+1,n_2-1} \\ \bar{\mathbf{A}}_{n_2-1}^{(n_2)} \end{bmatrix} \qquad (2.85)$$

$$\bar{\mathbf{A}}_{n_2-1}^{(n_2)} = \begin{bmatrix} a^{(n_2)}(n_2) & 0 & \cdots & 0 & 0 \\ a^{(n_2)}(n_2-1) & a^{(n_2)}(n_2) & \cdots & 0 & 0 \\ \vdots & \vdots & & \vdots & \vdots \\ a^{(n_2)}(2) & a^{(n_2)}(3) & \cdots & a^{(n_2)}(n_2) & 0 \\ a^{(n_2)}(1) & a^{(n_2)}(2) & \cdots & a^{(n_2)}(n_2-1) & a^{(n_2)}(n_2) \end{bmatrix}. \qquad (2.86)$$

Equality (2.85) is further transformed to the form

$$\substack{GL \\ k_0} \Sigma_k^{(n_1)} [\Delta^{n_2} f(k)]$$

$$= [a^{(-n_1)}(0)\ a^{(-n_1)}(1) \cdots a^{(-n_1)}(k-k_0)]\ \mathbf{A}_{k-k_0}^{(n_2)} \begin{bmatrix} f(k) \\ f(k-1) \\ \vdots \\ f(k_0+1) \\ f(k_0) \end{bmatrix}$$

$$+ [a^{(-n_1)}(k-k_0-n_2+1) \cdots a^{(-n_1)}(k-k_0)]\ \bar{\mathbf{A}}_{n_2-1}^{(n_2)} \begin{bmatrix} f(k_0-1) \\ f(k_0-2) \\ \vdots \\ f(k_0-n_2+1) \\ f(k_0-n_2) \end{bmatrix}.$$

$$(2.87)$$

The first term on the right-hand side of (2.87) equates to

$$\left[a^{(-n_1)}(0) \; a^{(-n_1)}(1) \; \cdots \; a^{(-n_1)}(k-k_0)\right] \mathbf{A}_{k-k_0}^{(n_2)} \begin{bmatrix} f(k) \\ f(k-1) \\ \vdots \\ f(k_0+1) \\ f(k_0) \end{bmatrix}$$

$$= \left[a^{(n_2-n_1)}(0) \; a^{(n_2-n_1)}(1) \; \cdots \; a^{(n_2-n_1)}(k-k_0)\right] \begin{bmatrix} f(k) \\ f(k-1) \\ \vdots \\ f(k_0+1) \\ f(k_0) \end{bmatrix}$$

$$= {}_{k_0}^{GL}\Delta_k^{(n_2-n_1)} f(k). \qquad (2.88)$$

To emphasize that the function $f(k)$ is defined over $[k_0, k]$, and its values $f(k_0 - 1), f(k_0 - 2), \ldots, f(k_0 - n_2)$ are treated as initial conditions, appropriate values will be further denoted as $f_{k_0-1}, k_{k_0-2}, \ldots, f_{k_0-n_2}$. Now, the second term on the right-hand side of (2.87) is denoted as a function of initial conditions

$$g_{(n_2-n_1)}(f_{k_0-1}, f_{k_0-2}, \ldots, f_{k_0-n_2})$$

$$= \left[a^{(-n_1)}(k-k_0-n_2+1) \; \cdots \; a^{(-n_1)}(k-k_0)\right] \bar{\mathbf{A}}_{n_2-1}^{(n_2)} \begin{bmatrix} f(k_0-1) \\ f(k_0-2) \\ \vdots \\ f(k_0-n_2+1) \\ f(k_0-n_2) \end{bmatrix}. \qquad (2.89)$$

Substituting results (2.88) and (2.89) into (2.87), one obtains

$$_{k_0}^{GL}\Sigma_k^{(n_1)} \left[\Delta^{n_2} f(k)\right] = {}_{k_0}^{GL}\Delta_k^{(n_2-n_1)} f(k) + g_{n_2-n_1}(f_{k_0-1}, f_{k_0-2}, \ldots, f_{k_0-n_2}). \qquad (2.90)$$

Hence,

$$_{k_0}^{GL}\Sigma_k^{(n_1)} \left[\Delta^{n_2} f(k)\right]$$

$$= \begin{cases} _{k_0}^{GL}\Delta_k^{(n_2-n_1)} f(k) + g_{n_2-n_1}(f_{k_0-1}, f_{k_0-2}, \ldots, f_{k_0-n_2}) & \text{for } n_2 > n_1 \\ f(k) + g_{n_2-n_1}(f_{k_0-1}, f_{k_0-2}, \ldots, f_{k_0-n_2}) & \text{for } n_1 = n_2 \\ _{k_0}^{GL}\Sigma_k^{(n_1-n_2)} f(k) + g_{n_2-n_1}(f_{k_0-1}, f_{k_0-2}, \ldots, f_{k_0-n_2}) & \text{for } n_1 > n_2 \end{cases} \qquad (2.91)$$

Analysis of formulas (2.80) and (2.91) show that for $f_{k_0-1} = f_{k_0} - 2 = \cdots = f_{k_0-n_2} = 0$ both formulas are equal. Analysis concerning the IOBS of the IOBD is illustrated by a numerical example.

Example 2.3. For a discrete-variable function $f(k)$ and $k_0 = 0$ evaluate: a) the BS of its second-order BD, b) the BS of its first-order BD, c) the two-fold BS of its first-order BD.

Solution. For part one a), $n_1 = 2, n_2 = 1$. One should first calculate the second order BD

$$\Delta^2 f(k) = f(k) - 2f(k-1) + f(k-2). \tag{2.92}$$

Next,

$$
\begin{aligned}
{}_0^{GL}\Sigma_k^{(1)}\left[\Delta^2 f(k)\right] &= {}_{k_0}^{GL}\Sigma_k^{(1)}\left[f(k) - 2f(k-1) + f(k-2)\right] \\
&= \sum_{i=0}^{k} f(i) - 2\sum_{i=0}^{k} f(i-1) + \sum_{i=0}^{k} f(i-2) \\
&= f(k) - f(k-1) - f_{-1} + f_{-2} = \Delta^1 f(k) - f_{-1} + f_{-2}.
\end{aligned} \tag{2.93}
$$

The result obtained above will be confirmed by checking equality (2.91) with

$$
{}_0^{GL}\Delta_k^{(n_2-n_1)} f(k) = {}_{k_0}^{GL}\Delta_k^{(1)} f(k)
$$

and

$$
\begin{aligned}
g_{(n_2-n_1)}\left(f_{k_0-1}, f_{k_0-2}, \ldots, f_{k_0-n_2}\right) &= g_{(-1)}\left(f_{k_0-1}, f_{k_0-2}\right) \\
&= \left[a^{(-1)}(k-1)\ a^{(-1)}(k)\right]\bar{\mathbf{A}}_1^{(2)} \begin{bmatrix} f_{-1} \\ f_{-2} \end{bmatrix}
\end{aligned} \tag{2.94}
$$

where

$$
\left[a^{(-1)}(k-1)\ a^{(-1)}(k)\right] = \begin{bmatrix} 1 & 1 \end{bmatrix} \tag{2.95}
$$

$$
\bar{\mathbf{A}}_1^{(2)} = \begin{bmatrix} a^{(2)}(2) & 0 \\ a^{(2)}(1) & a^{(2)}(2) \end{bmatrix} = \begin{bmatrix} 1 & 0 \\ -2 & 1 \end{bmatrix}. \tag{2.96}
$$

After a substitution of (2.95) and (2.96) into (2.94), one obtains

$$
g_{(-1)}\left(f_{k_0-1}, f_{k_0-2}\right) = \begin{bmatrix} 1 & 1 \end{bmatrix}\begin{bmatrix} 1 & 0 \\ -2 & 1 \end{bmatrix}\begin{bmatrix} f_{-1} \\ f_{-2} \end{bmatrix} = -f_{-1} + f_{-2}. \tag{2.97}
$$

For b), $n_1 = 1, n_1 = 1$ and

$$\Delta^1 f(k) = f(k) - f(k-1). \tag{2.98}$$

Then

$$
\begin{aligned}
{}_0^{GL}\Sigma_k^{(1)}\left[\Delta^{(1)} f(k)\right] &= {}_0^{GL}\Sigma_k^{(1)}\left[f(k) - f(k-1)\right] \\
&= \sum_{i=0}^{k} f(i) - \sum_{i=0}^{k} f(i-1) = f(k) - f_{-1}.
\end{aligned} \tag{2.99}
$$

In this example, equality (2.91) reduces to the form

$$_0^{GL} \Delta_k^{(n_2 - n_1)} f(k) =_{k_0}^{GL} \Delta_k^{(0)} f(k) = f(k),$$

and

$$g_{(n_2 - n_1)}(f_{k_0 - 1}, f_{k_0 - 2}, \ldots, f_{k_0 - n_2}) = g_{(0)}(f_{k_0 - 1})$$
$$= \left[a^{(-1)}(k) \right] \bar{\mathbf{A}}_0^{(1)} [f_{-1}] = [1][1][[f_{-1}]. \tag{2.100}$$

Hence, one gets the same form as (2.99). Realise that (2.99) has its counterpart in a continuous-variable function case

$$\int_0^t \left[\frac{df(x)}{dx} \right] dx = \int_0^t df(x) = f(x)|_0^t = f(t) - f(0). \tag{2.101}$$

Lastly, for part c), $n_1 = 2, n_2 = 1$ direct calculations show

$$_0^{GL} \Sigma_k^{(2)} \left[\Delta^1 f(k) \right] = \sum_{i=0}^{k} a^{(-2)}(i) \Delta f(k - i) \tag{2.102}$$

where by definition

$$a^{(-2)}(i) = \left. \frac{n(n + 1) \cdots (n + i - 1)}{i!} \right|_{n=2} = \frac{2 \cdot 3 \cdots (i + 1)}{1 \cdot 2 \cdots (i - 1)i} = i + 1. \tag{2.103}$$

Taking into account (2.104) equality (2.103) takes the form

$$_0^{GL} \Sigma_k^{(2)} \left[\Delta^1 f(k) \right] = \sum_{i=0}^{k} (i + 1) \left[f(k - i) - f(k - 1 - i) \right]$$
$$= \sum_{i=0}^{k} f(k - i) - (k + 1) f_{-1} =_0^{GL} \Sigma_k^{(1)} f(k) - (k + 1) f_{-1}. \tag{2.104}$$

An analogous result will be obtained by manipulating equality

$$_0^{GL} \Sigma_k^{(2)} \left[\Delta^1 f(k) \right] = \sum_{j_1=0}^{k} \sum_{j_2=0}^{j_1} \left[f(j_2) - f(j_2 - 1) \right]. \tag{2.105}$$

The result is also confirmed by an application of formula (2.91). The first right-hand term is obviously $_0^{GL} \Sigma_k^{(1)} f(k)$, whereas the second one equals

$$g_{(n_2 - n_1)}(f_{k_0 - 1}, f_{k_0 - 2}, \ldots, f_{k_0 - n_2}) = g_{(1)}(f_{k_0 - 1})$$
$$= \left[a^{(-2)}(k - 0 - 1 + 1) \right] \bar{\mathbf{A}}_0^{(1)} [f_{-1}] = [k + 1][1][[f_{-1}] = (k + 1) f_{-1}. \tag{2.106}$$

Finally

$$_{0}^{GL}\Sigma_{k}^{(2)}\left[\Delta^{1}f(k)\right] =_{0}^{GL}\Sigma_{k}^{(1)}f(k)-(k+1)f_{-1}. \qquad (2.107)$$

The last result is similar to the continuous-variable function case where

$$\int_{0}^{t}\left[\int_{0}^{\tau_{1}}df(\tau_{2})\right]d\tau_{1}=\int_{0}^{t}f(\tau_{1})d\tau_{1}-tf(0). \qquad (2.108)$$

An inversed concatenation: the IOBD of the IOBS is more simple due to the absence of initial conditions. This is due to the fact that both operations always are bounded by the same lower limit k_0.

Finally,

$$_{k_0}^{GL}\Delta_k^{(n_1)}\left[_{k_0}^{GL}\Sigma_k^{(n_2)}f(k)\right]$$

$$= _{k_0}^{GL}\Delta_k^{(n_1)}\left[_{k_0}^{GL}\Delta_k^{(-n_2)}f(k)\right]$$

$$= \left[a^{(n_1)}(0) \cdots a^{(n_1)}(k-k_0-1)\ a^{(n_1)}(k-k_0)\right]\begin{bmatrix} _{k_0}^{GL}\Sigma_k^{(n_2)}f(k) \\ _{k_0}^{GL}\Sigma_{k-1}^{(n_2)}f(k-1) \\ \vdots \\ _{k_0}^{GL}\Sigma_{k_0+1}^{(n_2)}f(k_0+1) \\ _{k_0}^{GL}\Sigma_{k_0}^{(n_2)}f(k_0) \end{bmatrix}$$

$$= \left[a^{(n_1)}(0) \cdots a^{(n_1)}(n_1)\ 0 \cdots 0\right]\mathbf{A}_{k-k_0}^{(-n_2)}\mathbf{f}(k)$$

$$= \left[a^{(n_1-n_2)}(0)\ a^{(n_1-n_2)}(1) \cdots a^{(n_1-n_2)}(k-k_0)\right]\mathbf{f}(k)$$

$$= \begin{cases} _{k_0}^{GL}\Sigma_k^{(n_2-n_1)}f(k) & \text{for } n_2 > n_1 \\ f(k) & \text{for } n_1 = n_2 \\ _{k_0}^{GL}\Delta_k^{(n_1-n_2)}f(k) & \text{for } n_1 > n_2 \end{cases} \qquad (2.109)$$

Here it is worth recalling that notation $_{k_0}^{GL}\Sigma_{k_0}^{(n_2)}f(k_0)$ means $_{k_0}^{GL}\Sigma_k^{(n_2)}f(k)\big|_{k=k_0}$ and $_{k_0}^{GL}\Sigma_k^{(n_2)}f(k)\neq_{k_0}^{GL}\Sigma_k^{(n_2)}f(k-1)$. In the next example the IOBD of the IOBS will be calculated.

Example 2.4. For a discrete-variable function $f(k)$ evaluate a) first-order BD of its two-fold sum, b) first-order BD of its BS and c) second-order BD of its BS. All operators evaluate over an interval $[0, k]$.

Solution. For part a), one should first calculate a two-forl sum of a discrete-variable function $f(k)$

$$_{0}^{GL}\Sigma_k^{(2)}f(k)=\sum_{i=0}^{k}a^{(-2)}(i)f(k-i)=\sum_{i=0}^{k}(i+1)f(k-i) \qquad (2.110)$$

$$
{}_0^{GL}\Sigma_{k-1}^{(2)}f(k-1) = \sum_{i=0}^{k-1} a^{(-2)}(i)f(k-1-i) = \sum_{i=0}^{k-1}(i+1)f(k-1-i). \quad (2.111)
$$

Then,

$$
{}_0^{GL}\Delta_k^{(1)}\left[{}_0^{GL}\Sigma_k^{(2)}f(k)\right] = {}_0^{GL}\Sigma_k^{(2)}f(k) - {}_0^{GL}\Sigma_{k-1}^{(2)}f(k-1)
$$

$$
\sum_{i=0}^{k}(i+1)f(k-i) - \sum_{i=0}^{k-1}(i+1)f(k-1-i)
$$

$$
\sum_{i=0}^{k}f(k-i) = \sum_{i=0}^{k}f(i) = {}_0^{GL}\Sigma_k^{(1)}f(k). \quad (2.112)
$$

Part b) is more simple because

$$
{}_0^{GL}\Delta_k^{(1)}\left[{}_0^{GL}\Sigma_k^{(1)}f(k)\right] = {}_0^{GL}\Sigma_k^{(1)}f(k) - {}_0^{GL}\Sigma_{k-1}^{(1)}f(k-1)
$$

$$
\sum_{i=0}^{k}f(k-i) - \sum_{i=0}^{k-1}f(k-1-i) = f(k). \quad (2.113)
$$

Lastly, for part c), one evaluates the second-order BD of the BS

$$
{}_0^{GL}\Delta_k^{(2)}\left[{}_0^{GL}\Sigma_k^{(1)}f(k)\right]
$$

$$
= {}_0^{GL}\Sigma_k^{(1)}f(k) - 2{}_0^{GL}\Sigma_{k-1}^{(1)}f(k-1) + {}_0^{GL}\Sigma_{k-2}^{(1)}f(k-2)
$$

$$
\sum_{i=0}^{k}f(k-i) - 2\sum_{i=0}^{k-1}f(k-1-i) + \sum_{i=0}^{k-2}f(k-2-i)
$$

$$
= f(k) - f(k-1) = {}_0^{GL}\Delta_k^{(1)}f(k). \quad (2.114)
$$

Now one considers the IOBDs of consecutive orders $n = 0, 1, 2, \ldots$ and collects them in one matrix-vector form

$$
\begin{bmatrix} \Delta^0 f(k) \\ \Delta^1 f(k) \\ \Delta^2 f(k) \\ \Delta^3 f(k) \\ \Delta^4 f(k) \\ \vdots \end{bmatrix} = \mathbf{B}_k \begin{bmatrix} f(k) \\ f(k-1) \\ f(k-2) \\ f(k-3) \\ f(k-4) \\ \vdots \end{bmatrix} \quad (2.115)
$$

where

$$\mathbf{B}_k = \begin{bmatrix} a^{(0)}(0) & 0 & 0 & 0 & 0 & 0 & \cdots \\ a^{(1)}(0) & a^{(1)}(1) & 0 & 0 & 0 & 0 & \cdots \\ a^{(2)}(0) & a^{(2)}(1) & a^{(2)}(2) & 0 & 0 & 0 & \cdots \\ a^{(3)}(0) & a^{(3)}(1) & a^{(3)}(2) & a^{(3)}(3) & 0 & 0 & \cdots \\ a^{(4)}(0) & a^{(4)}(1) & a^{(4)}(2) & a^{(4)}(3) & a^{(4)}(4) & 0 & \cdots \\ a^{(5)}(0) & a^{(5)}(1) & a^{(5)}(2) & a^{(5)}(3) & a^{(5)}(4) & a^{(5)}(5) & \cdots \\ \vdots & \vdots & \vdots & \vdots & \vdots & \vdots & \end{bmatrix}$$

$$= \begin{bmatrix} 1 & 0 & 0 & 0 & 0 & 0 & \cdots \\ 1 & -1 & 0 & 0 & 0 & 0 & \cdots \\ 1 & -2 & 1 & 0 & 0 & 0 & \cdots \\ 1 & -3 & 3 & -1 & 0 & 0 & \cdots \\ 1 & -4 & 6 & -4 & 1 & 0 & \cdots \\ 1 & -5 & 10 & -10 & 5 & -1 & \cdots \\ \vdots & \vdots & \vdots & \vdots & \vdots & \vdots & \end{bmatrix}. \tag{2.116}$$

Formula (2.115) is valid for $k = k_0$. This means that

$$\begin{bmatrix} \Delta^0 f(k) \\ \Delta^1 f(k) \\ \Delta^2 f(k) \\ \Delta^3 f(k) \\ \Delta^4 f(k) \\ \vdots \end{bmatrix}\Bigg|_{k=k_0} = \begin{bmatrix} \Delta^0 f(k_0) \\ \Delta^1 f(k_0) \\ \Delta^2 f(k_0) \\ \Delta^3 f(k_0) \\ \Delta^4 f(k_0) \\ \vdots \end{bmatrix} = \mathbf{B}_{k_0} \begin{bmatrix} f(k) \\ f(k-1) \\ f(k-2) \\ f(k-3) \\ f(k-4) \\ \vdots \end{bmatrix}\Bigg|_{k=k0} = \begin{bmatrix} f(k_0) \\ f(k_0-1) \\ f(k_0-2) \\ f(k_0-3) \\ f(k_0-4) \\ \vdots \end{bmatrix}. \tag{2.117}$$

Chapter 3

Fractional-order backward differ-sum

In the so-called fractional calculus one investigates derivatives and integrals of any finite real or complex orders [Ross (1974); Oldham and Spannier (1974); Oustaloup (1983); Nishimoto (1984, 1986, 1989, 1991a,b); Lubich (1986); Michalski (1993); Miller and Ross (1993); Samko *at al.* (1993); Kiryakova (1994); Oustaloup (1995); Rubin (1996); Gorenflo and Mainardi (1997a); Podlubny (1999a); Hilfer (2000); Kilbas *et al.* (2006a); Kaczorek (2011)]. There are several equivalent (under some specified conditions) definitions of the fractional-order (FO) derivatives (FOD) and integrals (FOI). Following the principles of classical calculus one may consider left-sided and right-sided FOD and FOI. Most of the literature concerning fractional calculus is focused on applications in a chosen field of science and technology [Oustaloup (1983, 1991, 1995); Oustaloup and Mathieu (1999); Kaczorek (2011); Luo and Chen (2012); Valério and Costa (2013a); Valerio *et al.* (2013b)], and is mainly related to continuous-time dynamic systems.

Over the last 60 years, information quantized horizontally (i.e. time) and vertically (i.e. voltage, speed, pressure, temperature etc.) has become more important. In practice, the quantization level (sampling time [Weeks (2011)] denoted usually by h or T_s) in a horizontal scale is much greater than in the vertical one (related mainly to the AD converter accuracy [Ifeachor and Jervis (1993)]). This is because with a satisfactory precision, one may assume its continuity. Hence, the analyses of the fractional-order backward difference (FOBD) and fractional-order backward-sum (FOBS) will be applied to the discrete-variable functions with values from a continuous set of real numbers. The FOBD and FOBS are counterpart of the FO derivative and integral, respectively.

Real-time calculations based on available current and previous information force the application of the backward difference and sum. This is related to causality of the analyzed physical or technical phenomena [Proakis (2007); Weeks (2011)]. Hence, there is a motivation why, in this book, only fractional-order backward-difference/sum (FOBD/S) will be considered. Assuming a quantization level as known and constant, one may omit the mentioned parameter in the analysis or put $h = 1$ in the discrete-variable function $f(kh)$, getting the more simpler notation $f(k)$.

In chapter five, equivalent forms (under some specified conditions) of the fractional-order backward-difference or sum (FOBD/S) are defined. In general, the FOBD/S in the Grünwald-Letnikov, Horner, Riemann-Liouville, Caputo, polynomial-matrix-like, and Laguerre-based forms are presented. Its equivalence is proved.

3.1 Grünwald-Letnikov form of the fractional-order backward difference

Consider a discrete-variable bounded real-valued function $f(k)$ defined over a discrete interval $[k_0, k]$. One starts with a fundamental one known as the Grünwald-Letnikov backward-difference (GL-FOBD) form.

Definition 3.1 (GL-FOBD form). The GL-FOBD of an order $\nu \in \mathbb{R}_+$ is defined as a finite sum

$$
{}^{GL}_{k_0}\Delta^{(\nu)}_k f(k) = \sum_{i=k_0}^{k} a^{(\nu)}(i - k_0)f(k + k_0 - i) = \sum_{i=0}^{k-k_0} a^{(\nu)}(i)f(k - i) \qquad (3.1)
$$

where the function $a^{(\nu)}(k)$ is defined by formula (1.11). The left and right subscripts in ${}^{GL}_{k_0}\Delta^{(\nu)}_k f(k)$ denote the FOBD evaluation range, respectively. The left and right superscripts stand for the form (Grünwald-Letnikov) and order ν, respectively. Parentheses surrounding ν stress the FOBD order, not an exponentiation operation.

It is evident that for $\nu = n \in \mathbb{Z}_+$ the GL-FOBD becomes the GL-IOBD discussed in Chapter 2. In many practical applications it is convenient to assume $k_0 = 0$. Consequently, the GL-FOBD may be expressed as a discrete convolution

$$
{}^{GL}_{0}\Delta^{(\nu)}_k f(k) = a^{(\nu)}(k) * f(k) = f(k) * a^{(\nu)}(k)
$$

$$
= \left[a^{(\nu)}(0)\ a^{(\nu)}(1) \cdots a^{(\nu)}(k-1)\ a^{(\nu)}(k) \right]
\begin{bmatrix} f(k) \\ f(k-1) \\ \cdots \\ f(1) \\ f(0) \end{bmatrix}. \qquad (3.2)
$$

To show the similarity between the GL-FOBD and the Grünwald-Letnikov FO left derivative its definition is recalled [Oustaloup (1995); Podlubny (1999a); Kilbas *et al.* (2006a)]

$$
{}^{GL}_{t_0}D^{(\nu)}_t f(t) = \lim_{\substack{h \to 0^+ \\ k = \left[\frac{t-t_0}{h}\right]}} \left\{ \frac{{}^{GL}_{k_0}\Delta^{(\nu)}_k f(k)}{h^\nu} \right\} \qquad (3.3)
$$

and $t_0 = k_0 h$. Analogously one defined the GL-FOBS.

Definition 3.2 (GL-FOBS form). The GL-FOBS of an order $\nu \in \mathbb{R}_+$ is defined as a finite sum

$$_{k_0}^{GL}\Sigma_k^{(\nu)} f(k) =_{k_0}^{GL} \Delta_k^{(-\nu)} f(k) = \sum_{i=k_0}^{k} a^{(-\nu)}(i - k_0)f(k + k_0 - i) \qquad (3.4)$$

where the function $a^{(-\nu)}(k)$ is defined by formula (1.11).

3.2 Horner form of the fractional-order backward difference

For a discrete-variable function $f(k)$ satisfying the same conditions as posed in Section 3.1., the Horner form of the FOBD (H-FOBD) is defined as below

Definition 3.3 (H-FOBD form). The H-FOBS of an order $\nu \in \mathbb{R}_+$ is defined by the formula

$$_{k_0}^{H}\Delta_k^{(\nu)} f(k) = f(k)$$
$$+ c^{(\nu)}(1) \left[f(k-1) + c^{(\nu)}(2) \left[f(k-2) + c^{(\nu)}(3) \left[f(k-3) + \cdots \right.\right.\right.$$
$$+ \left.\left.\left. c^{(\nu)}(k - k_0 - 1) \left[f(k_0 + 1) + c^{(\nu)}(k - k_0) \left[f(k_0) \right] \right] \cdots \right] \right] \right] \qquad (3.5)$$

where the function $c^{(\nu)}(k)$ is defined by formula (1.53). Realize that a sum of variables in functions $c^{(\nu)}(k - i)$ and $f(i)$ is always constant and equal to k. Direct calculations reveal immediately that

$$_{k_0}^{H}\Delta_k^{(\nu)} f(k) =_{k_0}^{GL} \Delta_k^{(\nu)} f(k). \qquad (3.6)$$

Hence, formula (3.5) is also valid for $k_0 = 0$ and integer orders $\nu = n \in \mathbb{Z}_+$

$$_{0}^{H}\Delta_k^{(n)} f(k) = \Delta^n f(k). \qquad (3.7)$$

In the following example a character of the Horner form will be considered.

Example 3.1. Express the IOBD and FOBD of orders $n = 4$ and $\nu = 4.1$ in the Horner form, respectively.

Solution. By the classical definition of the IOBD one has

$$\Delta^{(4)} f(k) = f(k) - 4f(k-1) + 6f(k-2) - 4f(k-3) + 1f(k-4). \qquad (3.8)$$

According to (1.53) one gets

$$c^{(4)}(0) = 1 \text{ and } c^{(4)}(k) = 1 - \frac{5}{k} \text{ for } k = 1, 2, \ldots. \qquad (3.9)$$

Table 3.1 Comparison of functions $a^{(4)}(k)$, $a^{(4.1)}(k)$, $c^{(4)}(k)$, $c^{(4.1)}(k)$ values.

k	$c^{(4)}(k)$	$a^{(4)}(k)$	$c^{(4.1)}(k)$	$a^{(4.1)}(k)$
0	1	1	1	1
1	-4	-4	-4.1	-4.1
2	-1.5000	6	-1.5500	6.3550
3	-0.6667	-4	-0.7000	-4.4485
4	-0.2500	1	-0.2750	1.2233
5	0	0	-0.0200	-0.0245
6	0.1667	0	0.1500	-0.0037
7	0.2857	0	0.2714	-0.0010
8	0.3750	0	0.3625	-0.0004
9	0.4444	0	0.4333	-0.0002
10	0.5000	0	0.4900	-0.0001

In Table 3.1, selected values of the functions $a^{(4)}(k), a^{(4.1)}(k), c^{(4)}(k), c^{(4.1)}(k)$ are collected.

Here a key role is played by the coefficient $c^{(4)}(5) = 0$ in the second column which cuts an influence to the H-IOBD calculation of all remaining terms, i.e.

$$
{}^{H}_{0}\Delta^{(4)}_{k} f(k)
$$

$$
= f(k) - 4\left[f(k-1) - \frac{3}{2}\left[f(k-2) - \frac{2}{3}\left[f(k-3) - \frac{1}{4}\left[f(k-4) \right] \right] \right] \right] \quad (3.10)
$$

which is contrary to the FO case where all $c^{(\nu)}(i) \neq 0$ for $i = 1, 2, \dots$.

3.3 Riemann-Liouville form of the fractional-order backward difference

Consider a discrete-variable function satisfying the same assumptions as those posed in the GL-FOBD. Regarding the order $\nu \in \mathbb{R}_{+} \setminus \mathbb{Z}_{+}$, one assumes that

$$
0 \leqslant n - 1 < \nu < n, \quad \text{where } n = \lfloor \nu \rfloor + 1. \quad (3.11)
$$

Remark 3.1. Realize that in light of (3.11), the following inequalities arise $0 < n - \nu \leqslant 1$.

The Riemann-Liouville FOBD (RL-FOBD) is defined as follows.

Definition 3.4 (RL-FOBD definition). The RL-FOBD of the discrete-variable function is defined as

$$
{}^{RL}_{k_0}\Delta^{(\nu)}_{k} f(k) = {}^{GL}_{k_0}\Delta^{(n)}_{k}\left[{}^{GL}_{k_0}\Sigma^{(n-\nu)}_{k} f(k) \right] = {}^{GL}_{k_0}\Delta^{(n)}_{k}\left[{}^{GL}_{k_0}\Delta^{(\nu-n)}_{k} f(k) \right]. \quad (3.12)
$$

To indicate that for $n \in \mathbb{Z}_+$ the defining formula (3.13) may be expressed as the classical n-th order BD of the FOBS of order $n - \nu$

$$\substack{RL\\k_0}\Delta_k^{(\nu)} f(k) = \Delta^n \left[\substack{GL\\k_0}\Sigma_k^{(n-\nu)} f(k) \right]. \tag{3.13}$$

The RL-FOBD has its continuous-function derivative counterpart known as the Riemann-Liouville fractional-order left-sided derivative of a continuous-variable function $f(t)$ defined over an interval $[t_0, t]$ [Kilbas *et al.* (2006a)]

$$\substack{RL\\t_0}D_t^{(\nu)} f(t) = \left(\frac{d}{dt} \right)^n \left[\frac{1}{\Gamma(n-\nu)} \int_{t_0}^{t} \frac{f(x)}{(t-x)^{\nu-n+1}} dx \right]. \tag{3.14}$$

3.4 Caputo form of the FOBD

The next definition of the FOBD is known as the Caputo form. For the same assumptions as those imposed above, the Caputo form of the FOBD (C-FOBD) is defined below.

Definition 3.5 (C-FOBD form). Let $f(k)$ be a discrete-variable function defined on a finite interval $[k_0, k]$ and let $\substack{GL\\k_0}\Delta_k^{(n)} f(k) = \Delta^n f(k)$ exists. Then for $n = \lfloor \nu \rfloor + 1$ the C-FOBD is defined as

$$\substack{C\\k_0}\Delta_k^{(\nu)} f(k) = \substack{GL\\k_0} \Delta_k^{(\nu-n)} [\Delta^n f(k)] = \substack{GL\\k_0} \Sigma_k^{(n-\nu)} [\Delta^n f(k)]. \tag{3.15}$$

This means that the Caputo FOBD is defined as the FOBD of the IOBD of the discrete-variable function $f(k)$. Similarly to the RL-FOBD there is a continuous-variable function FO left-sided derivative analogue. For a positive integer n

$$\substack{C\\t_0}D_t^{(\nu)} f(t) = \frac{1}{\Gamma(n-\nu)} \int_{t_0}^{t} \frac{f^{(n)}(x)}{(t-x)^{\nu-n+1}} dx. \tag{3.16}$$

Comment. The notation $f^{(n)}(t)$ justifies that used in the FOBD $\Delta^{(\nu)}$.

The equivalence conditions between the GL-FOBS, RL-FOBD and C-FOBD are established in the following theorem.

Theorem 3.1 (Equivalence of the GL-FOBD/S and H-FOBD/S). *The GL-FOBD/S is equivalent to the H-FOBD/S.*

Proof. There is a simple relation (1.54) between functions (1.11) and (1.53). Substitution of (1.54) into the H-FOBD/S yields, after simple rearrangements, the

GL-FOBD/S. Conversely, noting that

$$a^{(\nu)}(k) = \prod_{i=0}^{k} c^{(\nu)}(i) \tag{3.17}$$

and substituting it into the GL-FOBD/S after simple transformations gives the H-FOBD/S. $\qquad\square$

To establish a relation between the RL-FOBD, C-FOBD, and GL-FOBD defined over $[k_0, k]$, one needs the following Lemma.

Lemma 3.1. *The IOBDs of consecutive orders $n = 0, 1, 2, \ldots$ of the same discrete-variable function $f(k)$ and its shifted-back forms $f(k-i)$, for $i = 0, 1, 2, \ldots$ are related by a vector matrix equation with the Pascal involutory matrix (see Appendix, formula (A.6)*

$$\begin{bmatrix} \Delta^0 f(k) \\ \Delta^1 f(k) \\ \Delta^2 f(k) \\ \Delta^3 f(k) \\ \Delta^4 f(k) \\ \vdots \end{bmatrix} = \mathbf{P}_\infty \begin{bmatrix} f(k) \\ f(k-1) \\ f(k-2) \\ f(k-3) \\ f(k-4) \\ \vdots \end{bmatrix}. \tag{3.18}$$

Proof. Formula (3.18) is also valid for $k_0 - 1$, which means that

$$\begin{bmatrix} \Delta^0 f(k) \\ \Delta^1 f(k) \\ \Delta^2 f(k) \\ \Delta^3 f(k) \\ \Delta^4 f(k) \\ \vdots \end{bmatrix}_{k=k_0-1} = \begin{bmatrix} \Delta^0 f(k_0-1) \\ \Delta^1 f(k_0-1) \\ \Delta^2 f(k_0-1) \\ \Delta^3 f(k_0-1) \\ \Delta^4 f(k_0-1) \\ \vdots \end{bmatrix}$$

$$= \mathbf{P}_\infty \begin{bmatrix} f(k) \\ f(k-1) \\ f(k-2) \\ f(k-3) \\ f(k-4) \\ \vdots \end{bmatrix}_{k=k_0-1} = \mathbf{P}_\infty \begin{bmatrix} f(k_0-1) \\ f(k_0-2) \\ f(k_0-3) \\ f(k_0-4) \\ f(k_0-5) \\ \vdots \end{bmatrix} = \mathbf{P}_\infty \begin{bmatrix} f_{k_0} \\ f_{k_0-1} \\ f_{k_0-2} \\ f_{k_0-3} \\ f_{k_0-4} \\ \vdots \end{bmatrix}. \tag{3.19}$$

To emphasize that function values $f(k_0), f(k_0-1), \ldots$ are out of the IOBD range they are denoted as $f_{k_0}, f_{k_0-1}, \ldots$. Realize that vectors in (3.20) have an infinite number of elements. Values $f_{k_0}, f_{k_0-1}, \ldots$ will be further treated as initial conditions. The next Theorem states a relation between the three considered forms of the FOBDs. $\qquad\square$

Theorem 3.2. *Let $\nu \in \mathbb{R}_+ \setminus \mathbb{Z}_+$ and $n = \lfloor \nu \rfloor + 1$. If $\Delta f(k)$ exists, then the RL-FOBD can be expressed as*

$$
{}_{k_0}^{RL}\Delta_k^{(\nu)} f(k) = {}_{k_0}^{C}\Delta_k^{(\nu)} f(k) + \sum_{i=0}^{n-1} a^{(\nu-i)}(k - k_0)\Delta^{(i)} f(k_0 - 1). \tag{3.20}
$$

Proof. Three terms in formula (3.20) will be denoted as L, R_1, and R_2, respectively. So $L = R_1 + R_2$. By the C-FOBD definition formula (3.15), L is transformed successively as follows

$$
L = {}_{k_0}^{RL}\Delta_k^{(\nu)} f(k) = \Delta^n \left[{}_{k_0}^{GL}\Delta_k^{(\nu-n)} f(k) \right] = \Delta^n \left[{}_{k_0}^{GL}\Sigma_k^{(n-\nu)} f(k) \right]
$$

$$
= \left[a^{(n)}(0)\ a^{(n)}(1)\ \cdots\ a^{(n)}(n-1)\ a^{(n)}(n) \right]
\begin{bmatrix}
{}_{k_0}^{GL}\Sigma_k^{(n-\nu)} f(k) \\
{}_{k_0}^{GL}\Sigma_{k-1}^{(n-\nu)} f(k-1) \\
\vdots \\
{}_{k_0}^{GL}\Sigma_{k-n+1}^{(n-\nu)} f(k-n+1) \\
{}_{k_0}^{GL}\Sigma_{k-n}^{(n-\nu)} f(k-n)
\end{bmatrix}
$$

$$
= \left[a^{(n)}(0)\ a^{(n)}(1)\ \cdots\ a^{(n)}(n-1)\ a^{(n)}(n) \right]
$$

$$
\times
\begin{bmatrix}
a^{(\nu-n)}(0) & a^{(\nu-n)}(1) & \cdots & a^{(\nu-n)}(k-k_0) \\
0 & a^{(\nu-n)}(0) & \cdots & a^{(\nu-n)}(k-k_0-1) \\
\vdots & \vdots & & \vdots \\
0 & 0 & & a^{(\nu-n)}(k-n-k_0-1) \\
0 & 0 & \cdots & a^{(\nu-n)}(k-n-k_0)
\end{bmatrix}
\begin{bmatrix}
f(k) \\
f(k-1) \\
\vdots \\
f(k-k_0+1) \\
f(k-k_0)
\end{bmatrix}
$$

$$
= \left[a^{(\nu)}(0)\ a^{(\nu)}(1)\ \cdots\ a^{(\nu)}(k-k_0-1)\ a^{(\nu)}(k-k_0) \right]
\begin{bmatrix}
f(k) \\
f(k-1) \\
\vdots \\
f(k_0+1) \\
f(k_0)
\end{bmatrix}. \tag{3.21}
$$

The first term on the right-hand side of (3.21) equates to

$$
R_1 = {}_{k_0}^{GL}\Sigma_k^{(n-\nu)} \left[\Delta^n f(k) \right]
$$

$$
= \left[a^{(\nu-n)}(0)\ \cdots\ a^{(\nu-n)}(k-k_0-1)\ a^{(\nu-n)}(k-k_0) \right]
\begin{bmatrix}
\Delta^n f(k) \\
\Delta^n f(k-1) \\
\vdots \\
\Delta^n f(k_0+1) \\
\Delta^n f(k_0)
\end{bmatrix}
$$

$$
= \left[a^{(\nu-n)}(0)\ a^{(\nu-n)}(1)\ \cdots\ a^{(\nu-n)}(k-k_0-1)\ a^{(\nu-n)}(k-k_0) \right]
$$

$$\times \begin{bmatrix} a^{(n)}(0) \ a^{(n)}(1) \cdots & a^{(n)}(n) & 0 & \cdots & 0 & \cdots & 0 & 0 \\ 0 & a^{(n)}(0) \cdots a^{(n)}(n-1) \ a^{(n)}(n) & \cdots & 0 & \cdots & 0 & 0 \\ \vdots & \vdots & \vdots & \vdots & & \vdots & & \vdots & \vdots \\ 0 & 0 \cdots & 0 & 0 & \cdots a^{(n)}(0) \cdots & a^{(n)}(n) & 0 \\ 0 & 0 \cdots & 0 & 0 & \cdots & 0 & \cdots a^{(n)}(n-1) \ a^{(n)}(n) \end{bmatrix}$$

$$\times \begin{bmatrix} f(k) \\ f(k-1) \\ \vdots \\ f(k_0+1) \\ f(k_0) \\ f_{k_0-1} \\ f_{k_0-2} \\ \vdots \\ f_{k_0-n} \end{bmatrix}$$

$$= [a^{(\nu-n)}(0) \ a^{(\nu-n)}(1) \cdots a^{(\nu-n)}(k-k_0)] \, \mathbf{A}_{k-k_0}^{(n)} \begin{bmatrix} f(k) \\ f(k-1) \\ \vdots \\ f(k_0+1) \\ f(k_0) \end{bmatrix}$$

$$+ [a^{(\nu-n)}(0) \ a^{(\nu-n)}(1) \cdots a^{(\nu-n)}(k-k_0)] \, \mathbf{A}_{k-k_0,n-1}^{(n)} \begin{bmatrix} f_{k_0-1} \\ f_{k_0-2} \\ \vdots \\ f_{k_0-n} \end{bmatrix}.$$

$$(3.22)$$

Now one investigates the first addend in the latest form of R_1.

$$[a^{(\nu-n)}(0) \ a^{(\nu-n)}(1) \cdots a^{(\nu-n)}(k-k_0)] \, \mathbf{A}_{k-k_0}^{(n)} \begin{bmatrix} f(k) \\ f(k-1) \\ \vdots \\ f(k_0+1) \\ f(k_0) \end{bmatrix}$$

$$[a^{(\nu)}(0) \ a^{(\nu)}(1) \cdots a^{(\nu)}(k-k_0)] \begin{bmatrix} f(k) \\ f(k-1) \\ \vdots \\ f(k_0+1) \\ f(k_0) \end{bmatrix} = L. \qquad (3.23)$$

The second addend in (3.22) is transformed to the form

$$\left[a^{(\nu-n)}(0) \; a^{(\nu-n)}(1) \; \cdots \; a^{(\nu-n)}(k-k_0)\right] \mathbf{A}^{(n)}_{k-k_0,n-1} \begin{bmatrix} f_{k_0-1} \\ f_{k_0-2} \\ \vdots \\ f_{k_0-n} \end{bmatrix}$$

$$= \left[a^{(\nu-n)}(0) \; a^{(\nu-n)}(1) \; \cdots \; a^{(\nu-n)}(k-k_0)\right] \mathbf{A}^{(n)}_{k-k_0,n-1} \mathbf{P}_n \mathbf{P}_n \begin{bmatrix} f_{k_0-1} \\ f_{k_0-2} \\ \vdots \\ f_{k_0-n} \end{bmatrix}$$

$$= \left[a^{(\nu-n)}(0) \; \cdots \; a^{(\nu-n)}(k-k_0)\right] \mathbf{A}^{(n)}_{k-k_0,n-1} \mathbf{P}_n \begin{bmatrix} \Delta^0 f(k_0-1) \\ \Delta^1 f(k_0-1) \\ \vdots \\ \Delta^{n-1} f(k_0-1) \end{bmatrix}. \quad (3.24)$$

Straightforward but tedious calculations reveal that

$$\left[a^{(\nu-n)}(0) \; a^{(\nu-n)}(1) \; \cdots \; a^{(\nu-n)}(k-k_0)\right] \mathbf{A}^{(n)}_{k-k_0,n-1} \mathbf{P}_n$$

$$= -\left[a^{(\nu)}(k-k_0) \; a^{(\nu-1)}(k-k_0) \; \cdots \; a^{(\nu-n+1)}(k-k_0)\right]. \quad (3.25)$$

Substituting this result into (3.24) yields

$$\left[a^{(\nu-n)}(0) \; a^{(\nu-n)}(1) \; \cdots \; a^{(\nu-n)}(k-k_0)\right] \mathbf{A}^{(n)}_{k-k_0,n-1} \begin{bmatrix} f_{k_0-1} \\ f_{k_0-2} \\ \vdots \\ f_{k_0-n} \end{bmatrix}$$

$$= -\left[a^{(\nu)}(k-k_0) \; a^{(\nu-1)}(k-k_0) \; \cdots \; a^{(\nu-n+1)}(k-k_0)\right] \begin{bmatrix} \Delta^0 f(k_0-1) \\ \Delta^1 f(k_0-1) \\ \vdots \\ \Delta^{n-1} f(k_0-1) \end{bmatrix}$$

$$= -\sum_{i=0}^{n-1} a^{(\nu-i)}(k-k_0) \Delta^i f(k_0-1) = -R_2.$$

$$(3.26)$$

It was proved that $R_1 = L - R_2$, which confirms equality (3.20).

$\square$

Comment. Equality (3.20) can be also expressed in the form

$$
{}^{GL}_{k_0}\Sigma_k^{(n-\nu)}\left[\Delta^n f(k)\right]
$$

$$
= \Delta^n\left[{}^{GL}_{k_0}\Sigma_k^{(n-\nu)}f(k)\right] - \sum_{i=0}^{n-1} a^{(\nu)}(i)f(k_0 - i). \tag{3.27}
$$

The RL-FOBD and C-FOBD relation is similar to that valid for the Riemann-Liouville and Caputo FO left-sided derivative of a continuous-variable function $f(t)$

$$
{}^{RL}_{t_0}D_t^{(\nu)}f(t) = {}^{C}_{t_0}D_t^{(\nu)}f(t) + \sum_{i=0}^{n-1}\frac{(t-t_0)^{i-\nu}}{\Gamma(i+1-\nu)}f^{(i)}(t_0). \tag{3.28}
$$

Assuming that $\Delta^i f(k_0 - 1) = 0$ for $i = 0, 1, 2, ..., n-1$, all considered FOBDs are equal

$$
{}^{GL}_{k_0}\Delta_k^{(\nu)}f(k) = {}^{H}_{k_0}\Delta_k^{(\nu)}f(k) = {}^{RL}_{k_0}\Delta_k^{(\nu)}f(k) = {}^{C}_{k_0}\Delta_k^{(\nu)}f(k). \tag{3.29}
$$

3.5   Polynomial matrix like form of the FOBD (PML-BD)

The next form is strictly related to the GL-FOBD/S form (3.2) which is also valid for any $k-1, k-2, \ldots, k_0+1, k_0$. Adopting the notation used in the IOBD (2.15) in a matrix-vector form one gets an analogous description of the FOBD/S to that described by (2.15)–(2.17), where the integer order n is replaced by any real order $\nu \in \mathbb{R}_+$. Similar to the integer order example, to indicate an opposite action of the FOBD and FOBS, the latter will be denoted as the FOBD of a negative order.

$$
{}^{GL}_{k_0}\Delta_k^{(\nu)}\mathbf{f}(k) = {}^{GL}_{k_0}\Sigma_k^{(-\nu)}\mathbf{f}(k) \tag{3.30}
$$

and

$$
{}^{GL}_{k_0}\Delta_k^{(-\nu)}\mathbf{f}(k) = {}^{GL}_{k_0}\Sigma_k^{(\nu)}\mathbf{f}(k). \tag{3.31}
$$

The matrix $\mathbf{A}_{k-k_0}^{(\nu)}$ is always non-singular because all elements on the main diagonal $a^{(\nu)}(0) = 1$ for all orders. Therefore,

$$
\left[\mathbf{A}_{k-k_0}^{(\nu)}\right]^{-1}{}^{GL}_{k_0}\Delta_k^{(\nu)}\mathbf{f}(k) = \left[\mathbf{A}_{k-k_0}^{(-\nu)}\right]^{-1}\mathbf{A}_{k-k_0}^{(\nu)}\mathbf{f}(k) = \mathbf{f}(k). \tag{3.32}
$$

All elements of the set of matrices $\mathfrak{A} = \{\mathbf{A}_{k-k_0}^{(\nu)}, \nu \in \mathbb{R}\}$ satisfy the following properties

(a) For any two elements $\mathbf{A}_{k-k_0}^{(\nu_1)}$, $\mathbf{A}_{k-k_0}^{(\nu_2)} \in \mathfrak{A}$, the composition rule is satisfied (its product belongs to $\mathfrak{A}$)

$$\mathbf{A}_{k-k_0}^{(\nu_1)} \mathbf{A}_{k-k_0}^{(\nu_2)} = \mathbf{A}_{k-k_0}^{(\nu_1+\nu_2)} \in \mathfrak{A}. \tag{3.33}$$

(b) For any three elements $\mathbf{A}_{k-k_0}^{(\nu_i)} \in \mathfrak{A}$ for $i = 1, 2, 3$, the associative rule is satisfied

$$\left[\mathbf{A}_{k-k_0}^{(\nu_1)} \mathbf{A}_{k-k_0}^{(\nu_2)} \right] \mathbf{A}_{k-k_0}^{(\nu_3)} = \mathbf{A}_{k-k_0}^{(\nu_1)} \left[\mathbf{A}_{k-k_0}^{(\nu_2)} \mathbf{A}_{k-k_0}^{(\nu_3)} \right]. \tag{3.34}$$

(c) A set $\mathfrak{A}$ contains the unit element $\mathbf{A}_{k-k_0}^{(0)}$ such that

$$\mathbf{A}_{k-k_0}^{(\nu)} \mathbf{A}_{k-k_0}^{(0)} = \mathbf{A}_{k-k_0}^{(0)} \mathbf{A}_{k-k_0}^{(\nu)} = \mathbf{A}_{k-k_0}^{(\nu)}. \tag{3.35}$$

(d) For every element of the set $\mathfrak{A}$, there exists an inverse element such that

$$\mathbf{A}_{k-k_0}^{(\nu_1)} \mathbf{A}_{k-k_0}^{(\nu_2)} = \mathbf{A}_{k-k_0}^{(0)} \tag{3.36}$$

(evidently $\mathbf{A}_{k-k_0}^{(\nu_2)} = \mathbf{A}_{k-k_0}^{(-\nu_1)}$).

Comment. As a consequence of (3.36) one has

$$\left[\mathbf{A}_{k-k_0}^{(\nu)} \right]^{-1} = \mathbf{A}_{k-k_0}^{(-\nu)}. \tag{3.37}$$

and the set $\mathfrak{A}$ is a group [Bayin (2006)]. Contrary to matrix (2.18), in the matrix $\mathbf{A}_{k-k_0}^{(\nu)}$ all elements on and above the main diagonal are non-zero.

Using the above information, the polynomial matrix-like form of the FOBD/S (PML-BD/S) is defined as follows

Definition 3.6. The polynomial matrix-like form of the FOBS/S of a discrete-variable function having the GL-FOB/S is defined by a matrix-vector equation

$$\underset{k_0}{\overset{GL}{}}\Delta_k^{(\pm\nu)} \mathbf{f}(k) = \mathbf{A}_{k-k_0}^{(\pm\nu)} \mathbf{f}(k). \tag{3.38}$$

At this point, one should refer to the terminology used above. The proposed notation is very similar to that used in the so-called polynomial-matrix description of control systems [Kailath (1980)]. The similarity will be clear in subsequent chapters. The proposed form appears to be very useful in the FOBD equation solution.

3.6 Laguerre-based form of the FOBD

The Laguerre-based form of the FOBD (L-FOBD) of a discrete-variable function definition arises from the one-sided $\mathcal{Z}$-Transform FOBD expanded into an infinite series [Stanisławski (2013)]

$$\underset{k_0}{\overset{L}{}}\Delta_k^{(\nu)}\mathbf{f}(k) = \mathbf{f}(k) + \sum_{i=1}^{\infty} l^{(\nu)}(i)\mathbf{L}(k,i,p)\mathbf{f}(k) \tag{3.39}$$

where

$$l^{(\nu)}(i) = \lim_{k \to \infty} \langle \mathbf{a}^{(\nu)}(k), \mathbf{L}(k,i,p)\delta(k)\rangle \tag{3.40}$$

$$\mathbf{L}(k,i,p) = \sqrt{1-p^2}\,[\mathbf{E}(k,p)]^{-i}\,\mathbf{F}(k)\,[\mathbf{H}(k,p)]^{i-1} \tag{3.41}$$

$$\mathbf{E}(k,p) = \mathbf{1}_k + \begin{bmatrix} \mathbf{0}_{k-1} & -p\mathbf{1}_{k-1} \\ 0 & [\mathbf{0}_{k-1}]^\mathrm{T} \end{bmatrix} \tag{3.42}$$

$$\mathbf{F}(k) = \mathbf{1}_k + \begin{bmatrix} \mathbf{0}_{k-1} & \mathbf{1}_{k-1} \\ 0 & [\mathbf{0}_{k-1}]^\mathrm{T} \end{bmatrix} \tag{3.43}$$

$$\mathbf{H}(k,p) = -p\mathbf{1}_k + \begin{bmatrix} \mathbf{0}_{k-1} & \mathbf{1}_{k-1} \\ 0 & [\mathbf{0}_{k-1}]^\mathrm{T} \end{bmatrix} \tag{3.44}$$

and $\delta(k) = [0\ 0 \cdots 0\ 1]^\mathrm{T}$ is $k \times 1$ column vector and $p \in (-1,1)$ is a constant. Details of the parameter p selection are described in [Stanisławski (2013)]. The results formulated by following theorems serve as a useful tool for the L-FOBD numerical evaluation.

Theorem 3.3. *For the upper triangular band matrix (3.43) its i-th power is equal*

$$\mathbf{E}^i(k,p) = \begin{bmatrix} a^{(i)}(0)p^0 & a^{(i)}(1)p^1 & a^{(i)}(2)p^2 & \cdots & a^{(i)}(k-1)p^{k-1} & a^{(i)}(k)p^k \\ 0 & a^{(i)}(0)p^0 & a^{(i)}(1)p^1 & \cdots & a^{(i)}(k-2)p^{k-2} & a^{(i)}(k-1)p^{k-1} \\ 0 & 0 & a^{(i)}(0)p^0 & \cdots & a^{(i)}(k-3)p^{k-3} & a^{(i)}(k-2)p^{k-2} \\ \vdots & \vdots & \vdots & & \vdots & \vdots \\ 0 & 0 & 0 & \cdots & a^{(i)}(0)p^0 & a^{(i)}(1)p^1 \\ 0 & 0 & 0 & \cdots & 0 & a^{(i)}(0)p^0 \end{bmatrix}$$

$$\tag{3.45}$$

whereas the i-th power of its inverse is of the form

$$\left[\mathbf{E}^i(k,p)\right]^{-1} = \mathbf{E}^{-i}(k,p)$$

$$= \begin{bmatrix} a^{(-i)}(0)p^0 & a^{(-i)}(1)p^1 & a^{(-i)}(2)p^2 & \cdots & a^{(-i)}(k-1)p^{k-1} & a^{(-i)}(k)p^k \\ 0 & a^{(-i)}(0)p^0 & a^{(-i)}(1)p^1 & \cdots & a^{(-i)}(k-2)p^{k-2} & a^{(-i)}(k-1)p^{k-1} \\ 0 & 0 & a^{(-i)}(0)p^0 & \cdots & a^{(-i)}(k-3)p^{k-3} & a^{(-i)}(k-2)p^{k-2} \\ \vdots & \vdots & \vdots & & \vdots & \vdots \\ 0 & 0 & 0 & \cdots & a^{(-i)}(0)p^0 & a^{(-i)}(1)p^1 \\ 0 & 0 & 0 & \cdots & 0 & a^{(-i)}(0)p^0 \end{bmatrix}.$$

$$(3.46)$$

Proof. Performing consecutive multiplications of the considered matrices, one obtains the final result

$$\mathbf{E}(k,p)\mathbf{E}(k,p) = \mathbf{E}^2(k,p) = \begin{bmatrix} 1 & -p & 0 & \cdots & 0 & 0 \\ 0 & 1 & -p & \cdots & 0 & 0 \\ 0 & 0 & 1 & \cdots & 0 & 0 \\ \vdots & \vdots & \vdots & & \vdots & \vdots \\ 0 & 0 & 0 & \cdots & 1 & -p \\ 0 & 0 & 0 & \cdots & 0 & 1 \end{bmatrix} \begin{bmatrix} 1 & -p & 0 & \cdots & 0 & 0 \\ 0 & 1 & -p & \cdots & 0 & 0 \\ 0 & 0 & 1 & \cdots & 0 & 0 \\ \vdots & \vdots & \vdots & & \vdots & \vdots \\ 0 & 0 & 0 & \cdots & 1 & -p \\ 0 & 0 & 0 & \cdots & 0 & 1 \end{bmatrix}$$

$$= \begin{bmatrix} 1 & -2p & p^2 & 0 & \cdots & 0 & 0 \\ 0 & 1 & -2p & p^2 & \cdots & 0 & 0 \\ 0 & 0 & 1 & -2p & \cdots & 0 & 0 \\ \vdots & \vdots & \vdots & \vdots & & \vdots & \vdots \\ 0 & 0 & 0 & 0 & \cdots & 1 & -2p \\ 0 & 0 & 0 & 0 & \cdots & 0 & 1 \end{bmatrix}$$

$$= \begin{bmatrix} a^{(2)}(0) & a^{(2)}(1)p & a^{(2)}(2)p^2 & 0 & \cdots & 0 & 0 \\ 0 & a^{(2)}(0) & a^{(2)}(1)p & a^{(2)}(2)p^2 & \cdots & 0 & 0 \\ 0 & 0 & a^{(2)}(0) & a^{(2)}(1)p & \cdots & 0 & 0 \\ \vdots & \vdots & \vdots & \vdots & & \vdots & \vdots \\ 0 & 0 & 0 & 0 & \cdots & a^{(2)}(0) & a^{(2)}(1)p \\ 0 & 0 & 0 & 0 & \cdots & 0 & a^{(2)}(0) \end{bmatrix}$$

$$= \begin{bmatrix} a^{(2)}(0) & a^{(2)}(1)p & a^{(2)}(2)p^2 & a^{(2)}(3)p^3 & \cdots & a^{(2)}(k-1)p^{k-1} & a^{(2)}(k)p^k \\ 0 & a^{(2)}(0) & a^{(2)}(1)p & a^{(2)}(2)p^2 & \cdots & a^{(2)}(k-2)p^{k-2} & a^{(2)}(k-1)p^{k-1} \\ 0 & 0 & a^{(2)}(0) & a^{(2)}(1)p & \cdots & a^{(2)}(k-3)p^{k-3} & a^{(2)}(k-2)p^{k-2} \\ \vdots & \vdots & \vdots & \vdots & & \vdots & \vdots \\ 0 & 0 & 0 & 0 & \cdots & a^{(2)}(0) & a^{(2)}(1)p \\ 0 & 0 & 0 & 0 & \cdots & 0 & a^{(2)}(0) \end{bmatrix}.$$

$$(3.47)$$

In the last step all zeros above the main diagonal are substituted by $a^{(2)}(i) = 0$ for $i = 3, 4, \ldots, k-1, k$. Continuing multiplications one gets the final form. To prove formula (3.47) one realizes that

$$
\mathbf{E}^{-1}(k,p) =
\begin{bmatrix}
1 & p & p^2 & \cdots & p^{k-1} & p^k \\
0 & 1 & p & \cdots & p^{k-2} & p^{k-1} \\
0 & 0 & 1 & \cdots & p^{k-3} & p^{k-2} \\
\vdots & \vdots & \vdots & & \vdots & \vdots \\
0 & 0 & 0 & \cdots & 1 & p \\
0 & 0 & 0 & \cdots & 0 & 1
\end{bmatrix}.
\tag{3.48}
$$

This is true because

$$
\mathbf{E}(k,p)\mathbf{E}^{-1}(k,p) =
\begin{bmatrix}
1 & -p & 0 & \cdots & 0 & 0 \\
0 & 1 & -p & \cdots & 0 & 0 \\
0 & 0 & 1 & \cdots & 0 & 0 \\
\vdots & \vdots & \vdots & & \vdots & \vdots \\
0 & 0 & 0 & \cdots & 1 & -p \\
0 & 0 & 0 & \cdots & 0 & 1
\end{bmatrix}
\begin{bmatrix}
1 & p & p^2 & \cdots & p^{k-1} & p^k \\
0 & 1 & p & \cdots & p^{k-2} & p^{k-1} \\
0 & 0 & 1 & \cdots & p^{k-3} & p^{k-2} \\
\vdots & \vdots & \vdots & & \vdots & \vdots \\
0 & 0 & 0 & \cdots & 1 & p \\
0 & 0 & 0 & \cdots & 0 & 1
\end{bmatrix}
= \mathbf{1}_k.
\tag{3.49}
$$

Matrix (3.49) may be also expressed in the form

$$
\mathbf{E}^{-1}(k,p) =
\begin{bmatrix}
a^{(-1)}(0) & a^{(-1)}(1)p & a^{(-1)}(2)p^2 & \cdots & a^{(-1)}(k-1)p^{k-1} & a^{(-1)}(k)p^k \\
0 & a^{(-1)}(0) & a^{(-1)}(1)p & \cdots & a^{(-1)}(k-2)p^{k-2} & a^{(-1)}(k-1)p^{k-1} \\
0 & 0 & a^{(-1)}(0) & \cdots & a^{(-1)}(k-3)p^{k-3} & a^{(-1)}(k-2)p^{k-2} \\
\vdots & \vdots & \vdots & & \vdots & \vdots \\
0 & 0 & 0 & \cdots & a^{(-1)}(0) & a^{(-1)}(1)p \\
00 & 0 & 0 & \cdots & 0 & a^{(-1)}(0)
\end{bmatrix}.
\tag{3.50}
$$

Now one performs a multiplication

$$
\mathbf{E}^{-2}(k,p) = \mathbf{E}^{-1}(k,p)\mathbf{E}^{-1}(k,p)
$$

$$
=
\begin{bmatrix}
a^{(-2)}(0) & a^{(-2)}(1)p & a^{(-2)}(2)p^2 & \cdots & a^{(-2)}(k-1)p^{k-1} & a^{(-2)}(k)p^k \\
0 & a^{(-2)}(0) & a^{(-2)}(1)p & \cdots & a^{(-2)}(k-2)p^{k-2} & a^{(-2)}(k-1)p^{k-1} \\
0 & 0 & a^{(-2)}(0) & \cdots & a^{(-2)}(k-3)p^{k-3} & a^{(-2)}(k-2)p^{k-2} \\
\vdots & \vdots & \vdots & & \vdots & \vdots \\
0 & 0 & 0 & \cdots & a^{(-2)}(0) & a^{(-2)}(1)p \\
0 & 0 & 0 & \cdots & 0 & a^{(-2)}(0)
\end{bmatrix}.
\tag{3.51}
$$

Continuing with multiplications of matrices, leads to the result (3.46).

$\square$

The next theorem is also useful in numerical evaluation of the L-FOBD.

Theorem 3.4. *For an upper triangular band matrix (3.45), its i-th power is equal to*

$$\mathbf{H}^i(k,p)$$

$$= (-1)^i \begin{bmatrix} a^{(i)}(0)p^i & a^{(i)}(1)p^{i-1} & a^{(i)}(2)p^{i-2} & \cdots & a^{(i)}(k-10)p^{i-k+1} & a^{(i)}(k)p^{i-k} \\ 0 & a^{(i)}(0)p^i & a^{(i)}(1)p^{i-1} & \cdots & a^{(i)}(i-k+2)p^2 & a^{(i)}(i-k+1)p^1 \\ 0 & 0 & a^{(i)}(0)p^i & \cdots & a^{(i)}(k-3)p^{i-k+3} & a^{(i)}(k-2)p^{i-k+2} \\ \vdots & \vdots & \vdots & & \vdots & \vdots \\ 0 & 0 & 0 & \cdots & a^{(i)}(0)p^i & a^{(i)}(1)p^{i-1} \\ 0 & 0 & 0 & \cdots & 0 & a^{(i)}(0)p^i \end{bmatrix}$$

$$(3.52)$$

whereas the i-th power of its inverse has the form

$$\left[\mathbf{H}^i(k,p)\right]^{-1} = \mathbf{H}^{-i}(k,p)$$

$$= (p)^{-i} \begin{bmatrix} a^{(-i)}(0)p^0 & a^{(-i)}(1)p^{-1} & a^{(-i)}(2)p^{-2} & \cdots & a^{(-i)}(k-1)p^{1-k} & a^{(-i)}(k)p^{-k} \\ 0 & a^{(-i)}(0)p^0 & a^{(-i)}(1)p^{-1} & \cdots & a^{(-i)}(k-2)p^{2-k} & a^{(-i)}(k-1)p^{1-k} \\ 0 & 0 & a^{(-i)}(0)p^0 & \cdots & a^{(-i)}(k-3)p^{3-k} & a^{(-i)}(k-2)p^{2-k} \\ \vdots & \vdots & \vdots & & \vdots & \vdots \\ 0 & 0 & 0 & \cdots & a^{(-i)}(0)p^0 & a^{(-i)}(1)p^{-1} \\ 0 & 0 & 0 & \cdots & 0 & a^{(-i)}(0)p^0 \end{bmatrix}.$$

$$(3.53)$$

Proof. Matrix (3.45) can be expressed as

$$\mathbf{H}(k,p) = -p\mathbf{E}(k,p^{-1}). \qquad (3.54)$$

Hence,

$$\mathbf{H}^i(k,p) = (-p)^i \mathbf{E}^i(k,p^{-1})$$

$$= (-p)^i \begin{bmatrix} a^{(i)}(0)p^0 & a^{(i)}(1)p^{-1} & a^{(i)}(2)p^{-2} & \cdots & a^{(i)}(k-1)p^{1-k} & a^{(i)}(-k)p^{-k} \\ 0 & a^{(i)}(0)p^0 & a^{(i)}(1)p^{-1} & \cdots & a^{(i)}(k-2)p^{2-k} & a^{(i)}(k-1)p^{1-k} \\ 0 & 0 & a^{(i)}(0)p^0 & \cdots & a^{(i)}(k-3)p^{3-k} & a^{(i)}(k-2)p^{2-k} \\ \vdots & \vdots & \vdots & & \vdots & \vdots \\ 0 & 0 & 0 & \cdots & a^{(i)}(0)p^0 & a^{(i)}(1)p^{-1} \\ 0 & 0 & 0 & \cdots & 0 & a^{(i)}(0)p^0 \end{bmatrix}$$

$$= (-1)^i \begin{bmatrix} a^{(i)}(0)p^i & a^{(i)}(1)p^{i-1} & a^{(i)}(2)p^{i-2} & \cdots & a^{(i)}(k-1)p^{i+1-k} & a^{(i)}(-k)p^{i-k} \\ 0 & a^{(i)}(0)p^i & a^{(i)}(1)p^{i-1} & \cdots & a^{(i)}(k-2)p^{i+2-k} & a^{(i)}(k-1)p^{i+1-k} \\ 0 & 0 & a^{(i)}(0)p^i & \cdots & a^{(i)}(k-3)p^{i+3-k} & a^{(i)}(k-2)p^{i+2-k} \\ \vdots & \vdots & \vdots & & \vdots & \vdots \\ 0 & 0 & 0 & \cdots & a^{(i)}(0)p^i & a^{(i)}(1)p^{i-1} \\ 0 & 0 & 0 & \cdots & 0 & a^{(i)}(0)p^i \end{bmatrix}.$$

$$(3.55)$$

The second result of this theorem will be proved, noting that

$$[\mathbf{H}(k,p)]^{-1} = -p^{-1} \left[\mathbf{E}(k,p^{-1}) \right]^{-1} \tag{3.56}$$

and more generally

$$[\mathbf{H}(k,p)]^{-i} = (-p)^{-i} \left[\mathbf{E}(k,p^{-1}) \right]^{-i}. \tag{3.57}$$

Applying result (3.47) to (3.58) yields

$$[\mathbf{H}(k,p)]^{-i}$$

$$= (-p)^{-i} \begin{bmatrix} a^{(-i)}(0)p^0 & a^{(-i)}(1)p^{-1} & \cdots & a^{(-i)}(k-1)p^{1-k} & a^{(-i)}(k)p^{-k} \\ 0 & a^{(-i)}(0)p^0 & \cdots & a^{(-i)}(k-2)p^{2-k} & a^{(-i)}(k-1)p^{1-k} \\ 0 & 0 & \cdots & a^{(-i)}(k-3)p^{3-k} & a^{(-i)}(k-2)p^{2-k} \\ \vdots & \vdots & & \vdots & \vdots \\ 0 & 0 & \cdots & a^{(-i)}(0)p^0 & a^{(-i)}(1)p^{-1} \\ 0 & 0 & \cdots & 0 & a^{(-i)}(0)p^0 \end{bmatrix}$$

$$= (-1)^i \begin{bmatrix} a^{(-i)}(0)p^{-i} & a^{(-i)}(1)p^{-i-1} & \cdots & a^{(-i)}(-i+k-1)p^{1-k} & a^{(-i)}(k)p^{-i-k} \\ 0 & a^{(-i)}(0)p^{-i} & \cdots & a^{(-i)}(k-2)p^{-i+2-k} & a^{(-i)}(k-1)p^{-i+1-k} \\ 0 & 0 & \cdots & a^{(-i)}(k-3)p^{-i+3-k} & a^{(-i)}(k-2)p^{-i+2-k} \\ \vdots & \vdots & & \vdots & \vdots \\ 0 & 0 & \cdots & a^{(-i)}(0)p^{-i} & a^{(-i)}(1)p^{-i-1} \\ 0 & 0 & \cdots & 0 & a^{(-i)}(0)p^{-i} \end{bmatrix}.$$

$$(3.58)$$

□

3.7 Grünwald-Letnikov partial FOBD/S

Consider two discrete-variable function $f(k_1, k_2) \in \mathbb{R}$ with $k_1, k_2 \in \mathbb{Z}_+$. It is clear that for $k_2 = \text{const}$ one can evaluate the FOBD of order $\nu_1 \in \mathbb{R}_+$ with respect to the variable k_1 obtaining a new two discrete-variable function $g(k_1, k_2)$

$${}^{GL}_0 \Delta^{(\nu_2)}_{k_1} y(k_1, k_2) = \sum_{i_1=0}^{k_1} a^{(\nu_1)}(i_1) y(k_1 - i_1, k_2) = g(k_1, k_2). \tag{3.59}$$

To simplify the notation, one assumes here that $k_0 = 0$. In turn, for this new function one can further calculate the FOBD of order $\nu_2 \in \mathbb{R}_+$ with respect to a second discrete-variable k_2 (while treating discrete variable k_1 as a constant)

$$
\begin{aligned}
{}^{GL}_0\Delta^{(\nu_2)}_{k_2} g(k_1, k_2) &= {}^{GL}_0\Delta^{(\nu_2)}_{k_2} \left[{}^{GL}_0\Delta^{(\nu_1)}_{k_1} y(k_1, k_2) \right] \\
&= \sum_{i_2=0}^{k_2} a^{(\nu_2)}(i_2) \left[\sum_{i_1=0}^{k_1} a^{(\nu_1)}(i_1) y(k_1 - i_1, k_2 - i_2) \right] \\
&\qquad \sum_{i_1=0}^{k_1} \sum_{i_2=0}^{k_2} a^{(\nu_1)}(i_1) a^{(\nu_2)}(i_2) y(k_1 - i_1, k_2 - i_2).
\end{aligned}
\tag{3.60}
$$

Such a dual FOBD operation will be further denoted as

$$
{}^{GL}_{0,0}\Delta^{(\nu_1,\nu_2)}_{k_1,k_2} f(k_1, k_2) = \sum_{i_1=0}^{k_1} \sum_{i_2=0}^{k_2} a^{(\nu_1)}(i_1) a^{(\nu_2)}(i_2) y(k_1 - i_1, k_2 - i_2).
\tag{3.61}
$$

and called the fractional-order partial backward difference (FOPBD) . It can be alternatively expressed as

$$
\begin{aligned}
{}^{GL}_{0,0}\Delta^{(\nu_1,\nu_2)}_{k_1,k_2} f(k_1, k_2) &= {}^{GL}_0\Delta^{(\nu_2)}_{k_2} \left[{}^{GL}_0\Delta^{(\nu_1)}_{k_1} f(k_1, k_2) \right] \\
&= \sum_{i_2=0}^{k_2} a^{(\nu_2)}(i_2) \left[\sum_{i_1=0}^{k_1} a^{(\nu_1)}(i_1) f(k_1 - i_1, k_2 - i_2) \right] \\
&\qquad \sum_{i_1=0}^{k_1} \sum_{i_2=0}^{k_2} a^{(\nu_1)}(i_1) a^{(\nu_2)}(i_2) f(k_1 - i_1, k_2 - i_2).
\end{aligned}
\tag{3.62}
$$

A formal definition of the Grünwald-Letnikov partial FOBD (GL-FOPBD) is given below.

Definition 3.7 (GL-FOPBD). For a two discrete-variable function $f(k_1, k_2)$ (1.111) the Grünwald-Letnikov FO partial BD is defined as

$$
{}^{GL}_{0,0}\Delta^{(\nu_1,\nu_2)}_{k_1,k_2} f(k_1, k_2) = \sum_{i_1=0}^{k_1} \sum_{i_2=0}^{k_2} a^{(\nu_1,\nu_2)}(i_1, i_2) f(k_1 - i_1, k_2 - i_2)
\tag{3.63}
$$

where

$$
a^{(\nu_1,\nu_2)}(k_1, k_2) = a^{(\nu_1)}(k_1) a^{(\nu_2)}(k_2).
\tag{3.64}
$$

Note that the GL-FOPBD can be expressed equivalently as a discrete 2D convolution

$$
{}^{GL}_{0,0}\Delta^{(\nu_1,\nu_2)}_{k_1,k_2} f(k_1, k_2) = a^{(\nu_1,\nu_2)}(k_1, k_2) * f(k_1, k_2).
\tag{3.65}
$$

The double sum in (3.63) can also be represented as a matrix-vector product. Representing the fist sum in 3.59 as a vector-matrix product, introducing the unity matrix $\mathbf{N}_{k_1}\mathbf{N}_{k_1}$ and repeating the first operation for the second FOBD (of the FO ν_2) one obtains

$$
{}^{GL}_{0,0}\Delta^{(\nu_1,\nu_2)}_{k_1,k_2} f(k_1,k_2)
$$

$$
= \begin{bmatrix} a^{(\nu_1)}(0) \ a^{(\nu_1)}(1) \ \cdots \ a^{(\nu_1)}(k_1) \end{bmatrix}
\begin{bmatrix}
\sum_{i_2=0}^{k_2} a^{(\nu_2)}(i_2) f(k_1, k_2 - i_2) \\
\sum_{i_2=0}^{k_2} a^{(\nu_2)}(i_2) f(k_1 - 1, k_2 - i_2) \\
\vdots \\
\sum_{i_2=0}^{k_2} a^{(\nu_2)}(i_2) f(0, k_2 - i_2)
\end{bmatrix}
$$

$$
= \begin{bmatrix} a^{(\nu_1)}(0) \ a^{(\nu_1)}(1) \ \cdots \ a^{(\nu_1)}(k_1) \end{bmatrix} \mathbf{N}_{k_1}\mathbf{N}_{k_1}
\begin{bmatrix}
\sum_{i_2=0}^{k_2} a^{(\nu_2)}(i_2) f(k_1, k_2 - i_2) \\
\sum_{i_2=0}^{k_2} a^{(\nu_2)}(i_2) f(k_1 - 1, k_2 - i_2) \\
\vdots \\
\sum_{i_2=0}^{k_2} a^{(\nu_2)}(i_2) f(0, k_2 - i_2)
\end{bmatrix}
$$

$$
= \begin{bmatrix} a^{(\nu_1)}(k_1) \ a^{(\nu_1)}(k_1 - 1) \ \cdots \ a^{(\nu_1)}(0) \end{bmatrix}
\begin{bmatrix}
\begin{bmatrix} f(0,0) \ f(0,1) \ \cdots \ f(0,k_2) \end{bmatrix} \begin{bmatrix} a^{(\nu_2)}(k_2) \\ a^{(\nu_2)}(k_2 - 1) \\ \vdots \\ a^{(\nu_2)}(0) \end{bmatrix} \\
\begin{bmatrix} f(1,0) \ f(1,1) \ \cdots \ f(1,k_2) \end{bmatrix} \begin{bmatrix} a^{(\nu_2)}(k_2) \\ a^{(\nu_2)}(k_2 - 1) \\ \vdots \\ a^{(\nu_2)}(0) \end{bmatrix} \\
\vdots \\
\begin{bmatrix} f(k_1,0) \ f(k_1,1) \ \cdots \ f(k_1,k_2) \end{bmatrix} \begin{bmatrix} a^{(\nu_2)}(k_2) \\ a^{(\nu_2)}(k_2 - 1) \\ \vdots \\ a^{(\nu_2)}(0) \end{bmatrix}
\end{bmatrix}
$$

$$
= \begin{bmatrix} a^{(\nu_1)}(k_1) \ a^{(\nu_1)}(k_1 - 1) \ \cdots \ a^{(\nu_1)}(0) \end{bmatrix}
\begin{bmatrix}
f(0,0) & f(0,1) & \cdots & f(0,k_2) \\
f(1,0) & f(1,1) & \cdots & f(1,k_2) \\
\vdots & \vdots & & \\
f(k_1,0) & f(k_1,1) & \cdots & f(k_1,k_2)
\end{bmatrix}
\begin{bmatrix} a^{(\nu_2)}(k_2) \\ a^{(\nu_2)}(k_2 - 1) \\ \vdots \\ a^{(\nu_2)}(0) \end{bmatrix}
$$

$$
= \begin{bmatrix} a^{(\nu_1)}(k_1) \ a^{(\nu_1)}(k_1 - 1) \ \cdots \ a^{(\nu_1)}(0) \end{bmatrix} \mathbf{F}(k_1,k_2)
\begin{bmatrix} a^{(\nu_2)}(k_2) \\ a^{(\nu_2)}(k_2 - 1) \\ \vdots \\ a^{(\nu_2)}(0) \end{bmatrix}
\tag{3.66}
$$

where

$$
\mathbf{F}(k_1,k_2) = \begin{bmatrix}
f(0,0) & f(0,1) & \cdots & f(0,k_2) \\
f(1,0) & f(1,1) & \cdots & f(1,k_2) \\
\vdots & \vdots & & \\
f(k_1,0) & f(k_1,1) & \cdots & f(k_1,k_2)
\end{bmatrix}.
\tag{3.67}
$$

The GL-FOPBD may be also expressed in an equivalent form

$$\substack{GL \\ 0,0} \Delta_{k_1,k_2}^{(\nu_1,\nu_2)} f(k_1, k_2) = \mathbf{a}^{(\nu_1)}(k_1)\mathbf{N}_{k_1}\mathbf{F}(k_1, k_2)\mathbf{N}_{k_2}\left[\mathbf{a}^{(\nu_2)}(k_2)\right]^{\mathrm{T}}. \qquad (3.68)$$

In the following numerical example, one calculates the GL-FOPBD of the 2D discrete Dirac pulse and unit step function.

Example 3.2. Calculate the GL-FOPBD of orders ν_1 and ν_2 of the 2D discrete Dirac pulse and unit step function.

Solution. Substitution of (1.113) into (1.114) results in

$$\substack{GL \\ 0,0} \Delta_{k_1,k_2}^{(\nu_1,\nu_2)} \delta(k_1, k_2)$$

$$= \left[a^{(\nu_1)}(k_1)\; a^{(\nu_1)}(k_1 - 1) \; \cdots \; a^{(\nu_1)}(0)\right] \begin{bmatrix} 1 & 0 & \cdots & 0 \\ 0 & 0 & \cdots & 0 \\ \vdots & \vdots & & \ddots \\ 0 & 0 & \cdots & 0 \end{bmatrix} \begin{bmatrix} a^{(\nu_2)}(k_2) \\ a^{(\nu_2)}(k_2 - 1) \\ \vdots \\ a^{(\nu_2)}(0) \end{bmatrix}$$

$$= \left[a^{(\nu_1)}(k_1)\; a^{(\nu_1)}(k_1 - 1) \; \cdots \; a^{(\nu_1)}(0)\right] \begin{bmatrix} a^{(\nu_2)}(k_2) \\ 0 \\ \vdots \\ 0 \end{bmatrix} = a^{(\nu_1)}(k_1)a^{(\nu_2)}(k_2). \qquad (3.69)$$

A plot and an image of the 2D Dirac pulse GL-FOBD are given in Figs. 3.1(a) and 3.1(b), respectively.

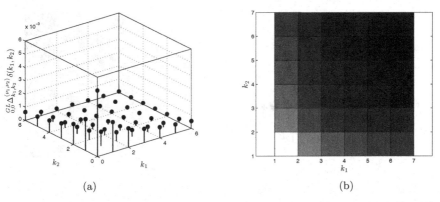

(a)   (b)

Fig. 3.1   3D plot and image of 2D GL-FOBD of orders $\nu_1 = 0.05$ and $\nu_2 = 0.1$ of $\delta(k_1, k_2)$.

In the case of a discrete unit step function, the GL-FOPBD is equal to

$$
{}^{GL}_{0,0}\Delta^{(\nu_1,\nu_2)}_{k_1,k_2}\delta(k_1,k_2)
$$

$$
= \begin{bmatrix} a^{(\nu_1)}(k_1)\ a^{(\nu_1)}(k_1-1)\ \cdots\ a^{(\nu_1)}(0) \end{bmatrix}
\begin{bmatrix} 1 & 1 & \cdots & 1 \\ 1 & 1 & \cdots & 1 \\ \vdots & \vdots & & \ddots \\ 1 & 1 & \cdots & 1 \end{bmatrix}
\begin{bmatrix} a^{(\nu_2)}(k_2) \\ a^{(\nu_2)}(k_2-1) \\ \vdots \\ a^{(\nu_2)}(0) \end{bmatrix}
$$

$$
= \begin{bmatrix} a^{(\nu_1)}(k_1)\ a^{(\nu_1)}(k_1-1)\ \cdots\ a^{(\nu_1)}(0) \end{bmatrix}
\begin{bmatrix} a^{(\nu_2-1)}(k_2) \\ a^{(\nu_2-1)}(k_2) \\ \vdots \\ a^{(\nu_2-1)}(k_2) \end{bmatrix}
$$

$$
= \begin{bmatrix} a^{(\nu_1)}(k_1)\ a^{(\nu_1)}(k_1-1)\ \cdots\ a^{(\nu_1)}(0) \end{bmatrix}
\begin{bmatrix} 1 \\ 1 \\ \vdots \\ 1 \end{bmatrix} a^{(\nu_2-1)}(k_2)
$$

$$
= a^{(\nu_1-1)}(k_1)a^{(\nu_2-1)}(k_2). \quad (3.70)
$$

A plot and an image of the 2D discrete unit step GL-FOBD are given in Figs. 3.2(a) and 3.2(b), respectively.

Definition 3.8 (2D GL-FOBS). For a two discrete-variable function $f(k_1,k_2)$ (1.111), the Grünwald-Letnikov 2D FOBS of orders $\nu_1,\nu_2 \in \mathbb{R}_+$ is defined as

$$
{}^{GL}_{0,0}\Sigma^{(\nu_1,\nu_2)}_{k_1,k_2}f(k_1,k_2) = {}^{GL}_{0,0}\Delta^{(-\nu_1,-\nu_2)}_{k_1,k_2}f(k_1,k_2)
$$

$$
= \sum_{i_1=0}^{k_1}\sum_{i_2=0}^{k_2} a^{(-\nu_1,-\nu_2)}(i_1,i_2)f(k_1-i_1,k_2-i_2) \quad (3.71)
$$

where $a^{(-\nu_1,-\nu_2)}$ is defined by formula (3.65).

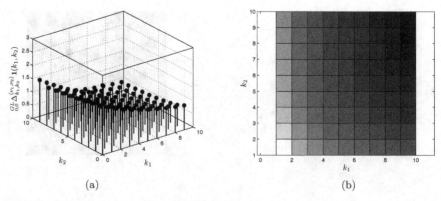

(a)   (b)

Fig. 3.2   3D plot and image of 2D GL-FOBD of orders $\nu_1 = 0.25$ and $\nu_2 = 0.5$ of $\mathbf{1}(k_1,k_2)$.

Performing analogous calculations one gets the following results

$$\begin{aligned}
{}_{0,0}^{GL}\Sigma_{k_1,k_2}^{(\nu_1,\nu_2)}\delta(k_1,k_2) &= {}_{0,0}^{GL}\Delta_{k_1,k_2}^{(-\nu_1,-\nu_2)}\delta(k_1,k_2) \\
&= a^{(-\nu_1)}(k_1)a^{(-\nu_2)}(k_2)
\end{aligned} \tag{3.72}$$

and

$$\begin{aligned}
{}_{0,0}^{GL}\Sigma_{k_1,k_2}^{(\nu_1,\nu_2)}\mathbf{1}(k_1,k_2) &= {}_{0,0}^{GL}\Delta_{k_1,k_2}^{(-\nu_1,-\nu_2)}\mathbf{1}(k_1,k_2) \\
&= a^{(-\nu_1-1)}(k_1)a^{(-\nu_2-1)}(k_2).
\end{aligned} \tag{3.73}$$

Related plots and images are given in Figs. 3.3(a), 3.3(b) and Figs. 3.4(a), 3.4(b), respectively.

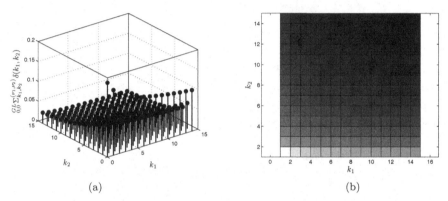

(a)   (b)

Fig. 3.3   3D plot and image of 2D GL-FOBS of orders $\nu_1 = 0.25$ and $\nu_2 = 0.5$ of $\delta(k_1,k_2)$.

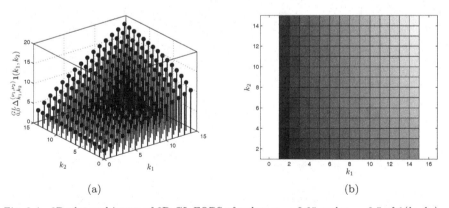

(a)   (b)

Fig. 3.4   3D plot and image of 2D GL-FOBS of orders $\nu_1 = 0.25$ and $\nu_2 = 0.5$ of $\mathbf{1}(k_1,k_2)$.

Chapter 4

The FOBD/S graphical interpretation

In this Chapter, an attempt of the FOBD and FOBS graphical interpretations is presented. Direct transfer of the IOBD and IOBS interpretations to their FO counterparts up to now is not effective. This is due to the fact that the IOBD is a local property of the discrete-variable function, whereas the FOBD is a global one [Ben Adda (1997); Moshrefi-Torbati and Hammond (1998); Podlubny (2002); Podlubny *at al.* (2007); Ostalczyk (2010a)]. In the FOBD/S case, the key role is played by the function $a^{(\pm\nu)}(k)$, which is considered in Section 1.3.3. Here, the considerations will be restricted to orders from a range $0 < \nu \leqslant 1$, whereas in the FO summation case one will consider orders satisfying inequalities $-1 \leqslant \nu < 1$. Here, upper and lower order ranges cover classical backward difference and sum, whose graphical interpretations are obvious and commonly understood. In this Chapter one demonstrates that in the FO difference operation, information about present and past events is compared (which is marked by "−" sign). In the FO summation operation present and past information is collected (which is marked by "+" sign).

4.1 FOBS graphical interpretation

To simplify considerations in this Chapter, it will be assumed that $k_0 = 0$. This assumption does not reduce the generality of the FOBD/S graphical interpretations. First, one will consider the FOBS. The GL-FOBS evaluated for $0 < \nu \leqslant 1$ can be equivalently expressed in the form

$$
{}_{0}^{GL}\Sigma_{k}^{(\nu)}f(k) = {}_{0}^{GL}\Delta_{k}^{(-\nu)}f(k) = f(k) + \sum_{i=0}^{k}\left|a^{(-\nu)}(i)\right|f(k-i). \tag{4.1}
$$

A general block diagram representing the FOBS evaluation due to formula (4.1) is depicted in Fig. 4.1.

The close-up block diagram given in Fig. 4.2 presents details of the FOBS calculation procedure. One treats a discrete-variable function value $f(k)$ as present information (a present sample is actual information about a dynamic system (for example: a financial system, a measured parameters of biological system,

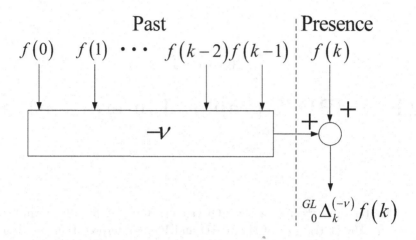

Fig. 4.1 Graphical block diagram of the FOBS realization.

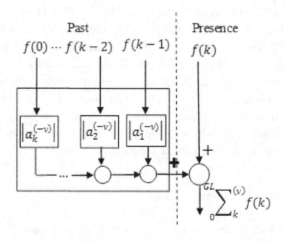

Fig. 4.2 Graphical block diagram of the FOBS realization.

a state of technical system or physical phenomena). The system past values $f(k-1), f(k-2), \ldots, 1, 0$ are treated as the considered system memory (or the history). One can ascertain that the FOBS is a collection of the presence and the past events. The consecutive events influence are diminished due to a distance from the FOBS evaluation moment. According to a natural loss of a memories (the memory engram decay) a past events influence is related to the distance from the presence. This memory decay level is represented by function $a^{(\nu)}(k-i)$ values (which can be treated as weighting factors) evaluated for $i = 1, 2, \ldots, k-1, k$. One can name the function $a^{(\nu)}(k-i)$ as a memory oblivion function.

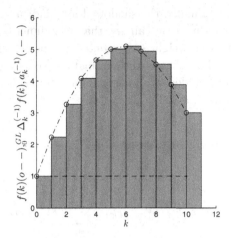

Fig. 4.3   Plot of function $f(k)$ (o–), its first-order sum ${}^{GL}_0\Delta^{(-1)}_k f(k)$ (■), and coefficients $a^{(-1)}(k)$ (. – –) in a reverse order.

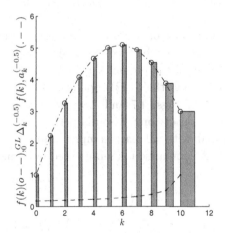

Fig. 4.4   Plot of function $f(k)$ (o- -), its FOBS ${}^{GL}_0\Delta^{(-0.5)}_k f(k)$ (■), and coefficients $a^{(-0.5)}(k)$ (. – –) in a reverse order.

In a classical first-order summation where $a^{(-1)}(k) = |a^{(-1)}(k)| = 1$ for $k = 0, 1, 2, \ldots$ the present function value $f(k)$ (a present sample) and all previous ones are collected with the same priority. The analyzed operation is depicted in Fig. 4.2.

One should note that the operation presented above is a classical rectangular integration of a continuous-time function (assuming a sampling period $h = 1$). Now, the summation of order $\nu = -0.5$ is depicted. In Fig. 4.4 the same function $f(k)$ as in Fig. 4.3 is plotted. There, the coefficients $a^{(-0.5)}(k)$ in a reverse order are plotted as well (. – –). The reverse order is due to the summation formula denoting the discrete convolution. Finally, the shadowed area indicating the value of the FOBS $_0^{GL}\Delta_k^{(-0.5)}f(k)$ is shown. One can see that past function samples are taken into account with decreasing weight (represented by the length of the rectangle base). Figures 4.3 and 4.4 show that in the summation (collection) procedure the present sample (experience) is added to all past samples diminished by appropriate oblivion coefficient $a^{(-\nu)}(k)$ $(0 < \nu \leqslant 1)$.

4.2 FOBD graphical interpretation

The FOBD of order $0 < \nu \leqslant 1$ can be expressed as

$$_0^{GL}\Delta_k^{(\nu)}f(k) = f(k) - \sum_{i=0}^{k}\left|a^{(\nu)}(i)\right|f(k-i). \tag{4.2}$$

A block diagram representing the FO backward differentiation idea is presented in Fig. 4.5 and a detailed block diagram is presented in Fig. 4.6.

A fundamental distinction between formulae 4.1 and 4.2 is determined by signs: "−" in the FOBD and "+" in the FOBS (except for different values of the same coefficients $a^{(\mp\nu)}(k)$).

The FOBD, as an inverse operation to the FOBS, is characterized by an opposite action. All past samples (experiences) diminished by appropriate oblivion coefficients are subtracted from the present sample. This is depicted in Figs. 4.7 and 4.8 where a first-order BD and the FOBD operations are considered, respectively.

Simple inspection of Figs. 4.7 and 4.8 reveals that in the discrete differentiation procedure an action of comparison is performed. The present function value (sample) is compared with all past samples diminished by appropriate oblivion coefficients $a^{(\nu)}(k)$. A comparison is represented by a subtraction operation. More-

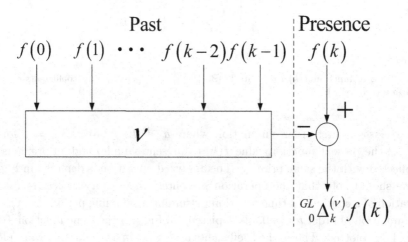

Fig. 4.5   Graphical block diagram of the FOBD realization.

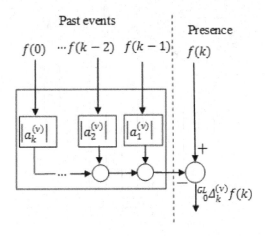

Fig. 4.6 Detailed block diagram of the FOBD realization.

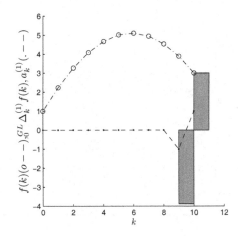

Fig. 4.7 Plot of function $f(k)$ (o–), its first-order sum ${}^{GL}_{0}\Delta^{(1)}_{k}f(k)$ (■), and ceofficients $a^{(1)}(k)$ (.– –) in a reverse order.

over, a comparison of Figs. 4.4 and 4.8 reveals a stronger action of the summation operation then the differentiation one (for the same FOs and functions). Quicker disappearance of coefficients $|a^{(\nu)}(k)|$ than $|a^{(-\nu)}(k)|$ can be seen in appropriate lengths of rectangular bases in the FOBD then in the FOBS. This is clarified by Fig. 4.9, where the decreasing intensities of $|a^{(\nu)}(k)|$ and $|a^{(-\nu)}(k)|$ are given.

Figure 4.4 contains three plots: function $f(k)$, a set of rectangles representing the FOBS ${}^{GL}_{0}\Delta^{(0.5)}_{k}f(k)$, and oblivion function $|a^{(-\nu)}(k)|$. These can be represented in the 3D plot given in Fig. 4.10.

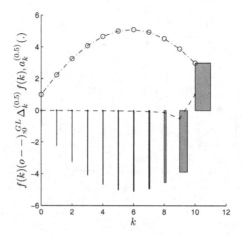

Fig. 4.8   Plot of function $f(k)$ (o- -), its FOBS $_0^{GL}\Delta_k^{(0.5)}f(k)$ (■), and coefficients $a^{(0.5)}(k)$ (.- - -) in a reverse order.

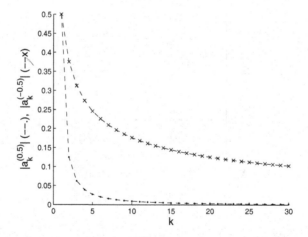

Fig. 4.9   Plot of functions $|a^{(\nu)(k)}|$ and $|a^{(-\nu)(k)}|$.

Noting that the length of k-th rectangular base is equal to $|a^{(-\nu)}(k)|$, all rectangles can be rotated 90 degrees clockwise. This operation is depicted in Fig. 4.11.

All the shadowed rectangles are equally spaced. Adding an additional constant factor h^ν representing a constant distance between consecutive rectangles, one gets a sum ob cubicoids described as

$$_0^{GL}\Sigma_k^{(\nu)}f(kh)h^\nu = h^\nu \sum_{i=0}^{k} a^{(-\nu)}(i)f\left[(k-i)h\right]. \tag{4.3}$$

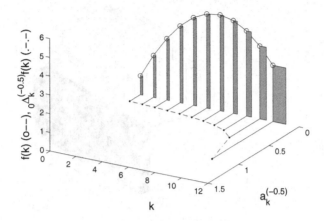

Fig. 4.10   FOBS 3D graphical representation for $\nu = 0.5$.

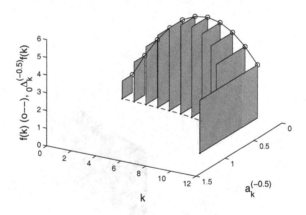

Fig. 4.11   FOBS 3D graphical representation for $\nu = 0.5$.

For $t = kh$ and $h \to 0^+$

$$\lim_{h \to 0^+} \left[{}_{0}^{GL}\Sigma_k^{(\nu)} f(kh) h^\nu \right] = \lim_{h \to 0^+} \sum_{i=0}^{\left\lfloor \frac{t}{h} \right\rfloor} a^{(-\nu)}(i) f\left[(k-i)h \right] h^\nu$$

$$= \frac{1}{\Gamma(1-\nu)} \int_0^t (t-\tau)^{\nu-1} f(\tau) d\tau. \qquad (4.4)$$

The sum of three factors' product: $a^{(\nu)}(i)$, $f(k-i)$ and h^ν in the discrete-variable function calculation represent the volume of a cuboid. This volume is shown in Fig. 4.12 for $\nu = 0.5$.

By extension, one may interpret the FOBD.

Discrete Fractional Calculus

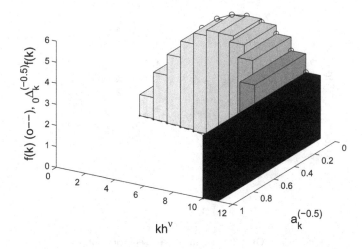

Fig. 4.12   FOBS 3D graphical representation for $\nu = 0.5$ and $h = 1$.

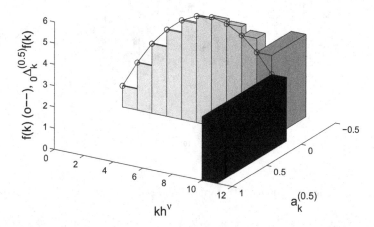

Fig. 4.13   FOBD 3D graphical representation for $\nu = 0.5$ and $h = 1$.

Chapter 5

The FOBD/S selected properties

In this Chapter, selected properties of the FOBD/S are discussed. They stem from the classical discrete-variable function analyses [Bayin (2006); Keisler (2013)] and are very useful in performing different operations on the FOBD/Ss. Here one considers the linearity, the FOBD/S concatenation rules and the FOBD of a product of two discrete-variable functions.

5.1 The FOBD/S linearity

First one considers two discrete-variable functions $f(k)$ and $g(k)$ and two real coefficients $a, b, \in \mathbb{R}$. One assumes that the FOBD/Ss of $f(k)$ and $g(k)$ exists. Accordingly, the following theorem holds true.

Theorem 5.1 (FOBD/S linearity). *For $\nu \in \mathbb{R}$ let $_{k_0}^{GL}\Delta_k^{(\nu)}f(k)$ and $_{k_0}^{GL}\Delta_k^{(\nu)}g(k)$ denote the FOBD/Ss of functions $f(k)$ and $g(k)$ respectively. The following equality holds*

$$a_{k_0}^{GL}\Delta_k^{(\nu)}f(k) + b_{k_0}^{GL}\Delta_k^{(\nu)}g(k) = _{k_0}^{GL}\Delta_k^{(\nu)}[af(k) + bg(k)]. \tag{5.1}$$

Proof. By definition of the FOBD/S

$$a_{k_0}^{GL}\Delta_k^{(\nu)}f(k) + b_{k_0}^{GL}\Delta_k^{(\nu)}g(k) \tag{5.2}$$

$$= a\begin{bmatrix} a^{(\nu)}(0)\ a^{(\nu)}(1) \cdots a^{(\nu)}(k) \end{bmatrix} \begin{bmatrix} f(k) \\ f(k-1) \\ \vdots \\ f(1) \\ f(0) \end{bmatrix} + b\begin{bmatrix} a^{(\nu)}(0)\ a^{(\nu)}(1) \cdots a^{(\nu)}(k) \end{bmatrix} \begin{bmatrix} g(k) \\ g(k-1) \\ \vdots \\ g(1) \\ g(0) \end{bmatrix}$$

$$+ \begin{bmatrix} a^{(\nu)}(0)\ a^{(\nu)}(1) \cdots a^{(\nu)}(k) \end{bmatrix} \left(a\begin{bmatrix} f(k) \\ k(k-1) \\ \vdots \\ f(1) \\ f(0) \end{bmatrix} + b\begin{bmatrix} g(k) \\ g(k-1) \\ \vdots \\ g(1) \\ g(0) \end{bmatrix} \right) = _{k_0}^{GL}\Delta_k^{(\nu)}[af(k) + bg(k)].$$

$$\square$$

This theorem can be generalized to more than two coefficients and functions. Moreover, it is important to note that the linearity property is valid for commensurate orders and the same differentiation ranges $[k_0, k]$.

5.2 The FOBD/S concatenation property

One defines two vectors for $\nu, \mu \in \mathbb{R}_+$

$$\mathbf{a}^{(\nu)}(k) = \left[a^{(\nu)}(0), a^{(\nu)}(1), a^{(\nu)}(2), \ldots, a^{(\nu)}(k) \right]^{\mathrm{T}}$$

$$\mathbf{a}^{(\mu)}(k) = \left[a^{(\mu)}(0), a^{(\mu)}(1), a^{(\mu)}(2), \ldots, a^{(\mu)}(k) \right]^{\mathrm{T}}. \tag{5.3}$$

Then

$$\begin{aligned}
{}^{GL}_{0}\Delta_k^{(\mu)} a^{(\nu)}(k) &= \begin{bmatrix} a^{(\mu)}(0) & a^{(\mu)}(1) & a^{(\mu)}(2) & \cdots & a^{(\mu)}(k) \end{bmatrix} \begin{bmatrix} a^{(\nu)}(k) \\ a^{(\nu)}(k-1) \\ \vdots \\ a^{(\nu)}(1) \\ a^{(\nu)}(0) \end{bmatrix} \\
&= \left[\mathbf{a}^{(\mu)}(k) \right]^{\mathrm{T}} \mathbf{N}_k \mathbf{a}^{(\nu)}(k). \tag{5.4}
\end{aligned}$$

where $\mathbf{N}_k$ is defined by (A.19). From formula (5.4) evaluated for $k = 0, 1, 2, \ldots$ one obtains a set of equalities

$$\left. {}^{GL}_{0}\Delta_k^{(\mu)} a^{(\nu)}(k) \right|_{k=0} = a^{(\mu)}(0) a^{(\nu)}(0) = 1 = a^{(\nu+\mu)}(0) \tag{5.5}$$

$$\left. {}^{GL}_{0}\Delta_k^{(\mu)} a^{(\nu)}(k) \right|_{k=1} = a^{(\mu)}(0) a^{(\nu)}(1) + a^{(\mu)}(1) a^{(\nu)}(0) = -\nu - \mu = a^{(\nu+\mu)}(1) \tag{5.6}$$

$$\left. {}^{GL}_{0}\Delta_k^{(\mu)} a^{(\nu)}(k) \right|_{k=2} = a^{(\mu)}(0) a^{(\nu)}(2) + a^{(\mu)}(1) a^{(\nu)}(1) + a^{(\mu)}(2) a^{(\nu)}(0)$$

$$= 1\frac{\nu(\nu-1)}{2!} + (-\nu)(-\mu) + \frac{\mu(\mu-1)}{2!}1 = \frac{(\nu+\mu)(\nu+\mu-1)}{2!} = a^{(\nu+\mu)}(2). \tag{5.7}$$

Continuing the calculations presented above, one finally gets a relation which is valid for all $\nu, \mu \in \mathbb{R}_+$ and $k = 0, 1, 2, \ldots$

$$ {}^{GL}_{0}\Delta_k^{(\mu)} a^{(\nu)}(k) = a^{(\nu+\mu)}(k). \tag{5.8}$$

Formula (5.8) *de facto* means that for $k = k_0, k_0 + 1, \ldots$

$$\left[a^{(\mu)}(0) \; a^{(\mu)}(1) \; a^{(\mu)}(2) \; \cdots \; a^{(\mu)}(k - k_0) \right] \begin{bmatrix} a^{(\nu)}(k) \\ a^{(\nu)}(k-1) \\ \vdots \\ a^{(\nu)}(k_0 + 1) \\ a^{(\nu)}(k_0) \end{bmatrix} = {}^{GL}_{k_0} \Delta^{(\mu)}_k a^{(\nu)}(k). \quad (5.9)$$

Collection of all such equalities evaluated for $k = 0, 1, \ldots$ in a matrix-vector form yields

$$\mathbf{A}^{(\mu)}_{k-k_0} \mathbf{A}^{(\nu)}_{k-k_0} = \mathbf{A}^{(\mu+\nu)}_{k-k_0} \quad (5.10)$$

where

$$\mathbf{A}^{(\mu)}_{k-k_0} = \begin{bmatrix} a^{(\mu)}(0) \; a^{(\mu)}(1) \; \cdots \; a^{(\mu)}(k - k_0 - 1) \; a^{(\mu)}(k - k_0) \\ 0 \quad a^{(\mu)}(0) \; \cdots \quad a^{(\mu)}(k - 2) \quad a^{(\mu)}(k - 1) \\ \vdots \quad \vdots \quad\quad\quad \vdots \quad\quad\quad \vdots \\ 0 \quad 0 \quad \cdots \quad a^{(\mu)}(0) \quad\quad a^{(\mu)}(1) \\ 0 \quad 0 \quad \cdots \quad 0 \quad\quad a^{(\mu)}(0) \end{bmatrix}$$

$$\mathbf{A}^{(\nu)}_{k-k_0} = \begin{bmatrix} a^{(\nu)}(0) \; a^{(\nu)}(1) \; \cdots \; a^{(\nu)}(k - k_0 - 1) \; a^{(\nu)}(k - k_0) \\ 0 \quad a^{(\nu)}(0) \; \cdots \quad a^{(\nu)}(k - 2) \quad a^{(\nu)}(k - 1) \\ \vdots \quad \vdots \quad\quad\quad \vdots \quad\quad\quad \vdots \\ 0 \quad 0 \quad \cdots \quad a^{(\nu)}(0) \quad\quad a^{(\nu)}(1) \\ 0 \quad 0 \quad \cdots \quad 0 \quad\quad a^{(\nu)}(0) \end{bmatrix}. \quad (5.11)$$

Simple calculations immediately prove that

$$\mathbf{A}^{(\mu)}_{k-k_0} \mathbf{A}^{(\nu)}_{k-k_0} = \mathbf{A}^{(\mu+\nu)}_{k-k_0} = \mathbf{A}^{(\nu+\mu)}_{k-k_0} = \mathbf{A}^{(\nu)}_{k-k_0} \mathbf{A}^{(\mu)}_{k-k_0}. \quad (5.12)$$

For a special case when $\nu = -\mu$, one obtains

$$\mathbf{A}^{(\mu)}_{k-k_0} \mathbf{A}^{(-\mu)}_{k-k_0} = \mathbf{A}^{(-\mu)}_{k-k_0} \mathbf{A}^{(\mu)}_{k-k_0} = \mathbf{A}^{(0)}_{k-k_0} = \mathbf{I}_{k-k_0}. \quad (5.13)$$

From the above equality one gets the following relation

$$[\mathbf{A}^{(\mu)}_{k-k_0}]^{-1} = \mathbf{A}^{(-\mu)}_{k-k_0} \quad \text{and} \quad [\mathbf{A}^{(-\mu)}_{k-k_)}]^{-1} = \mathbf{A}^{(\mu)}_{k-k_0}. \quad (5.14)$$

Now, one considers a concatenation operation of the two FOBDs. For this purpose one assumes the existence of the FOBD of the function $_{k_0}^{GL}\Delta_k^{(\nu)}f(k)$. Next,

$$
_{k_0}^{GL}\Delta_k^{(\nu)}\mathbf{f}(k) =
\begin{bmatrix}
_{0}^{GL}\Delta_k^{(\nu)}f(k) \\
_{0}^{GL}\Delta_{k-1}^{(\nu)}f(k-1) \\
\vdots \\
_{0}^{GL}\Delta_{k_0+1}^{(\nu)}f(1) \\
_{0}^{GL}\Delta_{k_0}^{(\nu)}f(0)
\end{bmatrix}
= \mathbf{A}_{k-k_0}^{(\nu)}\mathbf{f}(k)
\tag{5.15}
$$

$$
\mathbf{f}(k) =
\begin{bmatrix}
f(k) \\
f(k-1) \\
\vdots \\
f(k_0+1) \\
f(k_0)
\end{bmatrix}
\tag{5.16}
$$

and

$$
_{k_0}^{GL}\Delta_k^{(\mu)}\left[_{k_0}^{GL}\Delta_k^{(\nu)}f(k)\right] = \mathbf{A}_{k-k_0}^{(\mu)}\mathbf{A}_{k-k_0}^{(\nu)}\mathbf{f}(k) = \mathbf{A}_{k-k_0}^{(\mu+\nu)}\mathbf{f}(k).
\tag{5.17}
$$

A similar equality holds for $-\nu, -\mu \in \mathbb{R}_+$. This is a concatenation property for the FOBSs.

5.3 FOBD of the product of two functions (Fractional Leibnitz rule)

Now, the FOBD of the product of two discrete-variable functions $f(k)$ and $g(k)$ defined over interval k_0, k will be analyzed [Ostalczyk (2014b)]. The main result — The Leibnitz Rule for the FOBD is stated in the form of the following theorem.

Theorem 5.2. *The FOBD of the order $\nu > 0$ of the product of two functions $f(k)$ and $g(k)$ is equal*

$$
_{0}^{GL}\Delta_k^{(\nu)}[f(k)g(k)] = [_{0}^{GL}\Delta_k^{(\nu)}\mathbf{f}(k)]^{\mathrm{T}}\mathbf{M}_k^{(\nu)}{}_{0}^{GL}\Delta_k^{(\nu)}\mathbf{g}(k)
\tag{5.18}
$$

where

$$
\mathbf{g}(k) =
\begin{bmatrix}
g(0) \\
g(1) \\
\vdots \\
g(k-1) \\
g(k)
\end{bmatrix}, \quad
{}^{GL}\Delta^{(\nu)}\mathbf{g}(k) =
\begin{bmatrix}
_{0}^{GL}\Delta_k^{(\nu)}g(k) \\
_{0}^{GL}\Delta_{k-1}^{(\nu)}g(k-1) \\
\vdots \\
_{0}^{GL}\Delta_1^{(\nu)}g(1) \\
_{0}^{GL}\Delta_0^{(\nu)}g(0)
\end{bmatrix}
= \mathbf{A}_k^{(\nu)}\mathbf{N}_k\mathbf{g}(k)
\tag{5.19}
$$

$$
\mathbf{M}_k^{(\nu)} = [\mathbf{A}_k^{(-\nu)}]^T\mathbf{D}_k^{(\nu)}\mathbf{A}_k^{(\nu)},
\tag{5.20}
$$

$$\mathbf{D}_k^{(\nu)} = \begin{bmatrix} a_0^{(\nu)} & 0 & \cdots & 0 & 0 \\ 0 & a_1^{(\nu)} & \cdots & 0 & 0 \\ \vdots & \vdots & & \vdots & \vdots \\ 0 & 0 & \cdots & a_{k-1}^{(\nu)}0 & \\ 0 & 0 & \cdots & 0 & a_k^{(\nu)} \end{bmatrix} \qquad (5.21)$$

and $^{GL}\Delta^{(\nu)}\mathbf{g}(k)$ *is defined by formula (5.15). The FOBD* $^{GL}\Delta^{(\nu)}\mathbf{f}(k)$ *has analogous form.*

Proof. By Definition (3.1) formula (5.18) can be expressed in the form

$$_0^{GL}\Delta_k^{(\nu)}[f(k)g(k)] = \begin{bmatrix} f(k) \ f(k-1) \ \cdots \ f(1) \ f(0) \end{bmatrix} \mathbf{D}_k^{(\nu)} \begin{bmatrix} g(k) \\ g(k-1) \\ \vdots \\ g(1) \\ g(0) \end{bmatrix} \qquad (5.22)$$

$$= \begin{bmatrix} f(k) \ f(k-1) \ \cdots \ f(1) \ f(0) \end{bmatrix} [[\mathbf{A}_k^{(\nu)}]^{-1}\mathbf{A}_k^{(\nu)}]^{\mathrm{T}}\mathbf{D}_k^{(\nu)}[\mathbf{A}_k^{(\nu)}]^{-1}\mathbf{A}_k^{(\nu)} \begin{bmatrix} g(k) \\ g(k-1) \\ \vdots \\ g(1) \\ g(0) \end{bmatrix}$$

$$= \begin{bmatrix} f(k) \ f(k-1) \ \cdots \ f(1) \ f(0) \end{bmatrix} [\mathbf{A}_k^{(\nu)}]^{\mathrm{T}}[\mathbf{A}_k^{(-\nu)}]^{\mathrm{T}}\mathbf{D}_k^{(\nu)}\mathbf{A}_k^{(-\nu)}{}_0^{GL}\Delta_k^{(\nu)}\mathbf{g}(k)$$

$$= [\mathbf{A}_k^{(\nu)} \begin{bmatrix} f(k) \\ f(k-1) \\ \vdots \\ f(1) \\ f(0) \end{bmatrix}]^{\mathrm{T}}[\mathbf{A}_k^{(-\nu)}]^{\mathrm{T}}\mathbf{M}_k^{(\nu)}\mathbf{A}_k^{(-\nu)}{}_0^{GL}\Delta_k^{(\nu)}\mathbf{g}(k)$$

$$= [^{GL}\Delta_k^{(\nu)}\mathbf{f}(k)]^{\mathrm{T}}[\mathbf{A}_k^{(-\nu)}]^{\mathrm{T}}\mathbf{D}_k^{(\nu)}\mathbf{A}_k^{(-\nu)}{}_0^{GL}\Delta_k^{(\nu)}\mathbf{g}(k)$$

$$[_0^{GL}\Delta_k^{(\nu)}\mathbf{f}(k)]^{\mathrm{T}}\mathbf{M}_k^{(\nu)}{}_0^{GL}\Delta_k^{(\nu)}\mathbf{g}(k).$$

$\square$

The result proved above is valid for integer orders $n \in \mathbb{Z}_+$. One should only realize that in this case

$$
{}^{GL}\Delta^{(\nu)}\mathbf{f}(k) = \begin{bmatrix} {}^{GL}_{k-n}\Delta_k^{(n)}f(k) \\ {}^{GL}_{k-1-n}\Delta_{k-1}^{(n)}f(k-1) \\ \vdots \\ {}^{GL}_1\Delta_{n+1}^{(n)}f(n+1) \\ {}^{GL}_0\Delta_n^{(n)}f(n) \\ \vdots \\ {}^{GL}_0\Delta_0^{(n)}f(0) \end{bmatrix} = \mathbf{A}_k^{(n)}\mathbf{f}(k) \tag{5.23}
$$

$$
{}^{GL}\Delta^{(\nu)}\mathbf{g}(k) = \begin{bmatrix} {}^{GL}_{k-n}\Delta_k^{(n)}g(k) \\ {}^{GL}_{k-1-n}\Delta_{k-1}^{(n)}g(k-1) \\ \vdots \\ {}^{GL}_1\Delta_{n+1}^{(n)}g(n+1) \\ {}^{GL}_0\Delta_n^{(n)}g(n) \\ \vdots \\ {}^{GL}_0\Delta_0^{(n)}g(0) \end{bmatrix} = \mathbf{A}_k^{(n)}\mathbf{g}(k) \tag{5.24}
$$

$$
\mathbf{D}_k^{(n)} = \begin{bmatrix} a^{(n)}(0) & \cdots & 0 & \cdots & 0 \\ \vdots & & \vdots & & \vdots \\ 0 & \cdots & a^{(n)}(n) & \cdots & 0 \\ \vdots & & \vdots & & \vdots \\ 0 & \cdots & 0 & \cdots & 0 \end{bmatrix} \tag{5.25}
$$

with matrix $\mathbf{M}_k^{(\nu)}$ evaluated by formula (5.20). Below exemplary matrices calculated for $n = 1, \ldots, 5$ are given

$$
\mathbf{M}_k^{(1)} = \begin{bmatrix} 1 & 1 & 1 & 1 & \cdots \\ 1 & 0 & 0 & 0 & \cdots \\ 1 & 0 & 0 & 0 & \cdots \\ 1 & 0 & 0 & 0 & \cdots \\ \vdots & \vdots & \vdots & \vdots \end{bmatrix}, \quad \mathbf{M}_k^{(2)} = \begin{bmatrix} 1 & 2 & 3 & 4 & \cdots \\ 2 & 2 & 2 & 2 & \cdots \\ 3 & 2 & 2 & 2 & \cdots \\ 4 & 2 & 2 & 2 & \cdots \\ \vdots & \vdots & \vdots & \vdots \end{bmatrix}, \quad \mathbf{M}_k^{(3)} = \begin{bmatrix} 1 & 3 & 6 & 10 & 15 & \cdots \\ 3 & 6 & 9 & 12 & 15 & \cdots \\ 6 & 9 & 12 & 15 & 18 & \cdots \\ 10 & 12 & 15 & 18 & 21 & \cdots \\ \vdots & \vdots & \vdots & \vdots & \vdots \end{bmatrix} \tag{5.26}
$$

$$
\mathbf{M}_k^{(4)} = \begin{bmatrix} 1 & 4 & 10 & 20 & 35 & 56 & \cdots \\ 4 & 12 & 24 & 40 & 60 & 84 & \cdots \\ 10 & 24 & 42 & 64 & 90 & 120 & \cdots \\ 20 & 40 & 64 & 92 & 124 & 160 & \cdots \\ 35 & 60 & 90 & 124 & 162 & 204 & \cdots \\ 56 & 84 & 120 & 160 & 204 & 252 & \cdots \\ \vdots & \vdots & \vdots & \vdots & \vdots & \vdots \end{bmatrix}, \quad \mathbf{M}_k^{(5)} = \begin{bmatrix} 1 & 5 & 15 & 35 & 70 & 126 & \cdots \\ 5 & 20 & 50 & 100 & 60 & 84 & \cdots \\ 15 & 24 & 42 & 64 & 90 & 120 & \cdots \\ 35 & 40 & 64 & 92 & 124 & 160 & \cdots \\ 70 & 60 & 90 & 124 & 162 & 204 & \cdots \\ 126 & 84 & 120 & 160 & 204 & 252 & \cdots \\ \vdots & \vdots & \vdots & \vdots & \vdots & \vdots \end{bmatrix}.
$$

The correctness of the above formulae will be checked in the following numerical example.

Example 5.1. Verify the correctness of formula (5.18) for two integer orders $n = 1, 2$.

Solution. For $n = 1$ by (5.18) one gets

$$\begin{smallmatrix}GL\\k_0\end{smallmatrix}\Delta_k^{(1)}[f(k)g(k)] = [^{GL}\Delta^{(1)}\boldsymbol{f}(k)]^T \boldsymbol{M}_k^{(1)} {}^{GL}\Delta^{(1)}\boldsymbol{g}(k) \qquad (5.27)$$

$$= \begin{bmatrix} {}^{GL}_{k-1}\Delta_k^{(1)}f(k) \\ {}^{GL}_{k-2}\Delta_{k-1}^{(1)}f(k-1) \\ \vdots \\ {}^{GL}_1\Delta_2^{(1)}f(2) \\ {}^{GL}_0\Delta_1^{(1)}f(1) \\ \vdots \\ {}^{GL}_0\Delta_0^{(1)}f(0) \end{bmatrix}^T \begin{bmatrix} 1\;1\;1\;1\cdots \\ 1\;0\;0\;0\cdots \\ 1\;0\;0\;0\cdots \\ 1\;0\;0\;0\cdots \\ \vdots\;\vdots\;\vdots\;\vdots \end{bmatrix} \begin{bmatrix} {}^{GL}_{k-1}\Delta_k^{(1)}g(k) \\ {}^{GL}_{k-2}\Delta_{k-1}^{(1)}g(k-1) \\ \vdots \\ {}^{GL}_1\Delta_2^{(1)}g(2) \\ {}^{GL}_0\Delta_1^{(1)}g(1) \\ {}^{GL}_0\Delta_0^{(1)}g(0) \end{bmatrix}$$

$$= \begin{bmatrix} {}^{GL}_{k-1}\Delta_k^{(1)}f(k) \\ {}^{GL}_{k-2}\Delta_{k-1}^{(1)}f(k-1) \\ \vdots \\ {}^{GL}_1\Delta_2^{(1)}f(2) \\ {}^{GL}_0\Delta_1^{(1)}f(1) \\ {}^{GL}_0\Delta_0^{(1)}f(0) \end{bmatrix}^T \begin{bmatrix} \sum_{i=0}^k {}^{GL}_{i-1}\Delta_i^{(1)}g(i) \\ {}^{GL}_{k-1}\Delta_k^{(1)}g(k) \\ \vdots \\ {}^{GL}_{k-1}\Delta_k^{(1)}g(k) \\ {}^{GL}_{k-1}\Delta_k^{(1)}g(k)) \\ {}^{GL}_{k-1}\Delta_k^{(1)}g(k) \end{bmatrix}$$

$$= \begin{bmatrix} {}^{GL}_{k-1}\Delta_k^{(1)}f(k) \\ {}^{GL}_{k-2}\Delta_{k-1}^{(1)}f(k-1) \\ \vdots \\ {}^{GL}_1\Delta_2^{(1)}f(2) \\ {}^{GL}_0\Delta_1^{(1)}f(1) \\ {}^{GL}_0\Delta_0^{(1)}f(0) \end{bmatrix}^T \begin{bmatrix} g(k) \\ {}^{GL}_{k-1}\Delta_k^{(1)}g(k) \\ \vdots \\ {}^{GL}_{k-1}\Delta_k^{(1)}g(k) \\ {}^{GL}_{k-1}\Delta_k^{(1)}g(k)) \\ {}^{GL}_{k-1}\Delta_k^{(1)}g(k) \end{bmatrix}$$

$$= g(k){}^{GL}_{k-1}\Delta_k^{(1)}f(k) + \begin{bmatrix} {}^{GL}_{k-2}\Delta_{k-1}^{(1)}f(k) \\ {}^{GL}_{k-3}\Delta_{k-2}^{(1)}f(k-1) \\ \vdots \\ {}^{GL}_1\Delta_2^{(1)}f(2) \\ {}^{GL}_0\Delta_1^{(1)}f(1) \\ {}^{GL}_0\Delta_0^{(1)}f(0) \end{bmatrix}^T \begin{bmatrix} 1 \\ \vdots \\ 1 \\ 1 \\ 1 \end{bmatrix} {}^{GL}_{k-1}\Delta_k^{(1)}g(k)$$

Discrete Fractional Calculus

$$= g(k)_{k-1}^{GL}\Delta_k^{(1)}f(k) + f(k-1)_{k-1}^{GL}\Delta_k^{(1)}g(k)$$

$$= g(k)[f(k) - f(k-1)] + f(k-1)[g(k) - g(k-1)] = f(k)g(k) - f(k-1)g(k-1).$$

Comment. The last formula can be obtained by direct calculation of the first-order backward difference of the product of the two discrete-variable functions

$$_{k-1}^{GL}\Delta_k^{(1)}f(k)g(k) = f(k)g(k) - f(k-1)g(k-1)$$

$$= f(k)g(k) - f(k-1)g(k-1) + f(k-1)g(k) - f(k-1)g(k)$$

$$= g(k)\Delta^{(1)}f(k) + f(k-1)\Delta^{(1)}g(k). \qquad (5.28)$$

The above formula (5.28) is similar to the continuous function case known as the Leibnitz formula [Keisler (2013)]

$$\frac{d}{dt}[f(t)g(t)] = g(t)\frac{d}{dt}f(t) + f(t)\frac{d}{dt}g(t). \qquad (5.29)$$

Similar calculations performed for $n = 2$ yield

$$_0^{GL}\Delta_k^{(2)}[f(k)g(k)] = [_0^{GL}\Delta_k^{(2)}\mathbf{f}(k)]^T\mathbf{M}_k^{(2)}{}_0^{GL}\Delta_k^{(2)}\mathbf{g}(k) \qquad (5.30)$$

$$= \begin{bmatrix} _{k-1}^{GL}\Delta_k^{(2)}f(k) \\ _{k-2}^{GL}\Delta_{k-1}^{(2)}f(k-1) \\ \vdots \\ _1^{GL}\Delta_2^{(2)}f(2) \\ _0^{GL}\Delta_1^{(2)}f(1) \\ \vdots \\ _0^{GL}\Delta_0^{(2)}f(0) \end{bmatrix}^T \begin{bmatrix} 1\,2\,3\,4\,\cdots \\ 2\,2\,2\,2\,\cdots \\ 3\,2\,2\,2\,\cdots \\ 4\,2\,2\,2\,\cdots \\ \vdots\,\vdots\,\vdots\,\vdots \end{bmatrix} \begin{bmatrix} _{k-1}^{GL}\Delta_k^{(2)}g(k) \\ _{k-2}^{GL}\Delta_{k-1}^{(2)}g(k-1) \\ \vdots \\ _1^{GL}\Delta_2^{(2)}g(2) \\ _0^{GL}\Delta_1^{(2)}g(1) \\ _0^{GL}\Delta_0^{(2)}g(0) \end{bmatrix}$$

$$= \begin{bmatrix} _{k-1}^{GL}\Delta_k^{(2)}f(k) \\ _{k-2}^{GL}\Delta_{k-1}^{(2)}f(k-1) \\ \vdots \\ _1^{GL}\Delta_2^{(2)}f(2) \\ _0^{GL}\Delta_1^{(2)}f(1) \\ _0^{GL}\Delta_0^{(2)}f(0) \end{bmatrix}^T \begin{bmatrix} _0^{GL}\Delta_k^{(-2)}[_0^{GL}\Delta_k^{(2)}g(k)] \\ 2_0^{GL}\Delta_k^{(-1)}[_0^{GL}\Delta_k^{(2)}g(k)] \\ 2_{k-1}^{GL}\Delta_k^{(2)}g(k) \\ 2_{k-1}^{GL}\Delta_k^{(2)}g(k) \\ 2_{k-1}^{GL}\Delta_k^{(2)}g(k) \\ \vdots \end{bmatrix}$$

$$
= \begin{bmatrix} {}^{GL}_{k-1}\Delta^{(2)}_k f(k) \\ {}^{GL}_{k-2}\Delta^{(2)}_{k-1} f(k-1) \\ \vdots \\ {}^{GL}_1\Delta^{(2)}_2 f(2) \\ {}^{GL}_0\Delta^{(2)}_1 f(1) \\ {}^{GL}_0\Delta^{(2)}_0 f(0) \end{bmatrix}^T \begin{bmatrix} g(k) \\ 2{}^{GL}_{k-1}\Delta^{(1)}_k g(k) \\ 1{}^{GL}_{k-1}\Delta^{(2)}_k g(k) + 2{}^{GL}_{k-1}\Delta^{(1)}_k g(k) \\ 2{}^{GL}_{k-1}\Delta^{(2)}_k g(k) + 2{}^{GL}_{k-1}\Delta^{(1)}_k g(k) \\ 3{}^{GL}_{k-1}\Delta^{(2)}_k g(k) + 2{}^{GL}_{k-1}\Delta^{(1)}_k g(k) \\ 4{}^{GL}_{k-1}\Delta^{(2)}_k g(k) + 2{}^{GL}_{k-1}\Delta^{(1)}_k g(k) \\ \vdots \end{bmatrix}
$$

$$
= \begin{bmatrix} {}^{GL}_{k-1}\Delta^{(2)}_k f(k) \\ {}^{GL}_{k-2}\Delta^{(2)}_{k-1} f(k-1) \\ \vdots \\ {}^{GL}_1\Delta^{(2)}_2 f(2) \\ {}^{GL}_0\Delta^{(2)}_1 f(1) \\ {}^{GL}_0\Delta^{(2)}_0 f(0) \end{bmatrix}^T \left(\begin{bmatrix} g(k) - 2{}^{GL}_{k-1}\Delta^{(1)}_k g(k) \\ 0 \\ 1{}^{GL}_{k-1}\Delta^{(2)}_k g(k) \\ 2{}^{GL}_{k-1}\Delta^{(2)}_k g(k) \\ 3{}^{GL}_{k-1}\Delta^{(2)}_k g(k) \\ 4{}^{GL}_{k-1}\Delta^{(2)}_k g(k) \\ \vdots \end{bmatrix} + 2{}^{GL}_{k-1}\Delta^{(1)}_k g(k) \begin{bmatrix} 1 \\ 1 \\ 1 \\ 1 \\ \vdots \end{bmatrix} \right)
$$

$$
= ({}^{GL}_{k-1}\Delta^{(2)}_k f(k))[-g(k) + 2g(k-1)] + {}^{GL}_{k-1}\Delta^{(2)}_k g(k) \begin{bmatrix} {}^{GL}_{k-3}\Delta^{(2)}_{k-2} f(k-2) \\ {}^{GL}_{k-4}\Delta^{(2)}_{k-3} f(k-3) \\ \vdots \\ {}^{GL}_1\Delta^{(2)}_2 f(2) \\ {}^{GL}_0\Delta^{(2)}_1 f(1) \\ {}^{GL}_0\Delta^{(2)}_0 f(0) \end{bmatrix}^T \begin{bmatrix} 1 \\ 2 \\ 3 \\ 4 \\ \vdots \end{bmatrix}
$$

$$
+ 2{}^{GL}_{k-1}\Delta^{(1)}_k g(k) \begin{bmatrix} {}^{GL}_{k-2}\Delta^{(2)}_k f(k) \\ {}^{GL}_{k-3}\Delta^{(2)}_{k-1} f(k-1) \\ \vdots \\ {}^{GL}_1\Delta^{(2)}_2 f(2) \\ {}^{GL}_0\Delta^{(2)}_1 f(1) \\ {}^{GL}_0\Delta^{(2)}_0 f(0) \end{bmatrix}^T \begin{bmatrix} 1 \\ 1 \\ 1 \\ 1 \\ \vdots \end{bmatrix}
$$

$$
= {}^{GL}_{k-2}\Delta^{(2)}_k f(k)[-g(k) + 2g(k-1)] + {}^{GL}_{k-4}\Delta^{(2)}_{k-2} f(k-2){}^{GL}_{k-2}\Delta^{(2)}_k g(k)
$$

$$
+ 2{}^{GL}_{k-1}\Delta^{(1)}_k g(k){}^{GL}_{k-1}\Delta^{(1)}_k f(k)
$$

$$
= f(k)g(k) - 2f(k-1)g(k-1) + f(k-2)g(k-2).
$$

Corollary 5.1. *The FOBD of a product $f(k)g(k)$ is equal to the FOBD of a product $g(k)f(k)$.*

$$\,_{0}^{GL}\Delta_{k}^{(\nu)} f(k)g(k) =\,_{0}^{GL}\Delta_{k}^{(\nu)}g(k)f(k). \tag{5.31}$$

Proof. By Theorem 5.2

$$\,_{k_0}^{GL}\Delta_{k}^{(\nu)}[f(k)g(k)] = [\,^{GL}\Delta^{(\nu)}f(k)]^{T} M_{k}^{(\nu)\,GL}\Delta^{(\nu)}g(k) \tag{5.32}$$

$$= [[\,^{GL}\Delta^{(\nu)}f(k)]^{T} M_{k}^{(\nu)\,GL}\Delta^{(\nu)}g(k)]^{T}$$

$$= [\,^{GL}\Delta^{(\nu)}g(k)]^{T} M_{k}^{(\nu)}]^{T}[\,^{GL}\Delta^{(\nu)}f(k) = [\,^{GL}\Delta^{(\nu)}g(k)]^{T} M_{k}^{(\nu)\,GL}\Delta^{(\nu)}f(k)$$

$$=\,_{k_0}^{GL}\Delta_{k}^{(\nu)}[g(k)f(k)].$$

$\square$

Another one of the possible applications of the FO Leibnitz Rule is demonstrated in the numerical example given below.

Example 5.2. Consider a function $f(k) = k1(k)$. Knowing that $\,_{k-1}^{GL}\Delta_{k}^{(1)}f(k) =\,_{k-1}^{GL}\Delta_{k}^{(1)}k1(k) = 1(k) = 1(k)$ calculate $\,_{k-1}^{GL}\Delta_{k}^{(1)}k^2 1(k)$ using formula (5.18).

Solution. Denote $f(k) = g(k) = k1(k)$. Then for $\nu = n = 1$

$$\,_{k-1}^{GL}\Delta_{k}^{(1)}k^2 1(k) \tag{5.33}$$

$$= \begin{bmatrix} \,_{k-1}^{GL}\Delta_{k}^{(1)}f(k) \\ \,_{k-2}^{GL}\Delta_{k-1}^{(1)}f(k-1) \\ \vdots \\ \,_{1}^{GL}\Delta_{2}^{(1)}f(2) \\ \,_{0}^{GL}\Delta_{1}^{(1)}f(1) \\ \,_{0}^{GL}\Delta_{0}^{(1)}f(0) \end{bmatrix}^{T} \begin{bmatrix} 1\ 1\ 1\ 1\cdots \\ 1\ 0\ 0\ 0\cdots \\ 1\ 0\ 0\ 0\cdots \\ 1\ 0\ 0\ 0\cdots \\ \vdots\ \vdots\ \vdots\ \vdots \end{bmatrix} \begin{bmatrix} \,_{k-1}^{GL}\Delta_{k}^{(1)}g(k) \\ \,_{k-2}^{GL}\Delta_{k-1}^{(1)}g(k-1) \\ \vdots \\ \,_{1}^{GL}\Delta_{2}^{(1)}g(2) \\ \,_{0}^{GL}\Delta_{1}^{(1)}g(1) \\ \,_{0}^{GL}\Delta_{0}^{(1)}g(0) \end{bmatrix}$$

$$= [1\ 1\ 1\ \cdots\ 1\ 1] \begin{bmatrix} 1\ 1\ 1\ 1\cdots \\ 1\ 0\ 0\ 0\cdots \\ 1\ 0\ 0\ 0\cdots \\ 1\ 0\ 0\ 0\cdots \\ \vdots\ \vdots\ \vdots\ \vdots \end{bmatrix} \begin{bmatrix} 1 \\ 1 \\ \vdots \\ 1 \\ 1 \\ 1 \end{bmatrix}$$

$$= \begin{bmatrix} 1 & 1 & 1 & \cdots & 1 & 1 \end{bmatrix} \begin{bmatrix} k\mathbf{1}(k) \\ 1 \\ \vdots \\ 1 \\ 1 \\ 1 \end{bmatrix} = k\mathbf{1}(k) + (k-1)\mathbf{1}(k-1) = (2k-1)\mathbf{1}(k-1).$$

5.4 One-sided $\mathcal{Z}$-transform of the FOBD

In this Section the one-sided $\mathcal{Z}$-transform [Jury (1964); Ostalczyk (2002); Vich (1987)] of the FOBD is derived. Two distinct cases concerning the transformed discrete-time function $f(k)$ will be considered. One assumes that the one-sided $\mathcal{Z}$ transform of the function $f(k)$ exists and is denoted as $F(z)$. Subsequently

$$\mathcal{Z}\{^{GL}_0\Delta^{(\nu)}_k f(k)\} = \mathcal{Z}\{\sum_{i=0}^{k} a^{(\nu)}(i)f(k-i)\}$$

$$= \mathcal{Z}\{\sum_{i=0}^{k} a^{(\nu)}(i)f(k-i) + \sum_{i=k+1}^{+\infty} a^{(\nu)}(i)f(k-i) - \sum_{i=k+1}^{+\infty} a^{(\nu)}(i)f(k-i)\}$$

$$= \mathcal{Z}\{\sum_{i=0}^{+\infty} a^{(\nu)}(i)f(k-i)\} - \mathcal{Z}\{\sum_{i=k+1}^{+\infty} a^{(\nu)}(i)f(k-i)\}. \quad (5.34)$$

Next one examines the first right-hand term in 5.34.

$$\mathcal{Z}\{\sum_{i=0}^{+\infty} a^{(\nu)}(i)f(k-i)\} = \sum_{i=0}^{+\infty} a^{(\nu)}(i)\mathcal{Z}\{f(k-i)\}$$

$$= \sum_{i=0}^{+\infty} a^{(\nu)}(i)z^{-i}F(z) = (1-z^{-1})^{\nu}F(z). \quad (5.35)$$

The second term is slightly more complicated due to assumptions concerning the transofrmed function $f(k)$. By definition of the one-sided $\mathcal{Z}$-Transform

$$\mathcal{Z}\{\sum_{i=k+1}^{+\infty} a^{(\nu)}(i)f(ki)\} = \sum_{j=0}^{+\infty} \sum_{i=j+1}^{+\infty} a^{(\nu)}(i)f(j-i)$$

$$= \begin{bmatrix} z^0 & z^{-1} & z^{-2} & \cdots \end{bmatrix} \mathbf{H}^{(\nu)}_{\infty} \begin{bmatrix} f(-1) \\ f(-2) \\ f(-3) \\ \vdots \end{bmatrix}. \quad (5.36)$$

where

$$\mathbf{H}_\infty^{(\nu)} = \begin{bmatrix} a^{(\nu)}(1) \ a^{(\nu)}(2) \ a^{(\nu)}(3) \ \cdots \\ a^{(\nu)}(2) \ a^{(\nu)}(3) \ a^{(\nu)}(4) \ \cdots \\ a^{(\nu)}(3) \ a^{(\nu)}(4) \ a^{(\nu)}(5) \ \cdots \\ \vdots \qquad \vdots \qquad \vdots \end{bmatrix} \tag{5.37}$$

is so-called Hankel matrix. [Kailath (1980)]. To emphasize that funtion $f(k)$ values are related to negative arguments they will be denoted as $f_{-1}, f_{-2}, f_{-3}, \ldots$. Therefore,

$$\mathcal{Z}\{{}_0^{GL}\Delta_k^{(\nu)}f(k)\}$$

$$= (1 - z^{-1})^\nu F(z) - \begin{bmatrix} z^0 \ z^{-1} \ z^{-2} \cdots \end{bmatrix} \mathbf{H}_\infty^{(\nu)} \begin{bmatrix} f_{-1} \\ f_{-2} \\ f_{-3} \\ \vdots \end{bmatrix}. \tag{5.38}$$

The second term on the right-hand side will introduce the initial conditions in the solutions of the linear FO difference equations using the one-sided $\mathcal{Z}$-Transform method. It is useful to perform a further transformation of this term, as below

$$\begin{bmatrix} z^0 \ z^{-1} \ z^{-2} \cdots \end{bmatrix} [\mathbf{P}_\infty]^{\mathrm{T}} [\mathbf{P}_\infty]^{\mathrm{T}} \mathbf{H}_\infty^{(\nu)} [\mathbf{P}_\infty]^{\mathrm{T}} [\mathbf{P}_\infty]^{\mathrm{T}} \begin{bmatrix} f_{-1} \\ f_{-2} \\ f_{-3} \\ \vdots \end{bmatrix}$$

$$= \begin{bmatrix} (1 - z^{-1})^0 \ (1 - z^{-1})^1 (1 - z^{-1})^2 \cdots \end{bmatrix} [\mathbf{P}_\infty]^{\mathrm{T}} \mathbf{H}_\infty [\mathbf{P}_\infty]^{\mathrm{T}} \begin{bmatrix} f_{-1} \\ \Delta^1 f_{-1} \\ \Delta^2 f_{-1} \\ \vdots \end{bmatrix}. \tag{5.39}$$

Now, one considers a function $f(k)$ satisfying the condition $f(k) = 0$ for $k < 0$. Realize that such a function can be denoted as $f(k)\mathbf{1}(k)$. Then,

$$\mathcal{Z}\{{}_0^{GL}\Delta_k^{(\nu)}[f(k)\mathbf{1}(k)]\} = (1 - z^{-1})^\nu F(z). \tag{5.40}$$

Comment. Formulas (5.38) and (5.39) are also valid for integer orders $\nu = n \in \mathbb{Z}_+$.

The validity of formula (5.37) will be confirmed in the following numerical example concerning a solution of an integer-order difference equation (IODE).

Example 5.3. The solution to the second-order DE

$$y'(k' + 2) + (z_1 + z_2)y'(k' + 1) + z_1 z_2 y'(k') = 0, \tag{5.41}$$

subjected to two initial conditions $y'(0)$ and $y'(1)$ is of the form [Ogata (1987)]

$$y'(k') = \mathcal{Z}^{-1}\left\{Y'(z)\right\}$$

$$= \mathcal{Z}^{-1}\left\{\frac{z^2 y(0) + z(z_1 + z_2)y(0) + zy(1)}{z^2 + (z_1 + z_2)z + z_1 z_2}\right\} \tag{5.42}$$

$$= \frac{z_2 y(0) + y(1)}{z_2 - z_1}(-z_1)^{k'} - \frac{z_1 y(0) + y(1)}{z_2 - z_1}(-z_2)^{k'}.$$

One should find a solution using the one-sided $\mathcal{Z}$-Transform method applied to the modified equation (5.41) with $z_1 = -1/2$, $z_2 = -1/3$ and $y'(0) = 1, y'(1) = 2$. For the assumed values

$$Y'(z) = \frac{z^2 + \frac{7}{6}z}{z^2 - \frac{5}{6}z + \frac{1}{6}}, \tag{5.43}$$

$$y'(k) = 10\left(\frac{1}{2}\right)^k - 9\left(\frac{1}{3}\right)^k.$$

Here it is worth recalling that in a second-order difference equation to get a unique solution one needs two initial conditions.

Solution. First one should transform equation (5.41) into an equation containing only shifted back functions. A substitution

$$k = k' + 2, \tag{5.44}$$

yields

$$y(k) - \frac{5}{6}y(k-1) + \frac{1}{6}y(k-2) = 0. \tag{5.45}$$

Similarly, one should also transform the initial conditions

$$y_{-1} = y(1) = 2, \quad y_{-2} = y(0) = 1. \tag{5.46}$$

IODE (5.46) can be further transformed to an equivalent form, as follows

$$y(k) - \frac{5}{6}y(k-1) + \frac{1}{6}y(k-2) = \begin{bmatrix} 1 & -\frac{5}{6} & \frac{1}{6} \end{bmatrix}\begin{bmatrix} y(k) \\ y(k-1) \\ y(k-2) \end{bmatrix}$$

$$= \begin{bmatrix} 1 & -\frac{5}{6} & \frac{1}{6} \end{bmatrix}[\mathbf{P}_3]^{\mathrm{T}}[\mathbf{P}_3]^{\mathrm{T}}\begin{bmatrix} y(k) \\ y(k-1) \\ y(k-2) \end{bmatrix} = \begin{bmatrix} \frac{1}{6} & \frac{1}{2} & \frac{1}{3} \end{bmatrix}\begin{bmatrix} \Delta^2 y(k) \\ \Delta^1 y(k) \\ \Delta^0 y(k-2) \end{bmatrix}$$

$$= \Delta^2 y(k) + 3\Delta y(k) + y(k) = 0. \tag{5.47}$$

Multiplying both sides by 6 one gets the final form.

$$\Delta^2 y(k) + 3\Delta y(k) + y(k) = 0. \tag{5.48}$$

To solve this equation one applies the one-sided $\mathcal{Z}$-Transform method

$$\mathcal{Z}\left\{\Delta^2 y(k) + 3\Delta y(k) + 2y(k)\right\} = \mathcal{Z}\left\{\Delta^2 y(k)\right\} + 3\mathcal{Z}\left\{\Delta y(k)\right\} + 2y(k) = 0. \tag{5.49}$$

According to formula (5.38) the first term in (5.49) is equal to

$$\mathcal{Z}\left\{\Delta^2 y(k)\right\} = (1 - z^{-1})^2 + \begin{bmatrix} 1 & z^{-1} & z^{-2} & \cdots \end{bmatrix} \mathbf{H}_\infty^{(2)} \begin{bmatrix} y_{-1} \\ y_{-2} \\ y_{-3} \\ \vdots \end{bmatrix}. \tag{5.50}$$

In the considered example the Hankel matrix $\mathbf{H}_\infty^{(2)}$ has the form

$$\mathbf{H}_\infty^{(2)} = \begin{bmatrix} -2 & 1 & 0 & \cdots \\ 1 & 0 & 0 & \cdots \\ 0 & 0 & 0 & \cdots \\ \vdots & \vdots & \vdots & \end{bmatrix} \tag{5.51}$$

and this matrix "cuts out " all remaining initial conditions. This is a fundamental difference between the IODEs and its FO counterpart, where matrix $\mathbf{H}_\infty^{(\nu)}$ has all non-zero coefficients. Substituting (5.51) into (5.50) means

$$\mathcal{Z}\left\{\Delta^2 y(k)\right\} = (1 - z^{-1})^2 + \begin{bmatrix} 1 & z^{-1} & z^{-2} & \cdots \end{bmatrix} \begin{bmatrix} 2y_{-1} + y_{-2} \\ f_{-1} \\ 0 \\ \vdots \end{bmatrix}$$

$$= (1 - z^{-1})^2 + \begin{bmatrix} 1 & z^{-1} \end{bmatrix} \begin{bmatrix} 2y_{-1} + y_{-2} \\ f_{-1} \end{bmatrix} = (1 - z^{-1})^1 - 2y_{-1} + y_{-2} + y_{-1}z^{-1}. \tag{5.52}$$

The second term is more simple because of the Hankel matrix

$$\mathbf{H}_\infty^{(1)} = \begin{bmatrix} -1 & 0 & \cdots \\ 0 & 0 & \cdots \\ \vdots & \vdots & \end{bmatrix} \tag{5.53}$$

cutting all initial conditions except the first y_{-1} one. Hence,

$$\mathcal{Z}\left\{\Delta y(k)\right\} = (1 - z^{-1})^1 + \begin{bmatrix} 1 \, z^{-1} & \cdots \end{bmatrix} \begin{bmatrix} y_{-1} \\ y_{-2} \\ \vdots \end{bmatrix} = (1 - z^{-1})^1 - y_{-1}. \qquad (5.54)$$

Now one collects calculation effects (5.52) and (5.54) multiplied by appropriate coefficients. The result is of the form

$$(1 - z^{-1})^2 Y(z) - 2y_{-1} + y_{-2} + y_{-1} z^{-1}$$

$$+3 \left[(1 - z^{-1})^1 Y(z) - 2y_{-1} + y_{-2} + y_{-1} z^{-1} \right] + 2Y(z) = 0. \qquad (5.55)$$

After simple rearrangements of the above equation, one obtains the one-sided $\mathcal{Z}$-Transform of the solution to equation (5.46)

$$Y(z) = \frac{(5y_{-1} - y_{-2}) - y_{-1} z^{-1}}{6 - 5z^{-1} + z^{-2}} = \frac{9 - 2z^{-1}}{6 - 5z^{-1} + z^{-2}}$$

$$= \frac{\frac{3}{2} - \frac{1}{3} z^{-1}}{1 - \frac{5}{6} z^{-1} + \frac{1}{6} z^{-2}} = \frac{\frac{3}{2} z^2 - \frac{1}{3} z}{z^2 - \frac{5}{6} z + \frac{1}{6}}. \qquad (5.56)$$

To compare the obtained solutions it is sufficient to check its transforms (5.44) and (5.56). In fact

$$Y'(z) = 1 + 2z^{-1} + z^{-2} Y(z) = y(0) + y(1)z^{-1} + z^{-2} Y(z) \qquad (5.57)$$

which means that both solutions $y(k)$ and $y'(k)$ are of the same shape but $y(k)$ is shifted back two steps with respect to $y'(k)$. The solution $y(k)$ begins at $k = 0$ and its intial values y_{-1}, y_{-2} belong to the "past". Two initial conditions $y'(0)$ and $y'(1)$ are included in the solution $y'(k)$ which *de facto* begins at $k = 2$.

The one-sided $\mathcal{Z}$-Transform of the FOBD needs an infinitely large number of initial conditions $y_{-1}, y_{-2}, \ldots$. They carry information about the "past" — treated as the memory. Moreover, the function $a^{(\nu)}$ shape (which strongly diminishes) reveals a property that the influence of the initial condition y_{-i} on the value of the $\mathcal{Z}$-Transform strongly decreases due to an increasing distance $| - i|$.

Chapter 6

The FO dynamic system description

In this chapter, the FOBD fundamental application is considered. The FOBD is commonly used in the FO differential equation (FODE) [Diethelm (1995); Gorenflo and Mainardi (1997a); Diethelm (2010, 2011); Deng and Li (2012)], which describes the real dynamic plants or physical phenomena [Carlson and Halijak (1964); Caputo and Mainardi (1971); Bagley and Torvik (1984); Arena *et al.* (2000); Barkai (2002); Aoun *et al.* (2004); Caponetto *et al.* (2010)] that occupy the vast majority of the real world. A relation between differential equations and their discrete approximation in the simplest form using the GL-FOBD are presented. This means that the independent variable — the continuous time t — is replaced by the discrete one, $t = kh$ with constant sampling period h. For knowing h one can establish a simple relation between the continuous- and discrete-time scale and assume in considerations that $h = 1$. There are many types of differential and difference equations describing the aforementioned objects or physical phenomena: non-linear/linear, time-variant/time-invariant. A notion of commensurate/non-commensurate equations is discussed. An important class of systems, from a practical point of view, constitute so-called positive systems [Kaczorek (2004, 2009a,b, 2010, 2011a,b,c, 2014a,b)]. The first and the simplest approach in the non-linear system described by the non-linear FODE is the linearization procedure. The unique solution of the FODE is related to the known initial conditions and single- or multi-input signals. The initial conditions contain the FO system memory of dynamic system's past behavior. An important form of the FOE is the so-called state-space form [Raynaud and Zergaïnoh (2000); Dzieliński and Sierociuk (2005, 2006b, 2007a,b,c)]. Similar to classical discrete linear systems, the FO transfer function description is compact and elegant[Barbosa (2004a)]. Relations between selected forms of the FODSs are presented. In most cases without loss of generality there will be assumed an initial time instant $t_0 = k_0 h = 0$. Such an assumption suppresses technical considerations.

6.1 Linear time-invariant system description by the FO difference equation

Any classical differential equation (linear or non-linear, time-variant or time-invariant) that describes a single-input single-output (SISO) causal dynamical system [Kailath (1980); Kaczorek (1992); Khalil (2002)] with input signal $u(t)$ and output $y(t)$ is mainly expressed in the form

$$F\left[\frac{d^p y(t)}{dt^p}, \frac{d^{p-1} y(t)}{dt^{p-1}}, \dots, \frac{y(t)}{dt}, y(t), \frac{d^q u(t)}{dt^q}, \frac{d^{q-1} u(t)}{dt^{q-1}}, \dots, \frac{u(t)}{dt}, u(t), t\right] = 0 \quad (6.1)$$

where $p \geqslant q$ and an input signal $u(t)$ with appropriate derivatives is known for $t_0 \geqslant 0$. Its fractional-order counterpart contains FO derivatives

$$F\left[{}^{GL}_{t_0}D_t^{\nu_p} y(t), {}^{GL}_{t_0}D_t^{\nu_{p-1}} y(t), \dots, {}^{GL}_{t_0}D_t^{\nu_1} y(t), y(t), {}^{GL}_{t_0}D_t^{\mu_q} u(t), \dots, {}^{GL}_{t_0}D_t^{\mu_1} u(t), u(t), t\right]$$
$$= 0$$
$$(6.2)$$

where ${}^{GL}_{t_0}D_t^{\nu_i} y(t)$, ${}^{GL}_{t_0}D_t^{\mu_j} u(t)$ for $i = 1, 2, \dots, p$, $j = 1, 2, \dots, q$ denote the Grüwald-Letnikov fractional left derivative [Podlubny (1999a)]. Following the notation (6.1) one sorts the FOs

$$\nu_p > \nu_{p-1} > \dots > \nu_1 > 0, \ \mu_p > \mu_{p-1} > \dots > \mu_1 > 0. \quad (6.3)$$

Performing a substitution of the Grüwald-Letnikov fractional left derivative by the FOBD divided by a sampling time (which should be relatively small) in (6.2)

$$\substack{GL\\t_0}D_t^{(\nu_i)} y(t) \approx \frac{\substack{GL\\k_0h}\Delta_{kh}^{(\nu_i)} y(kh)}{h^{\nu_i}}, \ \substack{GL\\t_0}D_t^{(\mu_j)} u(t) \approx \frac{\substack{GL\\k_0h}\Delta_{kh}^{(\nu_j)} u(kh)}{h^{\mu_j}} \quad (6.4)$$

for $i = 1, 2, \dots, p$, $j = 1, 2, \dots, q$ one obtains the FO difference equation (FODE)

$$F\left[\substack{GL\\k_0h}\Delta_{kh}^{(\nu_p)} y(kh), \dots, \substack{GL\\k_0h}\Delta_{kh}^{(\nu_1)} y(kh), y(kh),\right.$$
$$\left. \substack{GL\\k_0h}\Delta_{kh}^{(\mu_q)} u(kh), \dots, \substack{GL\\k_0h}\Delta_{kh}^{(\mu_1)} u(kh), u(kh), kh\right] = 0. \quad (6.5)$$

For a known and constant sampling time $h = \text{const}$ there is used a simplified notation omitting h. Hence, the FODE (6.5) takes the form

$$F\left[\substack{GL\\k_0}\Delta_k^{(\nu_p)} y(k), \dots, \substack{GL\\k_0}\Delta_k^{(\nu_1)} y(k), y(k), \substack{GL\\k_0}\Delta_k^{(\mu_q)} u(k), \dots, \substack{GL\\k_0}\Delta_k^{(\mu_1)} u(k), u(k), k\right]$$
$$= 0$$
$$(6.6)$$

A formal definition of the FODE is given below.

Definition 6.1 (FODE). The DE is the FODE if at least one of the orders is an element of a set of positive non-integers i.e. there is $\nu_i, \mu_j \in \mathbb{R}_+ \setminus \mathbb{Z}_+$.

Definition 6.2 (FODE order). The FODE order is the highest order of a set $\nu_p, \nu_{p-1}, \dots, \nu_1$.

Comment. Recall that in a causal dynamical systems a soft inequality $\nu_p \geqslant \mu_q$ is always satisfied.

In the FODE (6.6), the unknown function is $y(k)$ (this is the FODE solution or the dynamical system output signal), whereas the function $u(k)$ together with its FOBDs is known for $k \geqslant k_0$. To obtain a unique solution of the FODE one should define a set of initial conditions. The initial condition problem will be discussed in details in Subsection 7.1.2. The FODE is called a homogenous equation when $u(k) = 0$ for $k \geqslant k_0$: otherwise it is referred to as a non-homogenous equation [Kailath (1980)]. If the function F in (6.6) is only a linear combination of all differences of $y(k)$ then it is called a linear FODE. This definition is less restrictive because the non-linear combinations of $u(k)$ are permitted. In dynamical system theory [Kailath (1980)], [O'Flynn and Moriarty (1987)], one imposes an additional assumption concerning linear combinations of input signal $u(k)$ and their differences.

6.1.1 *Commensurate and non-commensurate FODS*

Now one considers two separate classes of the FODEs - commensurate and non-commensurate FODEs. The notion is related to the orders ν_i and μ_j.

Definition 6.3 (Commensurate FODE). The FODE is a commensurate FODE if all orders $\nu_p > \nu_{p-1} > \cdots > \nu_1 > 0$, $\mu_p > \mu_{p-1} > \cdots > \mu_1 > 0$ are rational numbers i.e. $\nu_p, \ldots, \nu_1, \mu_q, \ldots, \mu_1 \in \mathbb{Q}_+$.

Definition 6.4 (Non-commensurate FODE). The FODE is a non-commensurate FODE if there is at least one order being an irrational number.

Modeling real dynamical systems or physical processes does not lead to its mathematical models described only by integer-orders difference equations (IODEs). Moreover, in a mentioned practical case one can state that with a probability equal to zero, there will be no irrational orders in the FODE description. So, due to this rather obvious assumption, further discussion will be concerned with the commensurate FODEs. Hence, the commensurate FODE of any rational orders can be transformed to the ordered form. Let all orders $\nu_p > \nu_{p-1} > \cdots > \nu_1 > 0$, $\mu_p > \mu_{p-1} > \cdots > \mu_1 > 0$ be expressed in the form

$$\nu_i = \frac{e_i}{d_i} \text{ for } i = 1, 2, \ldots, p, \text{ and} e_i, d_i \in \mathbb{Z}_+ \tag{6.7}$$

$$\mu_i = \frac{g_i}{f_i} \text{ for } j = 1, 2, \ldots, q, \text{ and} g_i, f_i \in \mathbb{Z}_+. \tag{6.8}$$

Let d be the least common denominator of fractions (6.7) and (6.8). Then the above orders take the following forms

$$\nu_i = \frac{n_i}{d} \text{ for } i = 1, 2, \ldots, p, \text{ and } e_i, d_i \in \mathbb{Z}_+ \tag{6.9}$$

$$\mu_i = \frac{m_i}{d} \text{ for } j = 1, 2, \ldots, q, \text{ and } g_i, f_i \in \mathbb{Z}_+ \tag{6.10}$$

where

$$\nu = \frac{1}{d}. \tag{6.11}$$

Next, introducing a notation

$$_{k_0h}^{GL}\Delta_{kh}^{(\frac{1}{d})}y(kh) = {}_{k_0h}^{GL}\Delta_{kh}^{(\nu)}y(kh) \tag{6.12}$$

$$_{k_0h}^{GL}\Delta_{kh}^{(\frac{1}{d})}u(kh) = {}_{k_0h}^{GL}\Delta_{kh}^{(\nu)}u(kh) \tag{6.13}$$

the appropriate FOBDs are as follows

$$_{k_0h}^{GL}\Delta_{kh}^{(\nu_i)}y(kh) = {}_{k_0h}^{GL}\Delta_{kh}^{(n_i\nu)}y(kh) = \left[\underbrace{{}_{k_0h}^{GL}\Delta_{kh}^{(\nu)}{}_{k_0h}^{GL}\Delta_{kh}^{(\nu)} \cdots {}_{k_0h}^{GL}\Delta_{kh}^{(\nu)}}_{n_i}\right]y(kh) \tag{6.14}$$

and

$$_{k_0h}^{GL}\Delta_{kh}^{(\mu_j)}u(kh) = {}_{k_0h}^{GL}\Delta_{kh}^{(m_j\nu)}u(kh) = \left[\underbrace{{}_{k_0h}^{GL}\Delta_{kh}^{(\nu)}{}_{k_0h}^{GL}\Delta_{kh}^{(\nu)} \cdots {}_{k_0h}^{GL}\Delta_{kh}^{(\nu)}}_{m_j}\right]u(kh). \tag{6.15}$$

In the light of equalities (6.14) and (6.15), the FODE transforms to a form of the IODE

$$F\left[{}_{k_0}^{GL}\Delta_{k}^{(n_p\nu)}y(k), {}_{k_0}^{GL}\Delta_{k}^{(n_p-1\nu)}y(k), \ldots, {}_{k_0}^{GL}\Delta_{k}^{(\nu)}y(k), y(k),\right.$$
$$\left.{}_{k_0}^{GL}\Delta_{k}^{(m_q\nu)}u(k), {}_{k_0}^{GL}\Delta_{k}^{(m_q-1\nu)}u(k), \ldots, u(k)\right] = 0. \tag{6.16}$$

In Table 6.1 some exemplary FODEs are given.

The linear time-invariant commensurate FODEs comprise the most important class. Therefore, in this book considerations will be focused on this type of the FOBDs. Evidently, all IODEs are of the commensurate type. The following numerical shows how to get a FODE of the commensurate type.

Example 6.1. Consider a linear FODE

$$_{0}^{GL}\Delta_{k}^{(1)}y(k) + a_2{}_{0}^{GL}\Delta_{k}^{(\frac{1}{2})}y(k) + a_1{}_{0}^{GL}\Delta_{k}^{(\frac{1}{3})}y(k) + a_0y(k) = b_0u(k).$$

Table 6.1   Examples of the FODEs.

FODE type	Example for $\nu = \frac{1}{d}, d \in a$
Non-linear, time-variant, non-homogenous	${}^{GL}_{0}\Delta_k^{(\nu)}y(k) + k^2 y(k) + y^2(k) = 0$
Non-linear, time-invariant, homogenous	${}^{GL}_{0}\Delta_k^{(\nu)}y(k) + 3y(k) + y^2(k) = 0$
Linear, time-invariant, homogenous, non-commensurate	${}^{GL}_{0}\Delta_k^{(\nu)}y(k) + {}^{GL}_{0}\Delta_k^{(\pi)}y(k) + 3y(k) + y(k) = 0$
Linear, time-invariant, homogenous, commensurate	${}^{GL}_{0}\Delta_k^{(3\nu)}y(k) + 2{}^{GL}_{0}\Delta_k^{(\nu)}y(k) + 3y(k) + y(k) = 0$
Linear, time-invariant, homogenous, commensurate	${}^{GL}_{0}\Delta_k^{(3\nu)}y(k) + 2{}^{GL}_{0}\Delta_k^{(\nu)}y(k) + 3y(k) = 4{}^{GL}_{0}\Delta_k^{(\nu)}u(k) + u(k)$

In this FODE

$$d = 6, \; \nu = \frac{1}{d} = \frac{1}{6}.$$

Hence,

$${}^{GL}_{0}\Delta_k^{(6\nu)}y(k) + a_2 {}^{GL}_{0}\Delta_k^{(3\nu)}y(k) + a_1 {}^{GL}_{0}\Delta_k^{(2\nu)}y(k) + a_0 y(k) = b_0 u(k).$$

The linear time-invariant FODE (6.16) can be equivalently expressed in the general form

$$\sum_{i=0}^{p} a_i {}^{GL}_{k_0}\Delta_k^{(i\nu)}y(k) = \sum_{j=0}^{q} b_j {}^{GL}_{k_0}\Delta_k^{(j\mu)}u(k) \qquad (6.17)$$

where a_i, b_j and $a_p = 1$ are constant coefficients, $p \geqslant p$, and $u(k)$ is a known input signal. For $d = 1$ (6.11) i.e. $\nu = 1$ a considered FOBD represents the classical integer-order difference equations (IOBD). Its most popular form can be easily obtained performing the following transformations. Substituting a considered value for $\nu = 1$, one obtains

$$\sum_{i=0}^{p} a_i \Delta^i y(k) = \sum_{j=0}^{q} b_j \Delta_k^j u(k) \qquad (6.18)$$

and further

$$\sum_{i=0}^{p} a_i \sum_{j=0}^{k} a^{(i)}(j)y(k-j) = \sum_{j=0}^{q} b_j \sum_{j=0}^{k} a^{(j)}(j)u(k-j) \qquad (6.19)$$

or equivalently in a vector form

$$
\begin{bmatrix} a_0 \; a_1 \; \cdots \; a_p \end{bmatrix}
\begin{bmatrix}
\sum_{j=0}^{k} a^{(0)}(j)y(k-j) \\
\sum_{j=0}^{k} a^{(1)}(j)y(k-j) \\
\vdots \\
\sum_{j=0}^{k} a^{(p-1)}(j)y(k-j) \\
\sum_{j=0}^{k} a^{(p)}(j)y(k-j)
\end{bmatrix}
= \begin{bmatrix} b_0 \; b_1 \; \cdots \; b_p \end{bmatrix}
\begin{bmatrix}
\sum_{j=0}^{k} a^{(0)}(j)u(k-j) \\
\sum_{j=0}^{k} a^{(1)}(j)u(k-j) \\
\vdots \\
\sum_{j=0}^{k} a^{(q-1)}(j)u(k-j) \\
\sum_{j=0}^{k} a^{(q)}(j)u(k-j)
\end{bmatrix}
$$
(6.20)

or

$$
\begin{bmatrix} a_0 \; a_1 \; \cdots \; a_p \end{bmatrix}
\begin{bmatrix}
a^{(0)}(0) & 0 & \cdots & 0 & 0 \\
a^{(0)}(1) & a^{(0)}(1) & \cdots & 0 & 0 \\
\vdots & \vdots & & \vdots & \vdots \\
a^{(0)}(p-1) & a^{(p-1)}(p-1) & \cdots & a^{(p-1)}(p-1) & 0 \\
a^{(0)}(p) & a^{(p-1)}(p) & \cdots & a^{(p)}(p-1) & a^{(p)}(p)
\end{bmatrix}
\begin{bmatrix}
y(k) \\
y(k-1) \\
\vdots \\
y(k-p+1) \\
y(k-p)
\end{bmatrix}
$$

$$
= \begin{bmatrix} b_0 \; b_1 \; \cdots \; b_p \end{bmatrix}
\begin{bmatrix}
a^{(0)}(0) & 0 & \cdots & 0 & 0 \\
a^{(0)}(1) & a^{(0)}(1) & \cdots & 0 & 0 \\
\vdots & \vdots & & \vdots & \vdots \\
a^{(0)}(q-1) & a^{(q-1)}(q-1) & \cdots & a^{(q-1)}(q-1) & 0 \\
a^{(0)}(p) & a^{(q-1)}(q) & \cdots & a^{(q)}(q-1) & a^{(q)}(q)
\end{bmatrix}
\begin{bmatrix}
u(k) \\
u(k-1) \\
\vdots \\
u(k-p+1) \\
u(k-q)
\end{bmatrix}.
$$
(6.21)

Now one defines new coefficients

$$
\begin{bmatrix} A_0 \; A_1 \; \cdots \; A_p \end{bmatrix}
$$

$$
= \begin{bmatrix} a_0 \; a_1 \; \cdots \; a_p \end{bmatrix}
\begin{bmatrix}
a^{(0)}(0) & 0 & \cdots & 0 & 0 \\
a^{(0)}(1) & a^{(0)}(1) & \cdots & 0 & 0 \\
\vdots & \vdots & & \vdots & \vdots \\
a^{(0)}(p-1) & a^{(p-1)}(p-1) & \cdots & a^{(p-1)}(p-1) & 0 \\
a^{(0)}(p) & a^{(p-1)}(p) & \cdots & a^{(p)}(p-1) & a^{(p)}(p)
\end{bmatrix}
$$
(6.22)

$$
\begin{bmatrix} B_0 \; B_1 \; \cdots \; B_q \end{bmatrix}
$$

$$
= \begin{bmatrix} b_0 \; b_1 \; \cdots \; b_q \end{bmatrix}
\begin{bmatrix}
a^{(0)}(0) & 0 & \cdots & 0 & 0 \\
a^{(0)}(1) & a^{(0)}(1) & \cdots & 0 & 0 \\
\vdots & \vdots & & \vdots & \vdots \\
a^{(0)}(p-1) & a^{(q-1)}(q-1) & \cdots & a^{(q-1)}(q-1) & 0 \\
a^{(0)}(p) & a^{(q-1)}(q) & \cdots & a^{(q)}(q-1) & a^{(q)}(q)
\end{bmatrix}
$$
(6.23)

An application of these new coefficients yields

$$\sum_{i=0}^{p} A_i y(k-i) = \sum_{j=0}^{q} B_j u(k-j) \tag{6.24}$$

where $A_0 \neq 0$. The condition arises from the assumption that the IODE is of order p. This is a commonly used form in computer simulations of its response evaluation.

6.2 Time-invariant FO system state-space description

Now, one describes a procedure which allows transformation of one non-linear, time invariant FODE of order $p\nu \in \mathbb{R}_+$ to $p \in \mathbb{Z}_+$ FODEs of order ν. Consider FODE (6.6) and assume that it can be expressed in the form

$$F\left[{}_{k_0}^{GL}\Delta_k^{(\nu p)} y(k), \, {}_{k_0}^{GL}\Delta_k^{(\nu(p-1))} y(k), \dots, \, {}_{k_0}^{GL}\Delta_k^{(\nu)} y(k), y(k), u(k) \right] = 0$$

$${}_{k_0}^{GL}\Delta_k^{(\nu p)} y(k)$$

$$= f_p\left[{}_{k_0}^{GL}\Delta_k^{(\nu(p-1))} y(k), \, {}_{k_0}^{GL}\Delta_k^{(\nu(p-2))} y(k), \dots, \, {}_{k_0}^{GL}\Delta_k^{(\nu)} y(k), y(k), u(k) \right]. \tag{6.25}$$

Comment. The vast majority of the DEs that describe real dynamical systems satisfy the above assumption. This applies to systems described by the FODE as well as the IODE.

Define a set of new functions

$$x_1(k) = y(k),$$
$$x_2(k) = {}_{k_0}^{GL}\Delta_k^{(\nu)} x_1(k) = {}_{k_0}^{GL}\Delta_k^{(\nu)} y(k),$$
$$x_3(k) = {}_{k_0}^{GL}\Delta_k^{(\nu)} x_2(k) = {}_{k_0}^{GL}\Delta_k^{(2\nu)} y(k),$$
$$\vdots \tag{6.26}$$
$$x_p(k) = {}_{k_0}^{GL}\Delta_k^{(\nu)} x_{p-1}(k) = {}_{k_0}^{GL}\Delta_k^{((p-1)\nu)} y(k).$$

Simple rearrangement of these equations and subsequent substitution into (6.17) yields

$${}_{k_0}^{GL}\Delta_k^{(\nu)} x_1(k) = x_2(k),$$
$${}_{k_0}^{GL}\Delta_k^{(\nu)} x_2(k) = x_3(k),$$
$$\vdots \tag{6.27}$$
$${}_{k_0}^{GL}\Delta_k^{(\nu)} x_{p-1}(k) = x_p(k),$$
$${}_{k_0}^{GL}\Delta_k^{(\nu)} x_p(k) = f_p\left[x_p(k), x_{p-1}(k), x_2(k), x_1(k), u(k) \right].$$

The set of equations given above can be alternatively expressed in a vector form

$$
\begin{bmatrix}
{}^{GL}_{k_0}\Delta_k^{(\nu)} x_1(k) \\
{}^{GL}_{k_0}\Delta_k^{(\nu)} x_2(k) \\
\vdots \\
{}^{GL}_{k_0}\Delta_k^{(\nu)} x_{p-1}(k) \\
{}^{GL}_{k_0}\Delta_k^{(\nu)} x_p(k)
\end{bmatrix}
= {}^{GL}_{k_0}\Delta_k^{(\nu)}
\begin{bmatrix}
x_1(k) \\
x_2(k) \\
\vdots \\
x_{p-1}(k) \\
x_p(k)
\end{bmatrix}
=
\begin{bmatrix}
x_2(k) \\
x_3(k) \\
\vdots \\
x_p(k) \\
f_p\left[x_p(k), x_{p-1}(k), x_2(k), x_1(k), u(k)\right]
\end{bmatrix}.
$$

(6.28)

The first equation in (6.26) denoting the FODE solution remains unchanged but is expressed in the opposite order

$$
y(k) = [x_1(k)].
$$

(6.29)

Equation (6.28) is commonly called the state equation whereas (6.29) is termed the output equation. The procedure described above will be now repeated for linear FODE

$$
\sum_{i=0}^{p} a_i \, {}^{GL}_{k_0}\Delta_k^{(\nu i)} y(k) = b_0 u(k)
$$

(6.30)

with $a_p = 1$. Analogous action gives

$$
\begin{bmatrix}
{}^{GL}_{k_0}\Delta_k^{(\nu)} x_1(k) \\
{}^{GL}_{k_0}\Delta_k^{(\nu)} x_2(k) \\
\vdots \\
{}^{GL}_{k_0}\Delta_k^{(\nu)} x_{p-1}(k) \\
{}^{GL}_{k_0}\Delta_k^{(\nu)} x_p(k)
\end{bmatrix}
= {}^{GL}_{k_0}\Delta_k^{(\nu)}
\begin{bmatrix}
x_1(k) \\
x_2(k) \\
\vdots \\
x_{p-1}(k) \\
x_p(k)
\end{bmatrix}
$$

$$
=
\begin{bmatrix}
x_2(k) \\
x_3(k) \\
\vdots \\
x_p(k) \\
-a_{p-1}x_p(k) - a_{p-2}x_{p-1}(k) \cdots - a_1 x_2(k) - a_0 x_1(k) + b_0 u(k)
\end{bmatrix}
$$

$$
=
\begin{bmatrix}
0 & 1 & 0 & \cdots & 0 & 0 \\
0 & 0 & 1 & \cdots & 0 & 0 \\
\vdots & \vdots & & \vdots & \vdots & \\
0 & 0 & 0 & \cdots & 0 & 1 \\
-a_0 & -a_1 & -a_2 & \cdots & -a_{p-2} & -a_{p-1}
\end{bmatrix}
\begin{bmatrix}
x_1(k) \\
x_2(k) \\
\vdots \\
x_{p-1}(k) \\
x_p(k)
\end{bmatrix}
+
\begin{bmatrix}
0 \\
0 \\
\vdots \\
0 \\
1
\end{bmatrix} u(k)
$$

(6.31)

and the output equation is the same as (6.29). The state and output equations

(6.28)–(6.29) and (6.31)–(6.29) are some special cases of general forms

$$
\begin{bmatrix}
{}^{GL}_{k_0}\Delta_k^{(\nu)}x_1(k) \\
{}^{GL}_{k_0}\Delta_k^{(\nu)}x_2(k) \\
\vdots \\
{}^{GL}_{k_0}\Delta_k^{(\nu)}x_{p-1}(k) \\
{}^{GL}_{k_0}\Delta_k^{(\nu)}x_p(k)
\end{bmatrix}
=
\begin{bmatrix}
f_1\left[x_p(k), x_{p-1}(k), x_2(k), x_1(k), u(k)\right] \\
f_2\left[x_p(k), x_{p-1}(k), x_2(k), x_1(k), u(k)\right] \\
\vdots \\
f_{p-1}\left[x_p(k), x_{p-1}(k), x_2(k), x_1(k), u(k)\right] \\
f_p\left[x_p(k), x_{p-1}(k), x_2(k), x_1(k), u(k)\right]
\end{bmatrix}
\tag{6.32}
$$

$$
y(k) = g\left[x_p(k), x_{p-1}(k), x_2(k), x_1(k), u(k)\right]
\tag{6.33}
$$

and in the linear system case

$$
\begin{bmatrix}
{}^{GL}_{k_0}\Delta_k^{(\nu)}x_1(k) \\
{}^{GL}_{k_0}\Delta_k^{(\nu)}x_2(k) \\
\vdots \\
{}^{GL}_{k_0}\Delta_k^{(\nu)}x_{p-1}(k) \\
{}^{GL}_{k_0}\Delta_k^{(\nu)}x_p(k)
\end{bmatrix}
=
\begin{bmatrix}
a_{11} & a_{12} & \cdots & a_{1,p-1} & a_{1,p} \\
a_{21} & a_{22} & \cdots & a_{2,p-1} & a_{2,p} \\
\vdots & \vdots & & \vdots & \vdots \\
a_{p-1,1} & a_{p-1,2} & \cdots & a_{p-1,p-1} & a_{p-1,p} \\
a_{p1} & a_{p2} & \cdots & a_{p,p-1} & a_{p,p}
\end{bmatrix}
\begin{bmatrix}
x_1(k) \\
x_2(k) \\
\vdots \\
x_{p-1}(k) \\
x_p(k)
\end{bmatrix}
+
\begin{bmatrix}
b_{1,1} \\
b_{2,1} \\
\vdots \\
b_{p-1,1} \\
b_{p,1}
\end{bmatrix}
u(k)
\tag{6.34}
$$

$$
y(k) = \begin{bmatrix} c_{1,1} & c_{1,2} & \cdots & c_{1,p-1} & c_{1,p} \end{bmatrix}
\begin{bmatrix}
x_1(k) \\
x_2(k) \\
\vdots \\
x_{p-1}(k) \\
x_p(k)
\end{bmatrix}
+ [d_{1,1}]u(k).
\tag{6.35}
$$

The state FODEs (6.32) and (6.34), as well as the algebraic output equations (6.33) and (6.35), are commonly expressed in more compact forms

$$
{}^{GL}_{k_0}\Delta_k^{(\nu)}\mathbf{x}(k) = \mathbf{f}\left[\mathbf{x}(k), \mathbf{u}(k)\right]
\tag{6.36}
$$

and

$$
\mathbf{y}(k) = \mathbf{g}\left[\mathbf{x}(k), \mathbf{u}(k)\right]
\tag{6.37}
$$

where

$$
\mathbf{x}(k) =
\begin{bmatrix}
x_1(k) \\
x_2(k) \\
\vdots \\
x_{p-1}(k) \\
x_p(k)
\end{bmatrix},\
\mathbf{u}(k) =
\begin{bmatrix}
u_1(k) \\
u_2(k) \\
\vdots \\
u_{q-1}(k) \\
u_q(k)
\end{bmatrix},\
\mathbf{y}(k) =
\begin{bmatrix}
y_1(k) \\
y_2(k) \\
\vdots \\
y_{r-1}(k) \\
y_r(k)
\end{bmatrix}
\tag{6.38}
$$

$$\mathbf{f}\left[\mathbf{x}(k), \mathbf{u}(k)\right] = \begin{bmatrix} f_1\left[\mathbf{x}(k), \mathbf{u}(k)\right] \\ f_2\left[\mathbf{x}(k), \mathbf{u}(k)\right] \\ \vdots \\ f_{p-1}\left[\mathbf{x}(k), \mathbf{u}(k)\right] \\ f_p\left[\mathbf{x}(k), \mathbf{u}(k)\right] \end{bmatrix} = \begin{bmatrix} f_1\left[x_p(k), \cdots, x_1(k), u_q(k) \ldots, u_1(k)\right] \\ f_2\left[x_p(k), \ldots, x_1(k), u_q(k) \ldots, u_1(k)\right] \\ \vdots \\ f_{p-1}\left[x_p(k), \ldots, x_1(k), u_q(k) \ldots, u_1(k)\right] \\ f_p\left[x_p(k), \ldots, x_1(k), u_q(k) \ldots, u_1(k)\right] \end{bmatrix}$$

$$\tag{6.39}$$

$$\mathbf{g}\left[\mathbf{x}(k), \mathbf{u}(k)\right] = \begin{bmatrix} g_1\left[\mathbf{x}(k), \mathbf{u}(k)\right] \\ g_2\left[\mathbf{x}(k), \mathbf{u}(k)\right] \\ \vdots \\ g_{q-1}\left[\mathbf{x}(k), \mathbf{u}(k)\right] \\ g_q\left[\mathbf{x}(k), \mathbf{u}(k)\right] \end{bmatrix} = \begin{bmatrix} g_1\left[x_p(k), \ldots, x_1(k), u_q(k) \ldots, u_1(k)\right] \\ g_2\left[x_p(k), \ldots, x_1(k), u_q(k) \ldots, u_1(k)\right] \\ \vdots \\ g_{r-1}\left[x_p(k), \ldots, x_1(k), u_q(k) \ldots, u_1(k)\right] \\ g_r\left[x_p(k), \ldots, x_1(k), u_q(k) \ldots, u_1(k)\right] \end{bmatrix}.$$

$$\tag{6.40}$$

The linear time-invariant FODE case is commonly denoted as

$$_{k_0}^{GL}\Delta_k^{(\nu)}\mathbf{x}(k) = \mathbf{A}\mathbf{x}(k) + \mathbf{B}\mathbf{u}(k), \tag{6.41}$$

and

$$\mathbf{y}(k) = \mathbf{C}\mathbf{x}(k) + \mathbf{D}\mathbf{u}(k), \tag{6.42}$$

where vectors $\mathbf{x}(k), \mathbf{u}(k), \mathbf{y}(k)$ are as in (6.38) and

$$\mathbf{A} = \begin{bmatrix} a_{11} & a_{12} & \cdots & a_{1,p-1} & a_{1,p} \\ a_{21} & a_{22} & \cdots & a_{2,p-1} & a_{2,p} \\ \vdots & \vdots & & \vdots & \vdots \\ a_{p-1,1} & a_{p-1,2} & \cdots & a_{p-1,p-1} & a_{p-1,p} \\ a_{p1} & a_{p2} & \cdots & a_{p,p-1} & a_{p,p} \end{bmatrix} \quad \mathbf{B} = \begin{bmatrix} b_{11} & \cdots & b_{1,q-1} & b_{1,q} \\ b_{21} & \cdots & b_{2,q-1} & b_{2,q} \\ \vdots & & \vdots & \vdots \\ b_{p-1,1} & \cdots & b_{p-1,q-1} & b_{p-1,q} \\ b_{p,1} & \cdots & b_{p,q-1} & b_{p,q} \end{bmatrix}$$

$$\mathbf{C} = \begin{bmatrix} c_{11} & \cdots & c_{1,p-1} & c_{1,p} \\ c_{21} & \cdots & c_{2,p-1} & c_{2,p} \\ \vdots & & \vdots & \vdots \\ c_{r-1,1} & \cdots & c_{r-1,p-1} & c_{r-1,p} \\ c_{r,1} & \cdots & c_{r,p-1} & c_{r,p} \end{bmatrix} \quad \mathbf{D} = \begin{bmatrix} d_{11} & d_{12} & \cdots & d_{1,q-1} & d_{1,q} \\ d_{21} & d_{22} & \cdots & d_{2,q-1} & d_{2,q} \\ \vdots & \vdots & & \vdots & \vdots \\ d_{r-1,1} & d_{r-1,2} & \cdots & d_{q-1,q-1} & d_{q-1,q} \\ d_{r,1} & d_{r,2} & \cdots & d_{r,q-1} & d_{r,q} \end{bmatrix}.$$

$$\tag{6.43}$$

The matrix $\mathbf{D}$ is usually a zero one.

6.3   Linearization of a time-invariant non-linear DE

Before starting a non-linear FODE linearization process, one should take note of the concept of a steady-state solution. A steady-state solution is usually obtained for constant external signal. Hence, one assumes that

$$\lim_{k \to +\infty} u(k) = u_s = \text{const.} \tag{6.44}$$

and for $k \to +\infty$ a solution achieves its steady-state

$$\lim_{k \to +\infty} \mathbf{x}(k) = \mathbf{x}_s = \text{const.} \tag{6.45}$$

The linearization procedure is based on the following preliminary assertion.

Lemma 6.1. *For $k \to +\infty$ the FOBD of order $\nu \in (0,1]$ of a constant function $x_{s,i}$ tends to zero.*

$$\substack{GL \\ k_0} \Delta_k^{(\nu)} x_{s,i} 1(k - k_0) = 0. \tag{6.46}$$

Proof. A steady-state is obtained for $k \to +\infty$ and is independent of k_0 so it may be put $k_0 = 0$. Hence,

$$\lim_{k \to +\infty} \left[\substack{GL \\ 0} \Delta_k^{(\nu)} x_{s,i} 1(k) \right] = \lim_{k \to +\infty} \left[\substack{GL \\ 0} \Delta_k^{(\nu)} 1(k) \right] x_{s,i}$$

$$= \lim_{k \to +\infty} \left[\sum_{j=0}^{k} a^{(\nu)}(j) \right] x_{s,i} = \left[\sum_{j=0}^{+\infty} a^{(\nu)}(j) \right] x_{s,i} = 0. \tag{6.47}$$

since by Theorem 1.4 the sum in brackets is equal to zero. This ends the proof.

$\square$

From Lemma 6.1 follows immediately the next result

$$\substack{GL \\ k_0} \Delta_k^{(\nu)} \mathbf{x}_s 1(k - k_0) = \mathbf{0}. \tag{6.48}$$

Now, one evaluates a steady-state solution of equation 6.36. A steady-state $\mathbf{x}_s$ is achieved for $k \to +\infty$. Thus from (6.36)

$$\lim_{k \to +\infty} \left[\substack{GL \\ k_0} \Delta_k^{(\nu)} \mathbf{x}_s \right] = \lim_{k \to +\infty} \mathbf{f}(\mathbf{x}_s, \mathbf{u}_s) \tag{6.49}$$

then

$$\mathbf{0} = \mathbf{f}(\mathbf{x}_s, \mathbf{u}_s). \tag{6.50}$$

A steady-state evaluation procedure may be performed in the following steps.

Algorithm 6.1.

(I) For a given constant u_s solve for $\mathbf{x}_s$ a non-linear algebraic equation $\mathbf{f}(\mathbf{x}_s, \mathbf{u}_s) = \mathbf{0}$

(II) For a given constant u_s and $\mathbf{x}_s$ find $y_s = \mathbf{g}(\mathbf{x}_s, \mathbf{u}_s)$.

Comment. In step I of the Algorithm 6.1, according to the algebraic equation $\mathbf{f}(\mathbf{x}_s, \mathbf{u}_s) = \mathbf{0}$ one can get a single unique solution $\mathbf{x}_s$ or multiple solutions $\mathbf{x}_{s,i}$, or none. The latter case means that there is no steady-state. In this case, for non-zero initial conditions or/and constant input signal, the output tends to $\pm\infty$ or achieves the so-called limit circle. Such a situation occurs in the non-stable systems considered in 7.3.

Comment. In the linear time-invariant FODE case described by (6.34) and (6.35) a steady- state solution is obtained by solving for $\mathbf{x}_s$ a linear algebraic equation

$$0 = \mathbf{A}\mathbf{x}_s + \mathbf{B}\mathbf{u}_s. \tag{6.51}$$

For an invertible matrix A, one always gets a single steady-state

$$\mathbf{x}_s = -\mathbf{A}^{-1}\mathbf{B}\mathbf{u}_s \tag{6.52}$$

and a steady-state system solution is as follows

$$\mathbf{y}_s = \mathbf{C}\mathbf{x}_s + \mathbf{D}\mathbf{u}_s = -\left(\mathbf{C}\mathbf{A}^{-1}\mathbf{B} + \mathbf{D}\right)\mathbf{u}_s. \tag{6.53}$$

In the following numerical example one explains the step of Algorithm 6.1 needed to get the steady-state solutions.

Example 6.2. For a non-linear time-invariant FODE

$$_{k_0}^{GL}\Delta_k^{(\nu)}y(k) + (1-a)y(k) + ay^2(k) = bu(k) \tag{6.54}$$

and orders from the range $0 < \nu = \frac{1}{d} \geqslant 1$, $a \neq 0$ find its steady-state solution.

Solution. Formally FODE equation (6.54) is transformed to the state-space equations form

$$_{k_0}^{GL}\Delta_k^{(\nu)}x_1(k) = (a-1)x_1(k) - ax_1^2(k) = bu(k)$$
$$y(k) = x_1(k). \tag{6.55}$$

Now, according to step I of Algorithm 6.1, for a known constant u_s one solves an algebraic equation

$$0 = (a-1)x_{s,1} - ax_{s,1}^2 + bu_s$$
$$y_s = x_{s,1}. \tag{6.56}$$

From (6.56), one obtains two real solutions

$$x_{s,1} = \frac{a-1+\sqrt{(1-a)^2+4abu_s}}{2a}$$

$$x_{s,2} = \frac{a-1-\sqrt{(1-a)^2+4abu_s}}{2a} \tag{6.57}$$

which are valid for

$$u_s \geqslant -\frac{(a-1)^2}{2ab}. \tag{6.58}$$

A pair of complex conjugate solutions is rejected, indicating an unstable FODE. For $u_s = 0$ one immediate obtains

$$x_{s,1} = 0 \text{ and } x_{s,2} = 1 - \frac{1}{a}. \tag{6.59}$$

The linearisation procedure of non-linear FODE assumes a "small" change (a small perturbation) around the known input function $\mathbf{u}(k)$. This perturbation will be denoted as

$$\mathbf{u}(k) = \mathbf{u}_s + \delta\mathbf{u}(k). \tag{6.60}$$

Here, $\delta\mathbf{u}(k)$ denotes a relatively "small" perturbation (positive or negative) around the steady-state value of the input signal $\mathbf{u}(k)$. Accordingly, equation (6.54) is "slightly" changed due to (6.60) and will be denoted as

$$\mathbf{x}(k) = \mathbf{x}_s + \delta\mathbf{x}(k), \ \mathbf{y}(k) = \mathbf{y}_s + \delta\mathbf{y}(k). \tag{6.61}$$

The expressions $\delta\mathbf{x}(k)$ and $\delta\mathbf{y}(k)$ represent induced "small" changes around the steady-state solutions $\mathbf{x}_s$ and $\mathbf{y}_s$ caused by a change of (6.60). Then, by (6.61),

$$\begin{aligned} {}^{GL}_{k_0}\Delta_k^{(\nu)}\mathbf{x}(k) &= {}^{GL}_{k_0}\Delta_k^{(\nu)}\left[\mathbf{x}_s + \delta\mathbf{x}(k)\right] \\ {}^{GL}_{k_0}\Delta_k^{(\nu)}\mathbf{x}_s + {}^{GL}_{k_0}\Delta_k^{(\nu)}\delta\mathbf{x}(k) &= {}^{GL}_{k_0}\Delta_k^{(\nu)}\delta\mathbf{x}(k). \end{aligned} \tag{6.62}$$

Additionally,

$$\begin{matrix}{GL}\\{k_0}\end{matrix}\Delta_k^{(\nu)}\mathbf{x}(k) = \mathbf{f}\left[\mathbf{x}_s + \delta\mathbf{x}(k), \mathbf{u}_s + \delta\mathbf{u}(k)\right] \tag{6.63}$$

$$\mathbf{y}_s + \delta y(k) = \mathbf{g}\left[\mathbf{x}_s + \delta\mathbf{x}(k), \mathbf{u}_s + \delta\mathbf{u}(k)\right]. \tag{6.64}$$

Now the vector function $\mathbf{f}\left[\mathbf{x}_s + \delta\mathbf{x}(k), \mathbf{u}_s + \delta\mathbf{u}(k)\right]$ elements are expanded into Taylor series around the steady-state solution for $i = 1, 2, \ldots, p$

$$f_i\left[\mathbf{x}(k), u(k)\right] = f_i\left[\mathbf{x}(k), u(k)\right]\Big|_{\substack{\mathbf{x}(k)\,=\,\mathbf{x}_s \\ \mathbf{u}(k)\,=\,\mathbf{u}_s}}$$

$$+\sum_{j=1}^{p}\frac{\partial f_i\left[\mathbf{x}(k), u(k)\right]}{\partial x_j}\Bigg|_{\substack{\mathbf{x}(k)\,=\,\mathbf{x}_s \\ \mathbf{u}(k)\,=\,\mathbf{u}_s}}[x_j(k)-x_{s,j}]$$

$$+\sum_{j=1}^{q}\frac{\partial f_i\left[\mathbf{x}(k), u(k)\right]}{\partial u_j}\Bigg|_{\substack{\mathbf{x}(k)\,=\,\mathbf{x}_s \\ \mathbf{u}(k)\,=\,\mathbf{u}_s}}[u_j(k)-u_{s,j}]$$

$$+\frac{1}{2}\sum_{j=1}^{p}\frac{\partial^2 f_i\left[\mathbf{x}(k), u(k)\right]}{\partial x_j^2}\Bigg|_{\substack{\mathbf{x}(k)\,=\,\mathbf{x}_s \\ \mathbf{u}(k)\,=\,\mathbf{u}_s}}[x_j(k)-x_{s,j}]^2$$

$$+\frac{1}{2}\frac{\partial^2 f_i\left[\mathbf{x}(k), u(k)\right]}{\partial u^2}\Bigg|_{\substack{\mathbf{x}(k)\,=\,\mathbf{x}_s \\ \mathbf{u}(k)\,=\,\mathbf{u}_s}}[u_j(k)-u_{s,j}]^2$$

$$+\sum_{i=1}^{p}\sum_{j=1}^{p}\frac{\partial^2 f_i\left[\mathbf{x}(k), u(k)\right]}{\partial x_i\partial x_j}\Bigg|_{\substack{\mathbf{x}(k)\,=\,\mathbf{x}_s \\ \mathbf{u}(k)\,=\,\mathbf{u}_s}}[x_i(k)-x_{i,s}]\,[x_j(k)-x_{j,s}]+\cdots \qquad (6.65)$$

where ∂ denotes a partial derivative with respect to appropriate variables $x_j(k), u_i(k)$. A substitution of (6.60) and (6.61) into (6.65) yields

$$f_i\left[\mathbf{x}(k), \mathbf{u}(k)\right] = f_i\left[\mathbf{x}_s+\delta\mathbf{x}(k),\, \mathbf{u}_s+\delta\mathbf{u}(k)\right]+f_i\left[\mathbf{x}(k), u(k)\right]\Big|_{\substack{\mathbf{x}(k)\,=\,\mathbf{x}_s \\ \mathbf{u}(k)\,=\,\mathbf{u}_s}}$$

$$+\sum_{j=1}^{p}\frac{\partial f_i\left[\mathbf{x}(k), \mathbf{u}(k)\right]}{\partial x_j}\Bigg|_{\substack{\mathbf{x}(k)\,=\,\mathbf{x}_s \\ \mathbf{u}(k)\,=\,\mathbf{u}_s}}\delta x_j(k)+\sum_{j=1}^{q}\frac{\partial f_i\left[\mathbf{x}(k), \mathbf{u}(k)\right]}{\partial u_j}\Bigg|_{\substack{\mathbf{x}(k)\,=\,\mathbf{x}_s \\ \mathbf{u_j}(k)\,=\,\mathbf{u}_s}}\delta u_j(k)$$

$$+\frac{1}{2}\sum_{j=1}^{p}\frac{\partial^2 f_i\left[\mathbf{x}(k), \mathbf{u}(k)\right]}{\partial x_j^2}\Bigg|_{\substack{\mathbf{x}(k)\,=\,\mathbf{x}_s \\ \mathbf{u}_j(k)\,=\,\mathbf{u}_s}}\delta x_j^2(k)+\frac{1}{2}\sum_{j=1}^{q}\frac{\partial^2 f_i\left[\mathbf{x}(k), \mathbf{u}(k)\right]}{\partial u_j^2}\Bigg|_{\substack{\mathbf{x}(k)\,=\,\mathbf{x}_s \\ \mathbf{u}(k)\,=\,\mathbf{u}_s}}\delta u_j^2(k)$$

$$+\sum_{i=1}^{p}\sum_{j=1}^{p}\frac{\partial^2 f_i\left[\mathbf{x}(k), \mathbf{u}(k)\right]}{\partial x_i\partial x_j}\Bigg|_{\substack{\mathbf{x}(k)\,=\,\mathbf{x}_s \\ \mathbf{u}(k)\,=\,\mathbf{u}_s}}\delta x_i(k)\delta x_j(k)+R\left[\delta\mathbf{x}(k),\delta\mathbf{u}(k)\right]$$

$$(6.66)$$

where $R\left[\delta\mathbf{x}(k), \delta u(k)\right]$ represents the infinite number of terms containing higher powers of $\delta x_j(k)$ and $\delta u_j(k)$ (including its combinations). By assumption $\delta x_j(k) \gg [\delta x_j(k)]^i$ and $\delta u_j(k) \gg [\delta u_j(k)]^i$ for $i = 2, 3, \ldots$, so ignoring all nonlinear terms containing $\delta x_j(k)$ and $\delta u_j(k)$ and taking into account (6.62), one obtains the following

$$\begin{aligned} {}_{k_0}^{GL}\Delta_k^{(\nu)}\delta x_i(k) = f_i\left[\mathbf{x}(k), \mathbf{u}(k)\right]\Big|_{\substack{\mathbf{x}(k) = \mathbf{x}_s \\ \mathbf{u}(k) = \mathbf{u}_s}} \\ + \sum_{j=1}^{p} \frac{\partial f_i\left[\mathbf{x}(k), \mathbf{u}(k)\right]}{\partial x_j}\Bigg|_{\substack{\mathbf{x}(k) = \mathbf{x}_s \\ \mathbf{u}(k) = \mathbf{u}_s}} \delta x_j(k) + \sum_{j=1}^{q} \frac{\partial f_i\left[\mathbf{x}(k), \mathbf{u}(k)\right]}{\partial u_j}\Bigg|_{\substack{\mathbf{x}(k) = \mathbf{x}_s \\ \mathbf{u}(k) = \mathbf{u}_s}} \delta u_j(k). \end{aligned}$$

$$(6.67)$$

In view of an equality 6.50,

$$f_i\left[\mathbf{x}(k), \mathbf{u}(k)\right]\Big|_{\substack{\mathbf{x}(k) = \mathbf{x}_s \\ \mathbf{u}(k) = \mathbf{u}_s}} = 0 \text{ for } i = 1, 2, \ldots, p. \qquad (6.68)$$

This simplifies the set of equations (6.67) to the form

$$\begin{aligned} {}_{k_0}^{GL}\Delta_k^{(\nu)}\delta x_i(k) \\ = \sum_{j=1}^{p} \frac{\partial f_i\left[\mathbf{x}(k), \mathbf{u}(k)\right]}{\partial x_j}\Bigg|_{\substack{\mathbf{x}(k) = \mathbf{x}_s \\ \mathbf{u}(k) = \mathbf{u}_s}} \delta x_j(k) + \sum_{j=1}^{q} \frac{\partial f_i\left[\mathbf{x}(k), \mathbf{u}(k)\right]}{\partial u_j}\Bigg|_{\substack{\mathbf{x}(k) = \mathbf{x}_s \\ \mathbf{u}(k) = \mathbf{u}_s}} \delta u_j(k). \end{aligned}$$

$$(6.69)$$

All partial derivatives in (6.69) will be further denoted as

$$a_{ij} = \frac{\partial f_i\left[\mathbf{x}(k), \mathbf{u}(k)\right]}{\partial x_j}\Bigg|_{\substack{\mathbf{x}(k) = \mathbf{x}_s \\ \mathbf{u}(k) = \mathbf{u}_s}}, \quad b_i = \frac{\partial f_i\left[\mathbf{x}(k), \mathbf{u}(k)\right]}{\partial u}\Bigg|_{\substack{\mathbf{x}(k) = \mathbf{x}_s \\ \mathbf{u}(k) = \mathbf{u}_s}} \qquad (6.70)$$

for $i = 1, 2, \ldots, p$. Thus, the set (6.69) simplifies to

$$ {}_{k_0}^{GL}\Delta_k^{(\nu)}\delta x_i(k) = \sum_{j=1}^{p} a_{ij}\delta x_j(k) + \sum_{j=1}^{q} b_i \delta u_j(k) \text{ for } i = 1, 2, \ldots, p. \qquad (6.71)$$

Collecting all equations considered above one obtains a matrix-vector form

$$ {}_{k_0}^{GL}\Delta_k^{(\nu)}\delta\mathbf{x}(k) = \mathbf{A}\delta\mathbf{x}(k) + \mathbf{B}\delta\mathbf{u}(k) \qquad (6.72)$$

where

$$
{}^{GL}_{k_0}\Delta_k^{(\nu)}\delta\mathbf{x}(k) =
\begin{bmatrix}
{}^{GL}_{k_0}\Delta_k^{(\nu)}\delta x_1(k) \\
{}^{GL}_{k_0}\Delta_k^{(\nu)}\delta x_2(k) \\
\vdots \\
{}^{GL}_{k_0}\Delta_k^{(\nu)}\delta x_p(k)
\end{bmatrix}
\tag{6.73}
$$

and the matrices $\mathbf{A}$ and $\mathbf{B}$ are as in formula (6.43). Now, expanding into Taylor series a non-linear function $\mathbf{g}\left[\mathbf{x}_s + \delta\mathbf{x}(k), \mathbf{u}_s + \delta\mathbf{u}(k)\right]$ in equation (6.64), one gets a linear algebraic equation

$$
\delta\mathbf{y}(k) = \mathbf{C}\delta\mathbf{x}(k) + \mathbf{D}\delta\mathbf{u}(k)
\tag{6.74}
$$

where matrices $\mathbf{C}$ and $\mathbf{D}$ are also defined in (6.43) with

$$
c_{ij} = \left.\frac{\partial g_i\left[\mathbf{x}(k), \mathbf{u}(k)\right]}{\partial x_j}\right|
\begin{aligned}
\mathbf{x}(k) &= \mathbf{x}_s \\
\mathbf{u}(k) &= \mathbf{u}_s
\end{aligned}
$$

$$
d_{ij} = \left.\frac{\partial g_i\left[\mathbf{x}(k), \mathbf{u}(k)\right]}{\partial u_j}\right|
\begin{aligned}
\mathbf{x}(k) &= \mathbf{x}_s \\
\mathbf{u}(k) &= \mathbf{u}_s
\end{aligned}
\tag{6.75}
$$

Remark 6.1. One should keep in mind that equations (6.72) and (6.74) are only approximations of the non-linear ones. The approximation accuracy depends on the input signal (6.60) perturbation range. For specified ranges of input vector signal perturbations $\delta\mathbf{u}(k) \to 0$ and $\delta\mathbf{x}(k) \to 0$. However, this is not always true. For some perturbations or even a special set of initial conditions, the linearized equation may be stable whereas the original one, the non-linear equation, may be unstable. The perturbation range assumed in the linearization procedure depends on the equation.

A linearization process will be performed in a numerical example.

Example 6.3. Linearize the FODE considered in Example 6.2 for $u_s = 0$.

Solution. Step I of the Algorithm was performed in Example 6.2. Hence, for $u_s = 0$ there are two steady-state solutions

$$
x_{s,1} = 0, \quad x_{s,2} = 1 - \frac{1}{a}.
\tag{6.76}
$$

In step II, one evaluated two sets of matrices A_i, B_i, C_i, D_i, $i = 1, 2$. In this simple example there is

$$
f_1\left[x_1(k), u_1(k)\right] = (a - 1)x_1(k) - ax_1^2(k) + bu_1(k)
\tag{6.77}
$$

$$
g_1\left[x_1(k), u_1(k)\right] = x_1(k).
\tag{6.78}
$$

Hence, for $x_{s,1} = 0$

$$\mathbf{A}_1 = [a_{11}] = \left[\left. \frac{\partial f_1 \left[x_1(k), u_1(k) \right]}{\partial x_1} \right|_{\substack{x_{s,1} = 0 \\ u_{s,1} = 0}} \right] = \left. [a - 1 - 2ax_{s,1}] \right|_{\substack{x_{s,1} = 0 \\ u_{s,1} = 0}} = [a - 1]$$

(6.79)

$$\mathbf{B}_1 = [b_1] = \left[\left. \frac{\partial f_1 \left[x_1(k), u_1(k) \right]}{\partial u_1} \right|_{\substack{x_{s,1} = 0 \\ u_{s,1} = 0}} \right] = [b]$$

(6.80)

$$\mathbf{C}_1 = [c_1] = \left[\left. \frac{\partial g_1 \left[x_1(k), u_1(k) \right]}{\partial x_1} \right|_{\substack{x_{s,1} = 0 \\ u_{s,1} = 0}} \right] = [1]$$

(6.81)

$$\mathbf{D}_1 = [d_1] = \left[\left. \frac{\partial g_1 \left[x_1(k), u_1(k) \right]}{\partial u_1} \right|_{\substack{x_{s,1} = 0 \\ u_{s,1} = 0}} \right] = [0].$$

(6.82)

For the second solution (6.76)

$$\mathbf{A}_2 = [a_{11}] = \left[\left. \frac{\partial f_1 \left[x_1(k), u_1(k) \right]}{\partial x_1} \right|_{\substack{x_{s,2} = 0 \\ u_{s,1} = 0}} \right] = \left. [a - 1 - 2ax_{s,2}] \right|_{\substack{x_{s,2} = 0 \\ u_{s,1} = 0}} = [1 - a]$$

(6.83)

and $\mathbf{B}_2 = \mathbf{B}_1$, $\mathbf{C}_2 = \mathbf{C}_1$, $\mathbf{D}_2 = \mathbf{D}_1$.

In the next numerical example differences between exact and approximated solutions of the FOBD (6.54) are plotted.

Example 6.4. For $a = 1.8$, $b = 1$, $u(k) = 0.011(k)$, $\nu = 0.5$ and zero initial conditions, plot the solution of equation (6.55) as well as its linearized version.

Solution. After simple transformation equation (6.55) gets the form

$$ay^2(k) + (2 - a)y(k) + \left[a^{(\nu)}(1) \; a^{(\nu)}(2) \; \cdots \; a^{(\nu)}(k) \right] \begin{bmatrix} y(k-1) \\ y(k-2) \\ \vdots \\ y(0) \end{bmatrix} - bu(k) = 0. \quad (6.84)$$

The linearized solution is as follows

$$(2-a)\delta y(k) + \left[a^{(\nu)}(1)\ a^{(\nu)}(2)\ \cdots\ a^{(\nu)}(k)\right]\begin{bmatrix}\delta y(k-1)\\ \delta y(k-2)\\ \vdots\\ \delta y(0)\end{bmatrix} - b\delta u(k) = 0. \qquad (6.85)$$

Having evaluated $y_s = 0$ (6.76), one gets a linearized equation solution $y_l(k) = \delta y(k)$. Both solutions are plotted in Fig. 6.1.

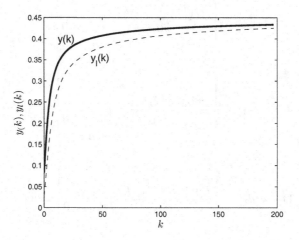

Fig. 6.1   Solutions of equations (6.84) and (6.85).

6.4   FODE initial condition problem

In this section a problem of initial conditions will be examined. Only homogenous DEs will be subjected to analysis. For better understanding of the relation between the initial conditions and the DE solution, the analysis starts with the fractional-order DEs. The analysis will concentrate on the linear time-invariant DEs. It is a well-known fact that to evaluate a unique solution of the n-th order differential equation, one should have n initial conditions. For simplicity, one considers only a case where the initial time is $k_0 = 0$. As an introduction to the problem, a numerical example will be analyzed.

Example 6.5. Evaluate the unique solution of a linear second-order differential equation of the form

$$\frac{d^2 y(t)}{dt^2} + a_1\frac{dy(t)}{dt} + a_0 y(t) = 0 \qquad (6.86)$$

with two initial conditions $y(0^-)$ and $\left.\dfrac{dy(t)}{dt}\right|_{t=0^-} = y'(0^-)$.

Solution. First, equation (6.86) will be solved by means of the one-sided Laplace transform. Applying to both sides the mentioned transform, one obtains

$$\mathcal{L}\left\{\frac{d^2 y(t)}{dt^2} + a_1 \frac{dy(t)}{dt} + a_0 y(t)\right\} = 0. \tag{6.87}$$

A routine procedure using the fundamental properties of the one-sided Laplace transform (linearity, a transform of the first-order derivative) ([O'Flynn and Moriarty (1987)]) yields

$$\left[s^2 Y(s) - sy(0^-) - y'(0^-)\right] + a_1 \left[sY(s) - y(0^-)\right] + a_0 Y(s) = 0 \tag{6.88}$$

where $Y(s) = \mathcal{L}\{y(t)\}$. Equation (6.88) is further transformed to the form

$$Y(s) = \frac{sy(0^-)}{s^2 + a_1 s + a_0} + \frac{a_1 y(0^-) + y'(0^-)}{s^2 + a_1 s + a_0}. \tag{6.89}$$

Now denoting

$$y_0(t) = \mathcal{L}^{-1}\left\{\frac{1}{s^2 + a_1 s + a_0}\right\} \tag{6.90}$$

where $\mathcal{L}^{-1}$ denotes an inverse Laplace transform, one gets

$$y(t) = \mathcal{L}^{-1}\{Y(s)\} = y(0^-)\frac{dy_0(t)}{dt} + \left[a_1 y(0^-) + y'(0^-)\right] y_0(t)$$

$$= y(0^-)e^{-\sigma t}\left[\cos\omega t + \frac{\sigma + \frac{y'(0^-)}{y(0^-)}}{\omega} \sin(\omega t)\right] \tag{6.91}$$

where $\sigma = 0.5a_1$, $\omega = \sqrt{a_0 - 0.25a_1^2}$. The solution given above will now be abandoned for a moment: derivatives in the equation will be approximated by classical backward differences.

$$\frac{dy(t)}{dt}\bigg|_{t=kh} \approx \frac{{}_{k-1}^{GL}\Delta_k^{(1)} y(kh)}{h} = \frac{y(kh) - y(kh - h)}{h} \tag{6.92}$$

$$\frac{d^2 y(t)}{dt^2}\bigg|_{t=kh} \approx \frac{{}_{k-2}^{GL}\Delta_k^{(2)} y(kh)}{h^2} = \frac{y(kh) - 2y(kh - h) + y(kh - 2h)}{h^2}. \tag{6.93}$$

Then,

$$\frac{y(hk) - 2y(kh - 1h) + y(kh - 2h)}{h^2} + a_1 \frac{y(kh) - y(kh - 1h)}{h} + a_0 y(kh) = 0 \tag{6.94}$$

and after simple rearrangements of similar terms the final form of the approximated DE is as follows

$$y(kh) + \frac{-2 - a_1 h}{1 + a_1 h + a_0 h^2} y(kh - 1h) + \frac{1}{1 + a_1 h + a_0 h^2} y(kh - 2h) = 0 \qquad (6.95)$$

where $1 + a_1 h + a_0 h^2 \neq 0$. Otherwise, the DE would be a first-order one. Now, one evaluates the initial conditions of DE (6.95). For $t = 0^-$ one obtains

$$y(0^-) = y_{-1} \qquad (6.96)$$

$$\left. \frac{dy(t)}{dt} \right|_{t=0^-} = y'(0^-) = \frac{y_{-1} - y_{-2}}{h}. \qquad (6.97)$$

From theses two linear algebraic equations one calculates initial conditions y_{-1} and y_{-2} as follows

$$y_{-1} = y(0^-) \qquad (6.98)$$

$$y_{-2} = y(0^-) - hy'(0^-). \qquad (6.99)$$

To simplify notation, one defines new coefficients

$$A_1 = \frac{-2 - a_1 h}{1 + a_1 h + a_0 h^2} y(kh - 1h), \ \ A_0 = \frac{1}{1 + a_1 h + a_0 h^2} \qquad (6.100)$$

and calculates step-by-step consecutive values of the solution,

$$y(0h) = -A_1 y_{-1} - A_2 y_{-2}, \text{ for } k = 0$$

$$y(1h) = -A_1 y(0h) - A_2 y_{-1} \text{ for } k = 1$$

$$y(kh) = -A_1 y(kh - 1h) - A_2 y(kh - 2h) \text{ for } k = 2, 3, \ldots. \qquad (6.101)$$

Comment. Two non-zero coefficients A_1 and A_2 determine that two initial conditions are needed to get the unique solution of the second-order DE . The first coefficient A_1 may be zero (this occurs for $a_1 h = -2$), but the second one A_0 is always non-zero.

For $a_1 = 2$, $a_0 = 101$, $y(0^-) = 2$, $y'(0^-) = 1$, $h = 2.5e-4$ solutions, (6.91) and (6.95) are plotted in Fig. 6.2. Both solutions practically overlap. To show differences, an error function $e(t) = y(t) - y(kh)$ is plotted in Fig. 6.3. The values of the error function can be reduced by decreasing the calculation step h.

For a unique solution of the n-th order differential equation, one should have n initial conditions $y(0^-), y'(0^-), \ldots, y^{(n-1)}(0^-)$. In its discrete approximation one should find n initial conditions $y_{-1}, y_{-2}, \ldots, y_{-n}$. In this calculation formula (3.20)

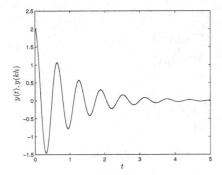

Fig. 6.2 Solutions $y(t)$ and $y(kh)$.

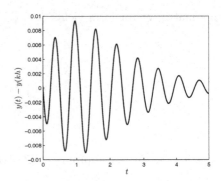

Fig. 6.3 Error function $e(t) = y(t) - y(kh)$.

is helpful. Multiplying its both sides by a matrix D_h

$$
\mathbf{D}_h = \begin{bmatrix} h^0 & 0 & 0 & \cdots \\ 0 & h^{-1} & 0 & \cdots \\ 0 & 0 & h^{-2} & \cdots \\ \vdots & \vdots & \vdots & \end{bmatrix}, \quad \mathbf{D}_h^{-1} = \begin{bmatrix} h^0 & 0 & 0 & \cdots \\ 0 & h^1 & 0 & \cdots \\ 0 & 0 & h^2 & \cdots \\ \vdots & \vdots & \vdots & \end{bmatrix} \tag{6.102}
$$

yields

$$
\begin{bmatrix} y(0^-) \\ y'(0^-) \\ y''(0^-) \\ \vdots \end{bmatrix} \approx \mathbf{D}_h \begin{bmatrix} \Delta^0 f(k) \\ \Delta^1 f(k) \\ \Delta^2 f(k) \\ \vdots \end{bmatrix}\Bigg|_{k=-1} = \mathbf{D}_h \mathbf{P}_\infty \begin{bmatrix} f_{-1} \\ f_{-2} \\ f_{-3} \\ \vdots \end{bmatrix}. \tag{6.103}
$$

Next multiplication of both sides of (6.103) by $[\mathbf{D}_h \mathbf{P}_\infty]^{-1}$ yields

$$
\begin{bmatrix} f_{-1} \\ f_{-2} \\ f_{-3} \\ \vdots \end{bmatrix} = \mathbf{P}_\infty \mathbf{D}_h^{-1} \begin{bmatrix} y(0^-) \\ y'(0^-) \\ y''(0^-) \\ \vdots \end{bmatrix} = \mathbf{P}_\infty \begin{bmatrix} y(0^-) \\ hy'(0^-) \\ h^2 y''(0^-) \\ \vdots \end{bmatrix}. \tag{6.104}
$$

For a relatively small h, the elements of the right-hand side vector rapidly decrease due to consecutive powers of h. This explains the decreasing influence of consecutive initial conditions.

Now, one considers the linear time-invariant FODE homogenous response. The relevant equation is of the form (6.17). Following the procedure performed for a special case $\nu \in \mathbb{Z}_+$ this time the solution is more sophisticated. Simple rearrangement

Discrete Fractional Calculus

to a matrix-vector form yields

$$
\begin{bmatrix} a_p\ a_{p-1}\ \cdots\ a_1\ a_0 \end{bmatrix}
\begin{bmatrix}
{}^{GL}_{k_0}\Delta_k^{(p\nu)}y(k) \\
{}^{GL}_{k_0}\Delta_k^{((p-1)\nu)}y(k) \\
\cdots \\
{}^{GL}_{k_0}\Delta_k^{(\nu)}y(k) \\
y(k)
\end{bmatrix} = 0.
\qquad (6.105)
$$

Substituting further appropriate FOBDs by vector sums one obtains

$$
\begin{bmatrix} a_p\ a_{p-1}\ \cdots\ a_1\ a_0 \end{bmatrix}
$$

$$
\times
\begin{bmatrix}
a^{(p\nu)}(0) & a^{(p\nu)}(1) & \cdots & a^{(p\nu)}(k-k_0+1) & a^{(p\nu)}(k-k_0) \\
a^{((p-1)\nu)}(0) & a^{((p-1)\nu)}(1) & \cdots & a^{((p-1)\nu)}(k-k_0+1) & a^{((p-1)\nu)}(k-k_0) \\
\vdots & \vdots & & \vdots & \vdots \\
a^{(\nu)}(0) & a^{(\nu)}(1) & \cdots & a^{(\nu)}(k-k_0+1) & a^{(\nu)}(k-k_0) \\
a^{(0)}(0) & 0 & \cdots & 0 & 0
\end{bmatrix}
$$

$$
\times
\begin{bmatrix}
y(k) \\
y(k-1) \\
\cdots \\
y(k_0+1) \\
y(k_0)
\end{bmatrix} = 0.
$$

$$
(6.106)
$$

Now, one introduces a notation

$$
\begin{bmatrix} a_p\ a_{p-1}\ \cdots\ a_1\ a_0 \end{bmatrix}
$$

$$
\times
\begin{bmatrix}
a^{(p\nu)}(0) & a^{(p\nu)}(1) & \cdots & a^{(p\nu)}(k-k_0+1) & a^{(p\nu)}(k-k_0) \\
a^{((p-1)\nu)}(0) & a^{((p-1)\nu)}(1) & \cdots & a^{((p-1)\nu)}(k-k_0+1) & a^{((p-1)\nu)}(k-k_0) \\
\vdots & \vdots & & \vdots & \vdots \\
a^{(\nu)}(0) & a^{(\nu)}(1) & \cdots & a^{(\nu)}(k-k_0+1) & a^{(\nu)}(k-k_0) \\
a^{(0)}(0) & 0 & \cdots & 0 & 0
\end{bmatrix}
$$

$$
= \begin{bmatrix} A_0\ A_1\ \cdots\ A_{k-k_0+1}\ A_{k-k_0} \end{bmatrix}.
$$

$$
(6.107)
$$

Note that by definition $a^{(i\nu)}(0) = 1$ for $i = 0, 1, \ldots, p$. Hence,

$$
A_0 = \sum_{i=0}^{p} a_i \neq 0.
\qquad (6.108)
$$

The case where $A_p = 0$ means that the FOBD is of an order less than $p\nu$. This case will not be considered. The analysis will concentrate on the FODE of order

$p\nu$. Furthermore, all other coefficients can be any. In light of equality (6.107), the considered FOBD takes the form

$$
\begin{bmatrix} A_0 \ A_1 \ \cdots \ A_{k-k_0+1} \ A_{k-k_0} \end{bmatrix}
\begin{bmatrix} y(k) \\ y(k-1) \\ \cdots \\ y(k_0+1) \\ y(k_0) \end{bmatrix}
= \sum_{i=0}^{k-k_0} A_i y(k-i) = 0. \tag{6.109}
$$

Now, the homogenous solution of (6.110) is analyzed. To evaluate a unique solution one must take into account initial conditions $y_{k_0-1}, y_{k_0-2}, \ldots$. Then, the solution is of the form

$$
A_0 y(k) + \begin{bmatrix} A_1 \ A_2 \ \cdots \ A_{k-k_0+1} \ A_{k-k_0} \end{bmatrix}
\begin{bmatrix} y(k-1) \\ y(k-2) \\ \cdots \\ y(k_0+1) \\ y(k_0) \end{bmatrix}
$$

$$
= - \begin{bmatrix} A_{k_0-1} \ A_{k_0-2} \ A_{k_0-3} \ \cdots \end{bmatrix}
\begin{bmatrix} y_{k_0-1} \\ y_{k_0-2} \\ y_{k_0-3} \\ \cdots \end{bmatrix}. \tag{6.110}
$$

and after simple rearrangements

$$
y(k) = -\frac{1}{A_0} \begin{bmatrix} A_1 \ A_2 \ \cdots \ A_{k-k_0+1} \ A_{k-k_0} \end{bmatrix}
\begin{bmatrix} y(k-1) \\ y(k-2) \\ \vdots \\ y(k_0+1) \\ y(k_0) \end{bmatrix}
$$

$$
= -\frac{1}{A_0} \begin{bmatrix} A_{k-k_0-1} \ A_{k-k_0-2} \ A_{k-k_0-3} \ \cdots \end{bmatrix}
\begin{bmatrix} y_{k_0-1} \\ y_{k_0-2} \\ y_{k_0-3} \\ \vdots \end{bmatrix}. \tag{6.111}
$$

Comment. The FODE particular solution evaluated for initial time instant k_0 requires an infinite set of initial conditions $y_{k_0-1}, k_{k_0-2}, \ldots$. To preserve the unity of the FODEs and IODEs this statement should also be regarded as true. The IODE solution as a special case of the FODE is also given by (6.111). One should only replace $\nu \in \mathbb{R}_+ \setminus \mathbb{Z}_+$ by $n \in \mathbb{Z}_+$. Then, contrary to the FO case where all $A_i \neq 0$, in the IO case only $A_i \neq 0$ for $i = 0, 1, \ldots, n-1, n$. All other coefficients are equal to zero i.e. $A_i \neq 0$ for $i = n+1, n+2, \ldots$. This explains the concept that in the FODE an infinite set of initial conditions should always be known. In the IODE case its zero coefficients for $i = n + 1, n + 2, \ldots$ cut out all initial conditions, except the n first ones. Such a property explains the FODE solution with its 'memory' taken into account.

Therefore, for $\nu = n \in \mathbb{Z}_+$ one has for $k - k_0 + 1 > n$

$$y(k) = -\frac{1}{A_0} \begin{bmatrix} A_1 & A_2 & \cdots & A_{n-1} & A_n \end{bmatrix} \begin{bmatrix} y(k-1) \\ y(k-2) \\ \vdots \\ y(k-n+1) \\ y(k-n) \end{bmatrix} \qquad (6.112)$$

(this is a classical representation of the IODE solution [Ralston (1965)]), and for $l = k - k_0 + 1 < n$,

$$y(k) = -\frac{1}{A_0} \begin{bmatrix} A_1 & A_2 & \cdots & A_{l-1} & A_l \end{bmatrix} \begin{bmatrix} y(k-1) \\ y(k-2) \\ \vdots \\ y(k-l+1) \\ y(k-l) \end{bmatrix}$$

$$-\frac{1}{A_0} \begin{bmatrix} A_{l+1} & A_{l+2} & \cdots & A_{n-1} & A_n \end{bmatrix} \begin{bmatrix} y_{k_l-1} \\ y_{k-l-2} \\ \vdots \\ y_{k-n} \end{bmatrix}. \qquad (6.113)$$

The above solution describes the system homogenous response. To show the impact of the initial conditions one substitutes the last vector of the right-hand side of (6.111) and (6.113) by relation (3.20). Then the general solution takes the form

$$y(k) = -\frac{1}{A_0} \begin{bmatrix} A_1 & A_2 & \cdots & A_{l-1} & A_l \end{bmatrix} \begin{bmatrix} y(k-1) \\ y(k-2) \\ \vdots \\ y(k-l+1) \\ y(k-l) \end{bmatrix}$$

$$-\frac{1}{A_0} \begin{bmatrix} A_{l+1} & A_{l+2} & \cdots & A_{n-1} & A_n \end{bmatrix} \mathbf{P}_\infty \begin{bmatrix} \Delta^0 y_{k_l-1} \\ \Delta^1 y_{k_l-1} \\ \vdots \\ \Delta^2 y_{k_l-1} \end{bmatrix}. \qquad (6.114)$$

The FODE initial conditions analysis will be clarified by the following numerical example.

Example 6.6. Plot the response of the FO system described by (6.105) for $k_0 = 0$ where $p = 1$, $a_0 = 1$. Consider several initial conditions and input signal cases

(a) $y_{-1} = 4, y_{-i} = 0$ for $i = 2, 3, \ldots, u(k) = 0$,

(b) $y_{-2} = 16, y_{-1} = 0, y_{-i} = 0$ for $i = 3, 4, \ldots, u(k) = 0$,

(c) $y_{-3} = 32, y_{-1} = 0, y_{-2} = 0, y_{-i} = 0$ for $i = 4, 5, \ldots, u(k) = \mathbf{1}(k - 1)$,

(d) $y_{-1} = 4, y_{-2} = 16, y_{-3} = 32, y_{-4} = 100, y_{-5} = 200, y_{-i} = 0$ for $i = 6, 7, \ldots, u(k) = \mathbf{1}(k - 1)$.

$u(k) = \mathbf{1}(k - 1)$ and $u(k) = 0$.

Solution. The solution is as follows

$$
y(k) = -\frac{1}{1 + a_0} \left[a^{(\nu)}(1) \; a^{(\nu)}(2) \; \cdots \; a^{(\nu)}(k) \right]
\begin{bmatrix} y(k-1) \\ y(k-2) \\ \vdots \\ y(0) \end{bmatrix}
+ \frac{b_0}{1 + a_0} u(k)
$$

$$
-\frac{1}{1 + a_0} \left[a^{(\nu)}(k+1) \; a^{(\nu)}(k+2) \; a^{(\nu)}(k+3) \; a^{(\nu)}(k+4) \right]
\begin{bmatrix} y_{-1} \\ y_{-2} \\ y_{-3} \\ y_{-4} \end{bmatrix}. \quad (6.115)
$$

Figures 6.4–6.7 present solutions to equation 6.115 for previously mentioned cases a)–d).

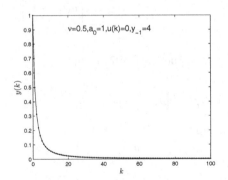

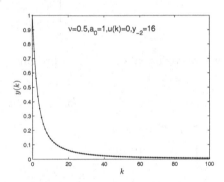

Fig. 6.4 Solution of (6.115) for $\nu = 0.5$, $a_0 = 1, b_0 = 1, u(k) = 0, y_1 = 4$.

Fig. 6.5 Solution of (6.115) for $\nu = 0.5$, $a_0 = 1, b_0 = 1, u(k) = 0, y_1 = 0, y_2 = 16$.

6.5 Linear time-invariant FODE in the polynomial-like matrix equation form

Now, one considers FODE in the form of formula (6.17) and assumes that $\nu = \mu = \frac{1}{d}$, where $d \in \mathbb{Z}_+$. Realize that for $d = 1$ one gets the classical IODE. Both sides of the

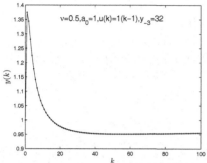
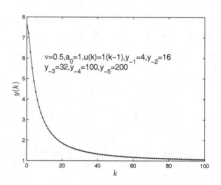

Fig. 6.6   Solution of (6.115) for $\nu = 0.5$, $a_0 = 1, b_0 = 1, u(k) = \mathbf{1}(k-1), y_1 = 0$, $y_2 = 0, y_3 = 32$.

Fig. 6.7   Solution of (6.115) for $\nu = 0.5$, $a_0 = 1, b_0 = 1, u(k) = \mathbf{1}(k-1), y_1 = 4$, $y_2 = 16, y_3 = 32, y_4 = 100, y_5 = 200$.

FODE are transformed to the vector-matrix in the way described in Section 6.4

$$
\begin{bmatrix} A_0 \ A_1 \ \cdots \ A_{k-k_0+1} \ A_{k-k_0} \end{bmatrix}
\begin{bmatrix} y(k) \\ y(k-1) \\ \vdots \\ y(k_0+1) \\ y(k_0) \end{bmatrix}
= \begin{bmatrix} B_0 \ B_1 \ \cdots \ B_{k-k_0+1} \ B_{k-k_0} \end{bmatrix}
\begin{bmatrix} u(k) \\ u(k-1) \\ \vdots \\ u(k_0+1) \\ u(k_0) \end{bmatrix}.
$$
(6.116)

where coefficients $B_0, B_1, \ldots$ are calculated by analogue to (6.107) formula

$$
\begin{bmatrix} B_0 \ B_1 \ \cdots \ B_{k-k_0+1} \ B_{k-k_0} \end{bmatrix} = \begin{bmatrix} b_q \ b_{q-1} \ \cdots \ b_1 \ b_0 \end{bmatrix}
$$

$$
\times \begin{bmatrix}
a^{(q\nu)}(0) & a^{(q\nu)}(1) & \cdots & a^{(q\nu)}(k-k_0+1) & a^{(q\nu)}(k-k_0) \\
a^{((q-1)\nu)}(0) & a^{((q-1)\nu)}(1) & \cdots & a^{((q-1)\nu)}(k-k_0+1) & a^{((q-1)\nu)}(k-k_0) \\
\vdots & \vdots & & \vdots & \vdots \\
a^{(\nu)}(0) & a^{(\nu)}(1) & \cdots & a^{(\nu)}(k-k_0+1) & a^{(\nu)}(k-k_0) \\
a^{(0)}(0) & 0 & \cdots & 0 & 0
\end{bmatrix}.
$$
(6.117)

Equation (6.116) is valid for $k, k-1, k-2, \ldots, k_0+1, k_0$. Collecting a set of such equations in vector-matrix form one obtains

$$
\mathbf{D}_{k-k_0}\mathbf{f}(k) = \mathbf{N}_{k-k_0}\mathbf{u}(k)
$$
(6.118)

where $\mathbf{f}(k)$ is defined by (5.17)

$$
\mathbf{D}_{k-k_0} = \begin{bmatrix} A_0 & A_1 & \cdots & A_{k-k_0-1} & A_{k-k_0} \\ 0 & A_0 & \cdots & A_{k-k_0-2} & A_{k-k_0-1} \\ \vdots & \vdots & & \vdots & \vdots \\ 0 & 0 & \cdots & A_0 & A_1 \\ 0 & 0 & \cdots & 0 & A_0 \end{bmatrix}, \mathbf{D}_{k-k_0} = \begin{bmatrix} B_0 & B_1 & \cdots & B_{k-k_0-1} & B_{k-k_0} \\ 0 & B_0 & \cdots & B_{k-k_0-2} & B_{k-k_0-1} \\ \vdots & \vdots & & \vdots & \vdots \\ 0 & 0 & \cdots & B_0 & B_1 \\ 0 & 0 & \cdots & 0 & B_0 \end{bmatrix}.
$$
(6.119)

I is worth recalling condition (6.108), which preserves the invertibility of the matrix $\mathbf{D}_{k-k_0}$ for $k \geq k_0$. On the matrix $\mathbf{D}_{k-k_0}$ there is no such condition imposed. Such condition forces mathematical modelling, where equation (6.118) describes a real dynamic system.

There is another way to determine matrices $\mathbf{D}_{k-k_0}$ and $\mathbf{N}_{k-k_0}$. Applying notation introduced by formula (2.15) and discussed in Section 3.5, it is possible to replace every FOBD in the considered FODE. Subsequently, the FODE form is as follows

$$
a_p \mathbf{A}_{k-k_0}^{(p\nu)} \mathbf{y}(k) + a_{p-1} \mathbf{A}_{k-k_0}^{((p-1)\nu)} \mathbf{y}(k) + \cdots + a_1 \mathbf{A}_{k-k_0}^{(1\nu)} \mathbf{y}(k) + a_0 \mathbf{A}_{k-k_0}^{(0\nu)} \mathbf{y}(k)
$$
$$
= b_q \mathbf{A}_{k-k_0}^{(q\nu)} \mathbf{u}(k) + b_{q-1} \mathbf{A}_{k-k_0}^{((q-1)\nu)} \mathbf{u}(k) + \cdots + b_1 \mathbf{A}^{(1\nu)}{}_{k-k_0}^{(1\nu)} \mathbf{u}(k) + b_0 \mathbf{A}_{k-k_0}^{(0\nu)} \mathbf{u}(k).
$$
(6.120)

The last form simplifies immediately because on both sides there exist the same vectors, $\mathbf{y}(k)$ and $\mathbf{u}(k)$

$$
\left[a_p \mathbf{A}_{k-k_0}^{(p\nu)} + a_{p-1} \mathbf{A}_{k-k_0}^{((p-1)\nu)} + \cdots + a_1 \mathbf{A}_{k-k_0}^{(1\nu)} + a_0 \mathbf{A}_{k-k_0}^{(0\nu)} \right] \mathbf{y}(k)
$$
$$
= \left[b_q \mathbf{A}_{k-k_0}^{(q\nu)} + b_{q-1} \mathbf{A}_{k-k_0}^{((q-1)\nu)} + \cdots + b_1 \mathbf{A}^{(1\nu)}{}_{k-k_0}^{(1\nu)} + b_0 \mathbf{A}_{k-k_0}^{(0\nu)} \right] \mathbf{u}(k).
$$
(6.121)

Comparison of equations (6.121) with (6.118) leads to a conclusion that

$$
\mathbf{D}_{k-k_0} = a_p \mathbf{A}_{k-k_0}^{(p\nu)} + a_{p-1} \mathbf{A}_{k-k_0}^{((p-1)\nu)} + \cdots + a_1 \mathbf{A}_{k-k_0}^{(1\nu)} + a_0 \mathbf{A}_{k-k_0}^{(0\nu)}
$$
$$
\mathbf{N}_{k-k_0} = b_q \mathbf{A}_{k-k_0}^{(q\nu)} + b_{q-1} \mathbf{A}_{k-k_0}^{((q-1)\nu)} + \cdots + b_1 \mathbf{A}^{(1\nu)}{}_{k-k_0}^{(1\nu)} + b_0 \mathbf{A}_{k-k_0}^{(0\nu)}.
$$
(6.122)

Evidently, $\mathbf{A}_{k-k_0}^{(0\nu)}$ in the above formulas there is the unit matrix of dimensions $(k + 1 - k_0) \times (k + 1 - k_0)$. The presented method seems to be more simple than that leading to formulas (6.106) and (6.117), especially when the resulting matrices will be represented in a form similar to the polynomial Horner form. Then, instead of a series of matrices $\mathbf{A}_{k-k_0}^{(p\nu)}$, $\mathbf{A}_{k-k_0}^{((p-1)\nu)}$ $\cdots$ and $\mathbf{A}_{k-k_0}^{(q\nu)}$, $\mathbf{A}_{k-k_0}^{((q-1)\nu)}$ $\cdots$ calculations one must calculate only one: $\mathbf{A}_{k-k_0}^{(\nu)}$.

$$\mathbf{D}_{k-k_0}$$

$$= a_0 \mathbf{1}_{k-k_0} + a_1 \left(\mathbf{A}_{k-k_0}^{(\nu)} + a_2 \left(\mathbf{A}_{k-k_0}^{(\nu)} + \cdots + \mathbf{A}_{k-k_0}^{(\nu)} \left(a_{p-1} + a_p \mathbf{A}_{k-k_0}^{(\nu)} \right) \cdots \right) \right) \tag{6.123}$$

$$\mathbf{N}_{k-k_0}$$

$$= b_0 \mathbf{1}_{k-k_0} + b_1 \left(\mathbf{A}_{k-k_0}^{(\nu)} + b_2 \left(\mathbf{A}_{k-k_0}^{(\nu)} + \cdots + \mathbf{A}_{k-k_0}^{(\nu)} \left(b_{q-1} + b_q \mathbf{A}_{k-k_0}^{(\nu)} \right) \cdots \right) \right). \tag{6.124}$$

There is a third equivalent method of evaluating the matrices $\mathbf{D}_{k-k_0}$ and $\mathbf{N}_{k-k_0}$. One starts with the FODE representation as a polynomial with a variable w defined as

$$w = {}_{k_0}^{GL} \Delta_k^{(\nu)} y(k). \tag{6.125}$$

Then, one can express equation (6.17) as a product of first degree polynomials

$$\prod_{i=1}^{p} [w - w_i] = \prod_{i=1}^{q} [w - z_i] \tag{6.126}$$

where w_i and z_i can be single, multiple, real or complex. In a complex value case there must also exist its complex conjugate counterpart. Returning to the FOBD one gets

$$\prod_{i=1}^{p} \left[{}_{k_0}^{GL} \Delta_k^{(\nu)} - w_i \right] y(k) = \prod_{i=1}^{q} \left[{}_{k_0}^{GL} \Delta_k^{(\nu)} - z_i \right] u(k). \tag{6.127}$$

A detailed description of the decomposition of (6.17) will be detailed in the next section.

Now, one introduces a notation

$$\mathbf{D}_{k-k_0,i} = \mathbf{A}_{k-k_0}^{(\nu)} + w_i \mathbf{1}_{k-k_0}, \ \ \mathbf{N}_{k-k_0,i} = \mathbf{A}_{k-k_0}^{(\nu)} + z_i \mathbf{1}_{k-k_0} \tag{6.128}$$

then

$$\mathbf{D}_{k-k_0} = \prod_{i=1}^{p} \mathbf{D}_{k-k_0,i}, \ \ \mathbf{N}_{k-k_0} = \prod_{i=1}^{q} \mathbf{N}_{k-k_0,i}. \tag{6.129}$$

Comment. The character of matrices $\mathbf{D}_{k-k_0,i}$ and $\mathbf{N}_{k-k_0,i}$ (upper-triangular band matrix) permits to note that they commute, i.e.

$$\mathbf{D}_{k-k_0,i} \mathbf{D}_{k-k_0,j} = \mathbf{D}_{k-k_0,j} \mathbf{D}_{k-k_0,i} \tag{6.130}$$

for $\mathbf{D}_{k-k_0,i}$ and $\mathbf{N}_{k-k_0,i}$ related to real w_i and w_j and a complex conjugate pair when $w_i = \bar{w}_j$.

The distribution algorithm will be summarized in the following steps.

Algorithm 6.2.

(I) For given coefficients $a_p, a_{p-1}, \ldots, p_0,\ b_q, b_{q-1}, \ldots, q_0$ and order ν find $w_1, \ldots, w_p$ and $z_1, \ldots, z_q$

(II) For $w_1, \ldots, w_p$ and $z_1, \ldots, z_q$ applying formula (6.128) evaluate matrices $\mathbf{D}_{k-k_0,i}$ and $\mathbf{N}_{k-k_0,i}$

(III) Evaluate the products of (6.129).

Decomposition (6.129) is a very important step among others in the dynamic system identification [Ljung (1999)] where in a decomposition procedure there may occur a case where $\mathbf{D}_{k-k_0,i} = \mathbf{N}_{k-k_0,j}$ or $\mathbf{D}_{k-k_0,i} \neq \mathbf{N}_{k-k_0,j}$. Performing the matrices $\mathbf{D}_{k-k_0,u}$ and $\mathbf{N}_{k-k_0,v}$, $u \in 1, 2, \ldots, p$, $v \in 1, 2, \ldots, q$ swapping operation, one obtains

$$\mathbf{D}_{k-k_0,u} \prod_{i=1,i\neq u}^{p} \mathbf{D}_{k-k_0,i}\mathbf{y}(k) = \mathbf{N}_{k-k_0,v} \prod_{i=1,i\neq v}^{q} \mathbf{N}_{k-k_0,i}\mathbf{u}(k). \qquad (6.131)$$

Multiplying both sides by an inverse matrix $[\mathbf{D}_{k-k_0,u}]^{-1}$ one gets a simpler FODE

$$\prod_{i=1,i\neq u}^{p} \mathbf{D}_{k-k_0,i}\mathbf{y}(k) = \prod_{i=1,i\neq v}^{q} \mathbf{N}_{k-k_0,i}\mathbf{u}(k) \qquad (6.132)$$

which after appropriate renaming is expressed as

$$\prod_{i=1}^{p-1} \mathbf{D}_{k-k_0,i}\mathbf{y}(k) = \prod_{i=1}^{q-1} \mathbf{N}_{k-k_0,i}\mathbf{u}(k). \qquad (6.133)$$

An inverse operation is much more complicated. For the given matrices (6.119) one should find its decomposition into products of matrices $\mathbf{D}_{k-k_0,i}$ and $\mathbf{N}_{k-k_0,i}$. This is related to the need for the $p+q+2$ non-linear algebraic equations solution. The problem is formulated as follows. For two given matrices $\mathbf{D}_{k-k_0}$ and $\mathbf{N}_{k-k_0}$ one should find p coefficients $a_{p-1}, a_{p-2}, \ldots, a_1, a_0$ (here one assumes $a_p = 1$), $q+1$ coefficients $b_q, b_{q-1}, \ldots, b_1, b_0$ and one FO ν. Knowledge of the matrix $\mathbf{D}_{k-k_0}$ means that a set of coefficients $A_0, A_1, \ldots$ is known. The non-linear equations are derived from a matrix equation

$$\begin{bmatrix} 1+w_1 & a^{(\nu)}(1) & a^{(\nu)}(2) & \cdots \\ 0 & 1+w_1 & a^{(\nu)}(1) & \cdots \\ 0 & 0 & 1+w_1 & \cdots \\ \vdots & \vdots & \vdots & \end{bmatrix} \begin{bmatrix} 1+w_2 & a^{(\nu)}(1) & a^{(\nu)}(2) & \cdots \\ 0 & 1+w_2 & a^{(\nu)}(1) & \cdots \\ 0 & 0 & 1+w_2 & \cdots \\ \vdots & \vdots & \vdots & \end{bmatrix} \cdots$$

$$\begin{bmatrix} 1+w_{p-1} & a^{(\nu)}(1) & a^{(\nu)}(2) & \cdots \\ 0 & 1+w_{p-1} & a^{(\nu)}(1) & \cdots \\ 0 & 0 & 1+w_{p-1} & \cdots \\ \vdots & \vdots & \vdots & \end{bmatrix} \begin{bmatrix} 1+w_p & a^{(\nu)}(1) & a^{(\nu)}(2) & \cdots \\ 0 & 1+w_p & a^{(\nu)}(1) & \cdots \\ 0 & 0 & 1+w_p & \cdots \\ \vdots & \vdots & \vdots & \end{bmatrix}$$

$$= \begin{bmatrix} A_0 & A_1 & A_2 & \cdots \\ 0 & A_0 & A_1 & \cdots \\ 0 & 0 & A_0 & \cdots \\ \vdots & \vdots & \vdots & \end{bmatrix}. \tag{6.134}$$

The few first equations have now been evaluated under the assumption that $w_i \neq 0$

$$(1+w_1)(1+w_2)\cdots(1+w_p) = A_0 \tag{6.135}$$

$$-\nu \left[\prod_{i=1}^{p}(1+w_i) \right] \left[\sum_{i=1}^{p} \frac{1}{1+w_i} \right] = A_1 \tag{6.136}$$

$$\left[\prod_{i=1}^{p}(1+w_i) \right] \begin{bmatrix} 1 & 0 & 0 \end{bmatrix} \prod_{i=1}^{p} \begin{bmatrix} 1 & -\frac{\nu}{1+w_i} & \frac{\nu(1-\nu)}{2(1+w_i)} \\ 0 & 1 & -\frac{\nu}{1+w_i} \\ 0 & 0 & 1 \end{bmatrix} \begin{bmatrix} 0 \\ 0 \\ 1 \end{bmatrix} = A_2 \tag{6.137}$$

$$\left[\prod_{i=1}^{p}(1+w_i) \right] \begin{bmatrix} 1 & 0 & 0 & 0 \end{bmatrix} \prod_{i=1}^{p} \begin{bmatrix} 1 & -\frac{\nu}{1+w_i} & \frac{\nu(1-\nu)}{2(1+w_i)} & -\frac{\nu(\nu-1)(\nu-2)}{3!(1+w_i)} \\ 0 & 1 & -\frac{\nu}{1+w_i} & \frac{\nu(1-\nu)}{2(1+w_i)} \\ 0 & 0 & 1 & -\frac{\nu}{1+w_i} \\ 0 & 0 & 0 & 1 \end{bmatrix} \begin{bmatrix} 0 \\ 0 \\ 0 \\ 1 \end{bmatrix} = A_3. \tag{6.138}$$

The procedure described above will be formulated in the form of Algorithm 6.3.

Algorithm 6.3.

(I) For given coefficients $A_0, A_1, \ldots$ evaluate equations (6.135)–(6.137)

(II) Solve a set of non-linear algebraic equations from Step I

(III) Having a set of solutions $w_1, w_2, \ldots, w_p$ and ν use the Vieta's equations to calculate a set $a_{p-1}, a_{p-2}, \ldots, p_1, p_0$

(IV) Repeat Steps I–III for $B_0, B_1, \ldots$ to get $b_q, b_{q-1}, \ldots, q_1, q_0$.

For $p = 1$ the solution is trivial. In the next numerical example the steps of Algorithm 6.3 will be checked for $p = 2$. This leads to three non-linear algebraic equations as well as its solution.

Example 6.7. First, the input data will be prepared. This gives a predicted solution. Then consider the linear time-invariant FODE

$$_{k_0}^{GL}\Delta_k^{(2\nu)}y(k) + a_1{}_{k_0}^{GL}\Delta_k^{(\nu)}y(k) + a_0y(k) = b_0u(k). \tag{6.139}$$

According to substitution (6.125) leading to polynomial (6.126), in the considered case one has

$$w^2 + a_1w + a_0 = (w - w_1)(w - w_2) = b_0 \tag{6.140}$$

with

$$a_1 = w_1 + w_2, \; a_0 = w_1w_2. \tag{6.141}$$

By formula (6.107)

$$A_0 = 1 + a_1 + a_0 \; \text{and} A_i = a^{(2\nu)}(i) + a^{(\nu)}(i) \; \text{for} i = 1, 2, \ldots. \tag{6.142}$$

These data are treated as input ones. Now, by equations (6.135)–(6.137)

$$(1 + w_1)(1 + w_2) = 1 + (w_1 + w_2) + w_1w_2 = 1 + a_1 + a_0 = A_0 \tag{6.143}$$

$$-\nu(1 + w_1)(1 + w_2)\left(\frac{1}{1 + w_1} + \frac{1}{1 + w_2}\right) = -\nu(2 + w_1 + w_2) = -\nu(2 + a_1) = A_2 \tag{6.144}$$

$$(1 + w_1)(1 + w_2)\begin{bmatrix}1 & 0 & 0\end{bmatrix}\begin{bmatrix}1 & -\frac{\nu}{1+w_1} & \frac{\nu(1-\nu)}{2(1+w_1)} \\ 0 & 1 & -\frac{\nu}{1+w_1} \\ 0 & 0 & 1\end{bmatrix}\begin{bmatrix}1 & -\frac{\nu}{1+w_2} & \frac{\nu(1-\nu)}{2(1+w_2)} \\ 0 & 1 & -\frac{\nu}{1+w_1} \\ 0 & 0 & 1\end{bmatrix}\begin{bmatrix}0 \\ 0 \\ 1\end{bmatrix}$$

$$= \begin{bmatrix}1 & 0 & 0\end{bmatrix}\begin{bmatrix}1 + w_1 & -\nu & \frac{\nu(1-\nu)}{2} \\ 0 & 1 + w_1 & -\nu \\ 0 & 0 & 1 + w_1\end{bmatrix}\begin{bmatrix}1 + w_2 & -\nu & \frac{\nu(1-\nu)}{2} \\ 0 & 1 + w_2 & -\nu \\ 0 & 0 & 1 + w_2\end{bmatrix}\begin{bmatrix}0 \\ 0 \\ 1\end{bmatrix} = A_3. \tag{6.145}$$

Having evaluated three non-linear algebraic equations with unknown a_1, a_0, ν the solution begins at (6.145). Simple calculations lead to the more simple form

$$\nu^2 - \frac{A_1}{2}\nu + \frac{A_1}{2} - A_2 = 0. \tag{6.146}$$

Then

$$\nu_1 = \frac{A_1 + \sqrt{\Delta}}{2}, \ \nu_2 = \frac{A_1 - \sqrt{\Delta}}{2} \tag{6.147}$$

$$\Delta = \frac{A_1^2}{4} - 2A_1 + 4A_2.$$

From ν_1, ν_2 one takes a positive solution. Assuming that ν is known from (6.144), one extracts a_1 and a_2

$$a_1 = -\frac{A_1}{\nu} \tag{6.148}$$

$$a_0 = A_0 - 1 - a_1. \tag{6.149}$$

The numerical example presented above reveals that the degree of complexity rapidly increases due to an increased number of the searched coefficients.

6.6 Discrete FO transfer function

An application of the one-sided $\mathcal{Z}$-Transform to both sides of equation (6.17) with zero initial conditions

$$\sum_{i=0}^{p} a_i \mathcal{Z} \left\{ {}_{k_0}^{GL}\Delta_k^{(i\nu)} y(k) \right\} = \sum_{j=0}^{q} b_j \mathcal{Z} \left\{ {}_{k_0}^{GL}\Delta_k^{(j\nu)} u(k) \right\} \tag{6.150}$$

gives

$$\sum_{i=0}^{p} a_i \left(1 - z^{-1}\right)^{i\nu} Y(z) = \sum_{j=0}^{q} b_j \left(1 - z^{-1}\right)^{j\nu} U(z). \tag{6.151}$$

where $Y(z)$ and $U(z)$ are the transforms of functions $y(k)$ and $u(k)$, respectively. Simple rearrangement gives the discrete FO transfer function (FOTF)

$$G(z) = \frac{Y(z)}{U(z)} = \frac{\sum_{j=0}^{q} b_j \left(1 - z^{-1}\right)^{j\nu}}{\sum_{i=0}^{p} a_i \left(1 - z^{-1}\right)^{i\nu}} \ \text{for } q \geq p. \tag{6.152}$$

On the basis of formula (5.35) the above discrete FOTF is expressed in the form

$$G(z) = \frac{Y(z)}{U(z)} = \frac{\sum_{j=0}^{q} b_j \sum_{i=0}^{+\infty} a^{(j\nu)}(i)z^{-i}}{\sum_{i=0}^{p} a_i \sum_{j=0}^{+\infty} a^{(i\nu)}(j)z^{-j}} \ \text{for } q \geq p \tag{6.153}$$

which is further simplified to

$$G(z) = \frac{Y(z)}{U(z)} = \frac{\sum_{i=0}^{+\infty} B_i z^{-i}}{\sum_{j=0}^{+\infty} A_j z^{-j}} \tag{6.154}$$

where coefficients A_i, B_i are defined by (6.107) and (6.117), respectively. This form of the FOTF is particularly convenient for expansion into an infinite series of Markov parameters ([Kailath (1980)]). Performing a long division of the numerator polynomial by the denominator one obtains

$$G(w) = \sum_{i=0}^{+\infty} g_i z^{-i}. \tag{6.155}$$

The elements of a set $\{g_0, g_1, g_2, \ldots\}$ create a so-called impulse response [Ogata (1987)]

Defining a new complex variable

$$w = (1 - z^{-1})^\nu \tag{6.156}$$

expression $\sum_{i=0}^{p} a_i \left(1 - z^{-1}\right)^{i\nu} = 0$ transforms to a polynomial of this new variable w

$$\sum_{i=0}^{p} a_i w^i = 0 \tag{6.157}$$

which may also be represented as a product of first and second degree polynomials

$$\prod_{i=1}^{p_r} (w - w_{r,i})^{r_i} \prod_{i=1}^{p_c} \left[w^2 - 2\Re(w_{c,i})w + |w_{c,i}|^2\right]^{s_i} = 0 \tag{6.158}$$

where

$w_{r,i}$ – i-th real pseudo-pole of $G(z)$
$w_{c,i}$ – i-th complex pseudo-pole of $G(z)$
$\underline{r}_i$ – i-th real pseudo-pole multiplicity
$\underline{s}_i$ – i-th complex pseudo-pole multiplicity
p_r – number of distinct real pseudo-poles of $G(z)$
p_c – number of distinct complex pseudo-poles of $G(z)$.

Analogously, one can treat a polynomial $\sum_{i=0}^{q} b_i \left(1 - z^{-1}\right)^{i\nu} = 0$, obtaining an equivalent form

$$\prod_{i=1}^{q_r} (w - v_{r,i})^{\bar{r}_i} \prod_{i=1}^{q_c} \left[w^2 - 2\Re(v_{c,i})w + |v_{c,i}|^2\right]^{\bar{s}_i} = 0 \tag{6.159}$$

where

$v_{r,i}$ – i-th real pseudo-zero of $G(z)$
$v_{c,i}$ – i-th complex pseudo-zero of $G(z)$
$\bar{r}_i$ – i-th real pseudo-zero multiplicity

$\bar{s}_i$ – i-th complex pseudo-zero multiplicity
q_r – number of distinct real pseudo-zeros of $G(z)$
q_c – number of distinct complex pseudo-zeros of $G(z)$.

Evidently,

$$\sum_{i=1}^{p_r} \underline{r}_i + 2\sum_{i=1}^{p_c} \underline{s}_i = p, \quad \sum_{i=1}^{q_r} \bar{r}_i + 2\sum_{i=1}^{q_c} \underline{s}_i = q. \tag{6.160}$$

Then,

$$G(z)\big|_{(1-z^{-1})^{\nu}=w} = G(w) \tag{6.161}$$

$$= \frac{\prod_{i=1}^{q_r} (w - v_{r,i})^{\bar{r}_i} \prod_{i=1}^{q_c} \left[w^2 - 2\Re(v_{c,i})w + |v_{c,i}|^2\right]^{\bar{s}_i}}{\prod_{i=1}^{p_r} (w - w_{r,i})^{\underline{r}_i} \prod_{i=1}^{p_c} \left[w^2 - 2\Re(w_{c,i})w + |w_{c,i}|^2\right]^{\underline{s}_i}}.$$

Here one should pay attention to the possible occurrence of common factors appearing when $w_{r,i} = v_{r,j}$ or $w_{c,i} = v_{c,j}$. Such factors should be cancelled. The FOTF $G(w)$ in the form (6.162) can be expressed as a product of discrete FO sub-transfer functions

$$G(w) = \prod_i G_i(w), \tag{6.162}$$

of types indicated below

$$G_i(w) = \begin{cases} \dfrac{1}{w - w_{r,1}} \\[2mm] \dfrac{w - v_{r,i}}{w - w_{r,1}} = 1 + \dfrac{w_{r,i} - v_{r,i}}{w - w_{r,1}} \\[2mm] \dfrac{1}{w^2 - 2\Re(w_{c,i})w + |w_{c,i}|^2} \\[2mm] \dfrac{w - v_{r,i}}{w^2 - 2\Re(w_{c,i})w + |w_{c,i}|^2} \\[2mm] \dfrac{(w - v_{r,i})(w - v_{r,j})}{w^2 - 2\Re(w_{c,i})w + |w_{c,i}|^2} = 1 + \dfrac{w[2\Re(w_{c,i}) - v_{r,i} - v_{r,j}] + v_{r,i}v_{r,j} - |w_{c,i}|^2}{w^2 - 2\Re(w_{c,i})w + |w_{c,i}|^2} \\[2mm] \dfrac{w^2 - 2\Re(v_{c,i})w + |v_{c,i}|^2}{w^2 - 2\Re(w_{c,i})w + |w_{c,i}|^2} = 1 + \dfrac{2(\Re\{w_{c,i}\} - \{v_{c,i}\})w + (|v_{c,i}|^2 - |w_{c,i}|^2)}{w^2 - 2\Re(w_{c,i})w + |w_{c,i}|^2} \end{cases} . \tag{6.163}$$

Recalling the assumption that $p \geq q$ in (6.162) there are possible elements with a numerator equal to 1.

To the FO discrete transfer function (6.152) one can also apply the partial fraction decomposition procedure

$$G(z)|_{(1-z^{-1})^\nu = w} = \frac{\sum_{j=0}^{q} b_j w^{j\nu}}{\prod_{i=1}^{p_r} (w - w_{r,i})^{r_i} \prod_{i=1}^{p_c} [w^2 - 2\Re\{w_{c,i}\} w + |w_{c,i}|^2]^{s_i}}$$

$$= \sum_{i=1}^{p_r} \sum_{j=1}^{r_i} \frac{C_{ij}}{(w - w_{r,i})^j} + \sum_{i=1}^{p_c} \sum_{j=1}^{s_i} \frac{D_{ij} w + D_{ir}}{(w^2 - 2\Re\{w_{c,i}\} w + |w_{c,i}|^2)^j}. \qquad (6.164)$$

One can note that in this equivalent form there are similar elements as indicated in formula (6.163). The most important ones are

$$G_i(z) = \frac{1}{(1-z^{-1})^\nu + w_{r,i}} \qquad (6.165)$$

$$G_i(z) = \frac{1}{(1-z^{-1})^{2\nu} - 2\Re\{w_{c,i}\}(1-z^{-1})^\nu + |w_{c,i}|^2}. \qquad (6.166)$$

6.7 Selected relations between the FODE, FOTF, FOSSE and FOPLE

In this section, some relations between considered equivalent forms of the FODE are discussed and summarized. Some of them have already been presented. The FODE equivalent descriptions are derived: FOSSE, PMLE, FOTF (6.150)–(6.153). The presented between the forms relations are illustrated in Fig. 6.8

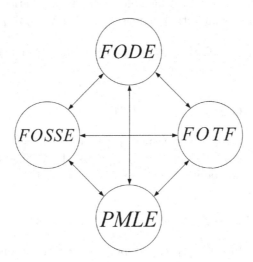

Fig. 6.8   Relations between different equivalent FO system descriptions.

In this section, the primary analysis will concentrate on the relations between the FOSSE and the FOTF, and vice versa. Hence, applying the one-sided $\mathcal{Z}$-Transform to both sides of equations (6.41) and (6.42) under the assumption of zero-initial conditions $\mathbf{0} = \mathbf{x}_{-1} = \mathbf{x}_{-2} = \cdots$ one gets

$$\mathcal{Z}\left\{{}_{k_0}^{GL}\Delta_k^{(\nu)}\mathbf{x}(k)\right\} = \mathbf{A}\mathcal{Z}\{\mathbf{x}(k)\} + \mathbf{B}\mathcal{Z}\{\mathbf{u}(k)\} \qquad (6.167)$$

and

$$\mathcal{Z}\{\mathbf{y}(k)\} = \mathbf{C}\mathcal{Z}\{\mathbf{x}(k)\} + \mathbf{D}\mathcal{Z}\{\mathbf{u}(k)\}. \qquad (6.168)$$

Hence,

$$(1 - z^{-1})^\nu \mathbf{X}(z) = \mathbf{A}\mathbf{X}(z) + \mathbf{B}\mathbf{U}(z) \qquad (6.169)$$

and

$$\mathbf{Y}(z) = \mathbf{C}\mathbf{X}(z) + \mathbf{D}\mathbf{U}(z) \qquad (6.170)$$

where $\mathcal{Z}\{\mathbf{x}(k)\} = \mathbf{X}(z)$, $\mathcal{Z}\{\mathbf{y}(k)\} = \mathbf{y}(z)$, $\mathcal{Z}\{\mathbf{u}(k)\} = \mathbf{U}(z)$.
Elementary transformations of (6.169) lead to

$$\left[(1 - z^{-1})^\nu \mathbf{1}_p - \mathbf{A}\right]\mathbf{X}(z) = \mathbf{B}\mathbf{U}(z) \qquad (6.171)$$

and further,

$$\mathbf{X}(z) = \left[(1 - z^{-1})^\nu \mathbf{1}_p - \mathbf{A}\right]^{-1}\mathbf{B}\mathbf{U}(z). \qquad (6.172)$$

Substituting (6.172) into (6.170) one gets the final relation between the $\mathcal{Z}$-Transforms of the input and output signals $\mathbf{U}(z)$ and $\mathbf{Y}(z)$, respectively.

$$\mathbf{Y}(z) = \left[\mathbf{C}\left[(1 - z^{-1})^\nu \mathbf{1}_p - \mathbf{A}\right]^{-1}\mathbf{B} + \mathbf{D}\right]\mathbf{U}(z) \qquad (6.173)$$

The matrix

$$\mathbf{G}(z) = \mathbf{C}\left[(1 - z^{-1})^\nu \mathbf{1}_p - \mathbf{A}\right]^{-1}\mathbf{B} + \mathbf{D} \qquad (6.174)$$

is called the FO matrix TF. For a special case when $\mathbf{C} = \mathbf{c}$ and $\mathbf{B} = \mathbf{b}$ are of dimensions $1 \times p$ and $p \times 1$ respectively, $\mathbf{G}(z)$ is a scalar FOTF. The output matrix $\mathbf{D} = [d]$. Hence a relation between the FOSSE and the FOTF has been established.
Now, one denotes

$$\mathbf{\Phi}(z) = \mathcal{Z}\{\mathbf{\Phi}(k)\} = \left[(1 - z^{-1})^\nu \mathbf{1}_p - \mathbf{A}\right]^{-1}. \qquad (6.175)$$

Matrix $\mathbf{\Phi}(k) = \mathcal{Z}^{-1}\{\mathbf{\Phi}(z)\}$ is a very important one in the linear time-invariant dynamic system analysis. The following definition formally defines this matrix.

Definition 6.5. The dynamic system described by the FOSSE (6.41) and (6.42) transient properties are characterized by its characteristic matrix $\mathbf{\Phi}(z)$ (6.175).

Comment. The characteristic matrix contains important information about the dynamic system transient behavior such as its stability and oscillatory or damping properties. The characteristic matrix can be expanded into an infinite series. The following result yields the formula for the characteristic matrix $\mathbf{\Phi}(z)$ expansion.

Theorem 6.1. *The characteristic matrix* $\mathbf{\Phi}(z)$ *can be represented as an infinite series*

$$\mathbf{\Phi}(z) = \sum_{i=0}^{+\infty} \mathbf{\Phi}(i) z^{-i} \tag{6.176}$$

where

$$\mathbf{\Phi}(0) = (\mathbf{1}_p - \mathbf{A})^{-1}$$
$$\mathbf{\Phi}(k) = -\mathbf{\Phi}(0) \sum_{i=0}^{k-1} a^{(\nu)}(k - 1 - i)\mathbf{\Phi}(i) \text{ for } k = 1, 2, \ldots.$$

Proof. A substitution into equality $[\mathbf{\Phi}(k)]^{-1}\mathbf{\Phi}(k) = \mathbf{1}_p$ formulas (6.175) and (6.176) yields the result

$$\left[(1 - z^{-1})^{\nu}\mathbf{1}_p - \mathbf{A}\right]\left[\mathbf{\Phi}(0) + \mathbf{\Phi}(1)z^{-1} + \mathbf{\Phi}(2)z^{-2} + \cdots\right] = \mathbf{1}_p. \tag{6.177}$$

Following result (1.17) the left factor in (6.177) can also be expanded into an infinite series

$$\left[(1 - z^{-1})^{\nu}\mathbf{1}_p - \mathbf{A}\right] = (\mathbf{1}_p - \mathbf{A}) + a^{(\nu)}(1)\mathbf{1}_p + a^{(\nu)}(2)\mathbf{1}_p + \cdots. \tag{6.178}$$

Hence, (6.177) is a product of two infinite polynomials in a variable z^{-1} with matrix coefficients. The multiplication result is equal to a unit matrix $\mathbf{1}_p$. Multiplying both polynomials term by term, one gets a polynomial for which all matrix coefficients equal to zero except the first one

$$(\mathbf{1}_p - \mathbf{A})\mathbf{\Phi}(0) + \left[(\mathbf{1}_p - \mathbf{A})\phi(1) + a^{(\nu)}(1)\mathbf{\Phi}(0)\right]z^{-1}$$
$$+ \left[(\mathbf{1}_p - \mathbf{A})\mathbf{\Phi}(2) + a^{(\nu)}(1)\mathbf{\Phi}(1) + a^{(\nu)}(2)\mathbf{\Phi}(0)\right]z^{-2}$$
$$+ \left[(\mathbf{1}_p - \mathbf{A})\mathbf{\Phi}(3) + a^{(\nu)}(1)\mathbf{\Phi}(2) + a^{(\nu)}(2)\mathbf{\Phi}(1) + a^{(\nu)}(3)\mathbf{\Phi}(0)\right]z^{-3} + \cdots$$
$$= \mathbf{1}_p + \mathbf{0}_p z^{-1} + \mathbf{0}_p z^{-2} + \mathbf{0}_p z^{-3} + \cdots. \tag{6.179}$$

Comparison of the left and right-hand sides of (6.179) leads to an infinite number of equations

$$(\mathbf{1}_p - \mathbf{A})\mathbf{\Phi}(0) = \mathbf{1}_p$$
$$(\mathbf{1}_p - \mathbf{A})\mathbf{\Phi}(1) + a^{(\nu)}(1)\mathbf{\Phi}(0) = \mathbf{0}_p$$
$$(\mathbf{1}_p - \mathbf{A})\mathbf{\Phi}(2) + a^{(\nu)}(1)\mathbf{\Phi}(1) + a^{(\nu)}(2)\mathbf{\Phi}(0) = \mathbf{0}_p$$
$$(\mathbf{1}_p - \mathbf{A})\mathbf{\Phi}(3) + a^{(\nu)}(1)\mathbf{\Phi}(2) + a^{(\nu)}(2)\mathbf{\Phi}(1) + a^{(\nu)}(3)\mathbf{\Phi}(0) = \mathbf{0}_p$$

$$\vdots$$

$$(\mathbf{1}_p - \mathbf{A})\mathbf{\Phi}(k) + \sum_{i=1}^{k-1} a^{(\nu)}(k - 1 - i)\mathbf{\Phi}(i) = \mathbf{0}_p. \qquad (6.180)$$

From the above equations one calculates consecutive matrices $\mathbf{\Phi}(k)$. $\qquad\square$

The above set of equalities may be expressed equivalently in one matrix equation

$$\begin{bmatrix} (\mathbf{1}_p - \mathbf{A}) & a^{(\nu)}(1)\mathbf{1}_p & \cdots & a^{(\nu)}(k-1)\mathbf{1}_p & a^{(\nu)}(k)\mathbf{1}_p \\ \mathbf{0}_p & (\mathbf{1}_p - \mathbf{A}) & \cdots & a^{(\nu)}(k-2)\mathbf{1}_p & a^{(\nu)}(k-1)\mathbf{1}_p \\ \vdots & \vdots & & \vdots & \vdots \\ \mathbf{0}_p & \mathbf{0}_p & \cdots & (\mathbf{1}_p - \mathbf{A}) & a^{(\nu)}(1)\mathbf{1}_p \\ \mathbf{0}_p & \mathbf{0}_p & \cdots & \mathbf{0}_p & (\mathbf{1}_p - \mathbf{A}) \end{bmatrix} \begin{bmatrix} \mathbf{\Phi}(k) \\ \mathbf{\Phi}(k-1) \\ \vdots \\ \mathbf{\Phi}(1) \\ \mathbf{\Phi}(0) \end{bmatrix} = \begin{bmatrix} \mathbf{0}_p \\ \mathbf{0}_p \\ \vdots \\ \mathbf{0}_p \\ \mathbf{1}_p \end{bmatrix}.$$

$$(6.181)$$

The square block matrix will be denoted as

$$\mathbf{D}_{kp} = \begin{bmatrix} (\mathbf{1}_p - \mathbf{A}) & a^{(\nu)}(1)\mathbf{1}_p & a^{(\nu)}(2)\mathbf{1}_p & \cdots & a^{(\nu)}(k-1)\mathbf{1}_p & a^{(\nu)}(k)\mathbf{1}_p \\ \mathbf{0}_p & (\mathbf{1}_p - \mathbf{A}) & a^{(\nu)}(1)\mathbf{1}_p & \cdots & a^{(\nu)}(k-2)\mathbf{1}_p & a^{(\nu)}(k-1)\mathbf{1}_p \\ \vdots & \vdots & \vdots & \cdots & \vdots & \vdots \\ \mathbf{0}_p & \mathbf{0}_p & \mathbf{0}_p & \cdots & (\mathbf{1}_p - \mathbf{A}) & a^{(\nu)}(1)\mathbf{1}_p \\ \mathbf{0}_p & \mathbf{0}_p & \mathbf{0}_p & \cdots & \mathbf{0}_p & (\mathbf{1}_p - \mathbf{A}) \end{bmatrix}. \quad (6.182)$$

By assumption, the matrix $\mathbf{1}_p - \mathbf{A}$ is non-singular. Therefore $\mathbf{D}_{pk}$ is invertible.

Corollary 6.1. *For a non-singular upper triangular block band matrix its inverse is also an upper triangular block band matrix.*

Proof. Let the upper triangular block matrix and its inverse be denoted, respectively as

$$\bar{\mathbf{A}}_k = \begin{bmatrix} \mathbf{A}_1 & \mathbf{A}_2 & \cdots & \mathbf{A}_{k-1} & \mathbf{A}_k \\ 0 & \mathbf{A}_1 & \cdots & \mathbf{A}_{k-2} & \mathbf{A}_{k-1} \\ \vdots & \vdots & & \vdots & \vdots \\ 0 & \mathbf{A}_1 & \cdots & \mathbf{A}_1 & \mathbf{A}_2 \\ 0 & 0 & \cdots & 0 & \mathbf{A}_1 \end{bmatrix}, \quad \bar{\mathbf{B}}_k = \begin{bmatrix} \mathbf{B}_1 & \mathbf{B}_2 & \cdots & \mathbf{B}_{k-1} & \mathbf{B}_k \\ 0 & \mathbf{B}_1 & \cdots & \mathbf{B}_{k-2} & \mathbf{B}_{k-1} \\ \vdots & \vdots & & \vdots & \vdots \\ 0 & 0 & \cdots & \mathbf{B}_1 & \mathbf{B}_2 \\ 0 & 0 & \cdots & 0 & \mathbf{B}_1 \end{bmatrix}. \quad (6.183)$$

This means that

$$\bar{\mathbf{A}}_k \bar{\mathbf{B}}_k = \begin{bmatrix} \mathbf{A}_1 & \mathbf{A}_2 & \cdots & \mathbf{A}_{k-1} & \mathbf{A}_k \\ 0 & \mathbf{A}_1 & \cdots & \mathbf{A}_{k-2} & \mathbf{A}_{k-1} \\ \vdots & \vdots & & \vdots & \vdots \\ 0 & \mathbf{A}_1 & \cdots & \mathbf{A}_1 & \mathbf{A}_2 \\ 0 & 0 & \cdots & 0 & \mathbf{A}_1 \end{bmatrix} \begin{bmatrix} \mathbf{B}_1 & \mathbf{B}_2 & \cdots & \mathbf{B}_{k-1} & \mathbf{B}_k \\ 0 & \mathbf{B}_1 & \cdots & \mathbf{B}_{k-2} & \mathbf{B}_{k-1} \\ \vdots & \vdots & & \vdots & \vdots \\ 0 & 0 & \cdots & \mathbf{B}_1 & \mathbf{B}_2 \\ 0 & 0 & \cdots & 0 & \mathbf{B}_1 \end{bmatrix}$$

$$= \begin{bmatrix} \mathbf{1}_p & \mathbf{0}_p & \cdots & \mathbf{0}_0 & \mathbf{0}_0 \\ \mathbf{0}_p & \mathbf{1}_p & \cdots & \mathbf{0}_0 & \mathbf{0}_p \\ \vdots & \vdots & & \vdots & \vdots \\ \mathbf{0}_p & \mathbf{0}_p & \cdots & \mathbf{1}_p & \mathbf{0}_p \\ \mathbf{0}_p & \mathbf{0}_p & \cdots & \mathbf{0}_p & \mathbf{1}_p \end{bmatrix}. \qquad (6.184)$$

From this matrix equation one derives k matrix equalities from which only three will be presented

$$\mathbf{A}_1 \mathbf{B}_1 = \mathbf{1}_p \ \Rightarrow \ \mathbf{B}_1 = (\mathbf{A}_1)^{-1} \qquad (6.185)$$

$$\mathbf{A}_1 \mathbf{B}_2 + \mathbf{A}_2 \mathbf{B}_1 = \mathbf{0}_p, \ \Rightarrow \mathbf{B}_2 = -(\mathbf{A}_1)^{-1} \mathbf{A}_2 \mathbf{B}_1 = -(\mathbf{A}_1)^{-1} \mathbf{A}_2 (\mathbf{A}_1)^{-1} \qquad (6.186)$$

$$\mathbf{A}_1 \mathbf{B}_3 + \mathbf{A}_2 \mathbf{B}_2 + \mathbf{A}_3 \mathbf{B}_1 = \mathbf{0}_p \ \Rightarrow \ \mathbf{B}_3 = -(\mathbf{A}_1)^{-1} (-\mathbf{A}_2 \mathbf{B}_2 + \mathbf{A}_3 \mathbf{B}_1)$$
$$= (\mathbf{A}_1^{-1} \mathbf{A}_2)^2 \mathbf{A}_1^{-1} - \mathbf{A}_1^{-1} \mathbf{A}_3 \mathbf{A}_1^{-1}. \qquad (6.187)$$

By formulas (7.14) and (7.15) the matrix $\bar{\mathbf{A}}_{k+1}$ and its inverse $\bar{\mathbf{B}}_{k+1}$ equal to

$$\bar{\mathbf{A}}_{k+1} = \begin{bmatrix} \mathbf{A}_1 & \mathbf{A}_2 \cdots \mathbf{A}_k & \mathbf{A}_{k+1} \\ 0 & & \\ \vdots & & \bar{\mathbf{A}}_k \\ 0 & & \end{bmatrix}, \ \bar{\mathbf{B}}_{k+1} = \begin{bmatrix} \mathbf{B}_1 & -\mathbf{A}_1^{-1}[\mathbf{A}_2 \cdots \mathbf{A}_{k+1}]\mathbf{B}_k \\ 0 & \\ \vdots & \bar{\mathbf{B}}_k \\ 0 & \end{bmatrix}.$$

$$(6.188)$$

It remains only to prove that

$$-\mathbf{A}_1^{-1} \begin{bmatrix} \mathbf{A}_2 & \mathbf{A}_3 & \cdots & \mathbf{A}_k & \mathbf{A}_{k+1} \end{bmatrix} \bar{\mathbf{B}}_k = \begin{bmatrix} \mathbf{B}_2 & \mathbf{B}_3 & \cdots \mathbf{B}_k & \mathbf{B}_{k+1} \end{bmatrix}. \qquad (6.189)$$

This equation is equivalent to

$$\begin{bmatrix} \mathbf{A}_2 & \mathbf{A}_3 & \cdots & \mathbf{A}_k & \mathbf{A}_{k+1} \end{bmatrix} = -\mathbf{A}_1 \begin{bmatrix} \mathbf{B}_2 & \mathbf{B}_3 & \cdots \mathbf{B}_k & \mathbf{B}_{k+1} \end{bmatrix} \bar{\mathbf{A}}_k$$

$$-\mathbf{A}_1 \begin{bmatrix} \mathbf{B}_2 & \mathbf{B}_3 & \cdots \mathbf{B}_k & \mathbf{B}_{k+1} \end{bmatrix} \begin{bmatrix} \mathbf{A}_1 & \mathbf{A}_2 & \cdots & \mathbf{A}_k & \mathbf{A}_{k+1} \\ 0 & \mathbf{A}_1 & \cdots & \mathbf{A}_{k-1} & \mathbf{A}_k \\ \vdots & \vdots & & \vdots & \vdots \\ 0 & \mathbf{A}_1 & \cdots & \mathbf{A}_1 & \mathbf{A}_2 \\ 0 & 0 & \cdots & 0 & \mathbf{A}_1 \end{bmatrix}. \qquad (6.190)$$

Performing step-by-step consecutive block matrix multiplications, one obtains the results represented by the first three matrix equations (6.185)-(6.187). This ends the proof. □

By Corollary, 6.1 the inverse of matrix (6.182) is also an upper triangular block band matrix. It will be denoted as

$$\mathbf{\Phi}_{kp} = \mathbf{D}_{kp}^{-1} = \begin{bmatrix} \mathbf{\Phi}(0) & \mathbf{\Phi}(1) & \mathbf{\Phi}(2) & \cdots & \mathbf{\Phi}(k-1) & \mathbf{\Phi}(k) \\ \mathbf{0}_p & \mathbf{\Phi}(0) & \mathbf{\Phi}(1) & \cdots & \mathbf{\Phi}(k-2) & \mathbf{\Phi}(k-1) \\ \vdots & \vdots & \vdots & \cdots & \vdots & \vdots \\ \mathbf{0}_p & \mathbf{0}_p & \mathbf{0}_p & \cdots & \mathbf{\Phi}(0) & \mathbf{\Phi}(1) \\ \mathbf{0}_p & \mathbf{0}_p & \mathbf{0}_p & \cdots & \mathbf{0}_p & \mathbf{\Phi}(0) \end{bmatrix}. \qquad (6.191)$$

Note that multiplying both sides of equation (6.181) by (6.191) leads to an identity.

The last equality in (6.180) is especially interesting because it can be immediately transformed to the form

$$\mathbf{1}_p \mathbf{\Phi}(k) + \sum_{i=1}^{k-1} a^{(\nu)}(k-1-i)\mathbf{\Phi}(i) = \mathbf{A}\mathbf{\Phi}(k)$$

$$\sum_{i=0}^{k} a^{(\nu)}(k-i)\mathbf{\Phi}(i) = \mathbf{A}\mathbf{\Phi}(k). \qquad (6.192)$$

Changing the summation variable in (6.192), one obtains

$$\sum_{i=0}^{k} a^{(\nu)}(k-i)\mathbf{\Phi}(i) = \sum_{i=0}^{k} a^{(\nu)}(i)\mathbf{\Phi}(k-i) = {}_0^{GL}\Delta_k^{(\nu)}\mathbf{\Phi}(k) = \mathbf{A}\mathbf{\Phi}(k). \quad (6.193)$$

The characteristic matrix can be equivalently expressed as an infinite series of powers of a state matrix A.

Theorem 6.2. *The FO system characteristic matrix $\mathcal{Z}$-Transform is equal to an infinite series of state matrix powers*

$$\mathbf{\Phi}(z) = \sum_{i=0}^{+\infty}(1-z^{-1})^{-(i+1)\nu}\mathbf{A}^i. \qquad (6.194)$$

Proof. The proof is very similar to the previous one. A substitution of (6.175) into (6.194) and multiplying both sides by $\mathbf{\Phi}^{-1}(z)$ gives

$$\left[(1-z^{-1})^{\nu}\mathbf{1}_p - \mathbf{A}\right]\sum_{i=0}^{+\infty}(1-z^{-1})^{-(i+1)\nu}\mathbf{A}^i$$

$$= \sum_{i=0}^{+\infty}(1-z^{-1})^{-i\nu}\mathbf{A}^i - \sum_{i=0}^{+\infty}(1-z^{-1})^{-(i+1)\nu}\mathbf{A}^{i+1}$$

$$\mathbf{1}_p + \sum_{i=1}^{+\infty}(1-z^{-1})^{-i\nu}\mathbf{A}^i - \sum_{i=0}^{+\infty}(1-z^{-1})^{-(i+1)\nu}\mathbf{A}^{i+1} = \mathbf{1}_p \qquad (6.195)$$

and relation (6.194) is proved because

$$\sum_{i=1}^{+\infty}(1-z^{-1})^{-i\nu}\mathbf{A}^i = \sum_{i=0}^{+\infty}(1-z^{-1})^{-(i+1)\nu}\mathbf{A}^{i+1}. \qquad (6.196)$$

$\square$

Definition 6.6. The characteristic polynomial of the dynamic system described by the FOSSE is defined as

$$det\left[(1-z^{-1})^{\nu}\mathbf{1}_p - \mathbf{A}\right]$$
$$= (1-z^{-1})^{p\nu} + a_{p-1}(1-z^{-1})^{(p-1)\nu} + \cdots + a_1(1-z^{-1})^{\nu} + a_0. \qquad (6.197)$$

Definition 6.7. The characteristic equation of the dynamic system described by the FOSSE is defined as

$$(1-z^{-1})^{p\nu} + a_{p-1}(1-z^{-1})^{(p-1)\nu} + \cdots + a_1(1-z^{-1})^{\nu} + a_0 = 0. \qquad (6.198)$$

Applying the inverse $\mathcal{Z}$ transforms to both sides of the characteristic equation one gets

$$_0^{GL}\Delta^{(p\nu)}y(k) + a_{p-1}{}_0^{GL}\Delta^{((p-1)\nu)}y(k) + \cdots + a_1{}_0^{GL}\Delta^{(\nu)}y(k) + a_0 y(k) = 0. \qquad (6.199)$$

The next Theorem states that the characteristic matrix is a solution of the FODE equation defined above.

Theorem 6.3. *The characteristic matrix* $\mathbf{\Phi}(k)$ *satisfies (6.199), i.e.*

$$_0^{GL}\Delta^{(p\nu)}\mathbf{\Phi}(k) + a_{p-1}{}_0^{GL}\Delta^{((p-1)\nu)}\mathbf{\Phi}(k) + \cdots + a_1{}_0^{GL}\Delta^{(\nu)}\mathbf{\Phi}(k) + a_0\mathbf{\Phi}(k) = \mathbf{0}_p. \qquad (6.200)$$

Proof. First, one recalls that the state matrix $\mathbf{A}$ satisfies its characteristic equation

$$\mathbf{A}^p + a_{p-1}\mathbf{A}^{p-1} + \cdots + a_1\mathbf{A}^1 + a_0\mathbf{1}_p = \mathbf{0}_p. \qquad (6.201)$$

From (6.193) one obtains a set of equations

$$_0^{GL}\Delta_k^{(2\nu)}\mathbf{\Phi}(k) = _0^{GL}\Delta_k^{(\nu)}\left[_0^{GL}\Delta_k^{(\nu)}\mathbf{\Phi}(k)\right]$$

$$= _0^{GL}\Delta_k^{(\nu)}\left[\mathbf{A}\mathbf{\Phi}(k)\right] = \mathbf{A}_0^{GL}\Delta_k^{(\nu)}\mathbf{\Phi}(k) = \mathbf{A}^2\mathbf{\Phi}(k)$$

$$_0^{GL}\Delta_k^{(3\nu)}\mathbf{\Phi}(k) = _0^{GL}\Delta_k^{(\nu)}\left[_0^{GL}\Delta_k^{(2\nu)}\mathbf{\Phi}(k)\right]$$

$$= _0^{GL}\Delta_k^{(\nu)}\left[\mathbf{A}^2\mathbf{\Phi}(k)\right] = \mathbf{A}^2{}_0^{GL}\Delta_k^{(\nu)}\mathbf{\Phi}(k) = \mathbf{A}^3\mathbf{\Phi}(k) \qquad (6.202)$$

$$\,_{0}^{GL}\Delta_k^{(4\nu)}\mathbf{\Phi}(k) = \mathbf{A}^4\mathbf{\Phi}(k)$$

$$\vdots$$ (6.203)

$$\,_{0}^{GL}\Delta_k^{(p\nu)}\mathbf{\Phi}(k) = \mathbf{A}^p\mathbf{\Phi}(k).$$

Substituting results (6.202) into (6.200) yields

$$\,_{0}^{GL}\Delta^{(p\nu)}\mathbf{\Phi}(k) + a_{p-1}\,_{0}^{GL}\Delta^{((p-1)\nu)}\mathbf{\Phi}(k) + \cdots + a_1\,_{0}^{GL}\Delta^{(\nu)}\mathbf{\Phi}(k) + a_0\mathbf{\Phi}(k)$$

$$\mathbf{A}^p\mathbf{\Phi}(k) + a_{p-1}\mathbf{A}^{p-1}\mathbf{\Phi}(k) + \cdots + a_1\mathbf{A}^1\mathbf{\Phi}(k) + a_0\mathbf{A}^0\mathbf{\Phi}(k)$$

$$= \left[\mathbf{A}^p + a_{p-1}\mathbf{A}^{p-1} + \cdots + a_1\mathbf{A}^1 + a_0\mathbf{A}^0\right]\mathbf{\Phi}(k) = \mathbf{0}_p\mathbf{\Phi}(k) = \mathbf{0}_p. \quad (6.204)$$

$$\square$$

The numerical example serves to support the theoretical analysis.

Example 6.8. Consider the FOSSE of the form

$$\begin{bmatrix} \,_{0}^{GL}\Delta^{(\nu)}x_1(k) \\ \,_{0}^{GL}\Delta^{(\nu)}x_2(k) \end{bmatrix} = \begin{bmatrix} -1.5 & 1 \\ -1.25 & 0.75 \end{bmatrix} \begin{bmatrix} x_1(k) \\ x_2(k) \end{bmatrix} + \begin{bmatrix} b_{1,1} \\ b_{2,1} \end{bmatrix} u(k) \quad (6.205)$$

$$y(k) = \begin{bmatrix} 1 & 0 \end{bmatrix} \begin{bmatrix} x_1(k) \\ x_2(k) \end{bmatrix} + \begin{bmatrix} 1 \\ -1 \end{bmatrix} u(k). \quad (6.206)$$

Find the characteristic polynomial and the FOTF.

Solution. From the characteristic matrix one obtains

$$det \begin{bmatrix} (1 - z^{-1})^\nu + 1.5 & -1 \\ 1.25 & (1 - z^{-1})^\nu - 0.75 \end{bmatrix} = (1 - z^{-1})^{2\nu} + 0.75(1 - z^{-1})^\nu + 0.125$$

$$= \left[(1 - z^{-1})^\nu + \frac{1}{2}\right]\left[(1 - z^{-1})^\nu + \frac{1}{4}\right].$$
(6.207)

To calculate the FOTF one needs the inverse of the characteristic matrix

$$\begin{bmatrix} (1 - z^{-1})^\nu + 1.5 & -1 \\ 1.25 & (1 - z^{-1})^\nu - 0.75 \end{bmatrix}^{-1}$$

$$= \frac{1}{(1 - z^{-1})^{2\nu} + 0.75(1 - z^{-1})^\nu + 0.125} \begin{bmatrix} (1 - z^{-1})^\nu - 0.75 & 1 \\ -1.25 & (1 - z^{-1})^\nu + 1.5 \end{bmatrix}.$$
(6.208)

Then by (6.175)

$$G(z)$$

$$= \frac{1}{(1 - z^{-1})^{2\nu} + \frac{3}{4}(1 - z^{-1})^{\nu} + \frac{1}{8}} \begin{bmatrix} 1 & 0 \end{bmatrix} \begin{bmatrix} (1 - z^{-1})^{\nu} - \frac{3}{4} & 1 \\ -\frac{5}{4} & (1 - z^{-1})^{\nu} + \frac{3}{2} \end{bmatrix} \begin{bmatrix} 1 \\ -1 \end{bmatrix}$$

$$= \frac{(1 - z^{-1})^{\nu} - \frac{7}{4}}{(1 - z^{-1})^{2\nu} + \frac{3}{4}(1 - z^{-1})^{\nu} + \frac{1}{8}} = \frac{(1 - z^{-1})^{\nu} - \frac{7}{4}}{\left[(1 - z^{-1})^{\nu} + \frac{1}{2}\right]\left[(1 - z^{-1})^{\nu} + \frac{1}{4}\right]}.$$

$$(6.209)$$

Chapter 7

Linear FO system analysis

In this Chapter, fundamental dynamical properties of the FO linear systems will be studied following the methodology commonly used in the classical IO dynamical system analysis [Kailath (1980)], [Ogata (1987)] [Kaczorek (1992)], [Franklin *et al.* (1998)], [Ifeachor and Jervis (1993)] [Dorf and Bishop (2005)]. After the FO system transient responses are explored, its discrete frequency characteristics are presented. Related to the state-space description, there are very important notions: reachability and observability [Klamka (1991)]. Finally, the FO system stability conditions are given. Considerations are supported by numerical examples. As well, all given results are incorporated into the classical discrete systems where the FO system with FOs close to the integer values is also close to the integer order system dynamical behavior. This property emphasizes that the FO systems bridge the gap between the systems of the IOs.

7.1 Linear time-invariant FO system response

The transient system response is the first important property characterizing the fractional dynamic. Evaluation of the FO system response is reduced to the solution of the FODE. In this Section solutions of the FODE in several equivalent forms are considered.

7.1.1 *Response of the linear time-invariant FOS described by the FODE*

One takes into account the FODE in a form of commensurate equation (6.17) (i.e. with the assumption that $\nu = \mu$). Also, a permanent assumption of $p \geq q$ is posed. A left-hand side of this equation was transformed to the form (6.110). When performing analogous operations on the right-hand side of the equation, one gets

$$\sum_{i=0}^{q} b_i {}^{GL}_{k_0}\Delta_k^{(i\nu)} u(k) = \begin{bmatrix} b_p & b_{p-1} & \cdots & b_1 & b_0 \end{bmatrix} \begin{bmatrix} {}^{GL}_{k_0}\Delta_k^{(p\nu)} u(k) \\ {}^{GL}_{k_0}\Delta_k^{((p-1)\nu)} u(k) \\ \cdots \\ {}^{GL}_{k_0}\Delta_k^{(\nu)} u(k) \\ u(k) \end{bmatrix}$$

$$= \begin{bmatrix} B_0 & B_1 & \cdots & B_{k-k_0+1} & B_{k-k_0} \end{bmatrix} \begin{bmatrix} u(k) \\ u(k-1) \\ \cdots \\ u(k_0+1) \\ u(k_0) \end{bmatrix} \qquad (7.1)$$

where

$$\begin{bmatrix} B_0 & B_1 & \cdots & B_{k-k_0+1} & B_{k-k_0} \end{bmatrix} = \begin{bmatrix} b_q & b_{q-1} & \cdots & b_1 & b_0 \end{bmatrix}$$

$$\times \begin{bmatrix} a^{(q\nu)}(0) & a^{(q\nu)}(1) & \cdots & a^{(q\nu)}(k-k_0+1) & a^{(q\nu)}(k-k_0) \\ a^{((q-1)\nu)}(0) & a^{((q-1)\nu)}(1) & \cdots & a^{((q-1)\nu)}(k-k_0+1) & a^{((q-1)\nu)}(k-k_0) \\ \vdots & \vdots & & \vdots & \vdots \\ a^{(\nu)}(0) & a^{(\nu)}(1) & \cdots & a^{(\nu)}(k-k_0+1) & a^{(\nu)}(k-k_0) \\ a^{(0)}(0) & 0 & \cdots & 0 & 0 \end{bmatrix}. \qquad (7.2)$$

Furthermore, collecting both sides of the equation (6.110) and (7.1), one arrives at the solution

$$y(k) = -\frac{1}{A_0} \begin{bmatrix} A_1 & A_2 & \cdots & A_{k-k_0+1} & A_{k-k_0} \end{bmatrix} \begin{bmatrix} y(k-1) \\ y(k-2) \\ \vdots \\ y(k_0+1) \\ y(k_0) \end{bmatrix}$$

$$+ \frac{1}{A_0} \begin{bmatrix} B_0 & B_1 & \cdots & B_{k-k_0+1} & B_{k-k_0} \end{bmatrix} \begin{bmatrix} u(k) \\ u(k-1) \\ \vdots \\ u(k_0+1) \\ u(k_0) \end{bmatrix}$$

$$- \frac{1}{A_0} \begin{bmatrix} A_{k-k_0+1} & A_{k-k_0+2} & A_{k-k_0+3} & \cdots \end{bmatrix} \begin{bmatrix} y_{k_0-1} \\ y_{k_0-2} \\ y_{k_0-3} \\ \vdots \end{bmatrix}. \qquad (7.3)$$

The formula given above is also valid for integer orders $\nu = n \in \mathbb{Z}_+$, as well as for linear time-variant DEs. In the case of the last-mentioned coefficients, a_i and

b_j are functions of independent variable k, being denoted as $a_i(k)$ and $b_j(k)$. As a consequence, all coefficients A_i and B_j will also be functions of k. The numerical example clarifies the FODE solution evaluation procedure.

Example 7.1. Solve the linear time variant FODE

$$\substack{GL\\k_0}\Delta_k^{(\nu)}y(k) + a_0(k)y(k) = b_0 u(k) \tag{7.4}$$

where $\nu = 0.5$, $k_0 = 0$, $a_0(k) = 0.1\,(1 - exp(-0.001k)\sin(0.02k))$, $b_0 = 1$, $u(k) = 1(k-1)$. Assume the initial conditions $y_1 = -10$ and $y_i = 0$ for $i = 2, 3, \ldots$.

Solution. First, one should find coefficients A_i and B_j. According to (6.107) and (7.2), one has

$$\begin{bmatrix} 1 & a_0(k) \end{bmatrix} \begin{bmatrix} a^{(1\nu)}(0) & a^{(\nu)}(1) & a^{(\nu)}(2) & \cdots \\ a^{(0\nu)}(0) & 0 & 0 & \cdots \end{bmatrix}$$
$$= \begin{bmatrix} 1 + a_0(k) & a^{(\nu)}(1) & a^{(\nu)}(2) & \cdots \end{bmatrix} = \begin{bmatrix} A_0 & A_1 & A_2 & \cdots \end{bmatrix} \tag{7.5}$$

$$\begin{bmatrix} b_0 \end{bmatrix} \begin{bmatrix} a^{(0\nu)}(0) & 0 & 0 & \cdots \end{bmatrix} = \begin{bmatrix} b_0 & 0 & 0 & \cdots \end{bmatrix} = \begin{bmatrix} B_0 & B_1 & B_2 & \cdots \end{bmatrix}. \tag{7.6}$$

For these coefficients, the solution has the form

$$y(k) = -\frac{1}{1 + a_0(k)} \begin{bmatrix} a^{(\nu)}(1) & a^{(\nu)}(2) & \cdots & a^{(\nu)}(k-1) & a^{(\nu)}(k) \end{bmatrix} \begin{bmatrix} y(k-1) \\ y(k-2) \\ \vdots \\ y(1) \\ y(0) \end{bmatrix}$$

$$+ \frac{1}{1 + a_0(k)} \begin{bmatrix} b_0 \end{bmatrix} \begin{bmatrix} u(k) \end{bmatrix} - \frac{1}{1 + a_0(k)} \begin{bmatrix} a^{(\nu)}(k+1) \end{bmatrix} \begin{bmatrix} y_{-1} \end{bmatrix}. \tag{7.7}$$

The plots of the coefficient function $a_0(k)$ and the FODE response are presented in Figs. 7.1 and 7.2, respectively.

A steady-state solution is evaluated according to Algorithm 6.1. For a constant input signal one obtains a steady-state solution

$$\lim_{k\to+\infty} \left[\substack{GL\\k_0}\Delta_k^{(\nu)} y_s + a_0(k)y_s \right] = \lim_{k\to+\infty} \begin{bmatrix} b_0 u_s \end{bmatrix}. \tag{7.8}$$

The FOBD of a constant function tends towards zero. Hence, the corresponding equation (7.8) is

$$\lim_{k\to+\infty} \begin{bmatrix} a_0(k) \end{bmatrix} y_s = \lim_{k\to+\infty} (b_0 u_s) = b_0 u_s \tag{7.9}$$

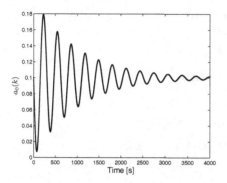
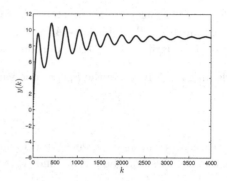

Fig. 7.1 Plot of the coefficient function $a_0(k)$.

Fig. 7.2 Plot of the FODE (7.7) solution.

and

$$y_s = \frac{b_0 u_s}{\lim_{k \to +\infty} [a_0(k)]} = \frac{1}{0.1} = 10. \qquad (7.10)$$

The plot presented in Fig. 7.2 confirms the above result.

The correctness of the solution method can be confirmed when assuming a constant coefficient $a_0(k) = \lim_{k \to +\infty} a_0(k) = 0.1(1 - e^{-0.0001k} \sin(0.02k)) = 0.1$. Under the aforementioned new conditions the FODE response simulation result is given in Fig. 7.3. In further simplifications, one assumes an integer order $\nu = n = 1$. The related plot of the equation response is given in Fig. 7.4. This is the typical response of the first-order linear time-invariant DE. In the last considered case, the steady-state solution level leaves no doubt.

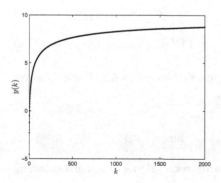
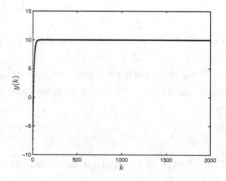

Fig. 7.3 Plot of the FODE (7.7) solution for $a_0(k) = 0.1$.

Fig. 7.4 Plot of the FODE (7.7) solution for $a_0(k) = 0.1$ and $\nu = n = 1$.

7.1.2 Response of the linear time-invariant FOS described by the PLE

The solution of the FODE in the PLE, described by equations (6.118) with (6.119), is very simple. One should add a matrix that introduces the initial conditions

$$\mathbf{D}_{k-k_0}\mathbf{y}(k) + \mathbf{I}_{ic}\mathbf{y}_{ic} = \mathbf{N}_{k-k_0}\mathbf{u}(k) \tag{7.11}$$

into the solution where

$$\mathbf{I}_{ic} = \begin{bmatrix} A_{k-k_0+1} & A_{k-k_0+2} & A_{k-k_0+3} & \cdots \\ A_{k-k_0} & A_{k-k_0+1} & A_{k-k_0+2} & \cdots \\ \vdots & \vdots & \vdots & \\ A_2 & A_3 & A_4 & \cdots \\ A_1 & A_2 & A_3 & \cdots \end{bmatrix} \quad \mathbf{y}_{ic} = \begin{bmatrix} y_{k_0-1} \\ y_{k_0-2} \\ y_{k_0-3} \\ \vdots \end{bmatrix} \tag{7.12}$$

and $\mathbf{y}_{ic}$ is a vector of the initial conditions. By assumption, (6.108) matrix D_{k-k_0} is non-singular. Therefore, multiplying both sides of the equation (7.11) by the inverse of D_{k-k_0}, yields

$$\mathbf{y}(k) = [\mathbf{D}_{k-k_0}]^{-1}\mathbf{N}_{k-k_0}\mathbf{u}(k) - [\mathbf{D}_{k-k_0}]^{-1}\mathbf{I}_{ic}\mathbf{y}_{ic}. \tag{7.13}$$

A special form of the matrix $\mathbf{D}_{k-k_0}$, may be used in its inverse calculation. Note that it may be expressed as a block matrix

$$\mathbf{D}_{k-k_0} = \begin{bmatrix} A_0 & [A_1 \; A_2 \cdots A_{k-k_0}] \\ \mathbf{0}_{k-k_0,1} & \mathbf{D}_{k-k_0-1} \end{bmatrix}. \tag{7.14}$$

Then

$$\mathbf{D}_{k-k_0}^{-1} = \begin{bmatrix} \frac{1}{A_0} & -\frac{1}{A_0}[A_1 \; A_2 \cdots A_{k-k_0}]\mathbf{D}_{k-k_0-1}^{-1} \\ \mathbf{0}_{k-k_0,1} & \mathbf{D}_{k-k_0-1}^{-1} \end{bmatrix}. \tag{7.15}$$

Solution (6.13) is also valid in the IODE case. The next numerical example reveals some characteristics of the considered solution.

Example 7.2. Consider the FODE

$$_{k_0}^{GL}\Delta_k^{(2\nu)}y(k) + a_1{}_{k_0}^{GL}\Delta_k^{(1\nu)}y(k) + a_0{}_{k_0}^{GL}\Delta_k^{(0\nu)}y(k) = b_0 u(k) \tag{7.16}$$

where $\nu = n = 1$, $k_0 = 0$ and $\mathbf{y}_{ic} = [y_{-1}y_{-2}\cdots]^{\mathrm{T}}$ is $\infty \times 1$ vector of initial conditions. For this assumption, this is a classical IODE. Following the procedure described in

Section 6.5, the matrix $\mathbf{D}_k$ has a form

$$\mathbf{D}_k = \begin{bmatrix} 1+a_1+a_0 & -2-a_1 & 1 & 0\cdots & 0 & 0 \\ 0 & 1+a_1+a_0 & -2-a_1 & 1\cdots & 0 & 0 \\ \vdots & \vdots & \vdots & \vdots & \vdots & \vdots \\ 0 & 0 & 0 & 0\cdots 1+a_1+a_0 & -2-a_1 \\ 0 & 0 & 0 & 0\cdots & 0 & 1+a_1+a_0 \end{bmatrix}, \ \mathbf{N}_k = b_0\mathbf{1}_k.$$

$$(7.17)$$

In the considered case, the matrix $\mathbf{D}_k$ is an upper triangular band matrix with two super-diagonals.

$$\mathbf{I}_{ic} = \begin{bmatrix} 0 & 0\ 0\cdots \\ 0 & 0\ 0\cdots \\ \vdots & \vdots\ \vdots \\ 1 & 0\ 0\cdots \\ -2-a_1 & 1\ 0\cdots \end{bmatrix}. \qquad (7.18)$$

The $(k+1) \times (\inf)$ matrix $\mathbf{I}_{ic}$ is multiplied by $(\infty \times 1)$ vector of initial conditions $\mathbf{y}_{ic}$

$$\mathbf{I}_{ic}\mathbf{y}_{ic} = \begin{bmatrix} 0 & 0\ 0\cdots \\ 0 & 0\ 0\cdots \\ \vdots & \vdots\ \vdots \\ 1 & 0\ 0\cdots \\ -2-a_1 & 1\ 0\cdots \end{bmatrix}\begin{bmatrix} y_{-1} \\ y_{-2} \\ y_{-3} \\ \vdots \end{bmatrix} = \begin{bmatrix} 0 & 0 \\ 0 & 0 \\ \vdots & \vdots \\ 1 & 0 \\ -2-a_1 & 1 \end{bmatrix}\begin{bmatrix} y_{-1} \\ y_{-2} \end{bmatrix} = \mathbf{I}_{ic,2}\mathbf{y}_{ic,2}. \qquad (7.19)$$

This result is particularly important because the above simplification confirms the results given in Section 6.4, where it states that to solve the FODE one needs an infinite number of initial conditions. One may find that the same is true in the case of the IODEs, but the matrix $\mathbf{I}_{ic}$, being in its special form $\mathbf{I}_{ic,2}$, "cuts" an impact into the solution of all initial conditions except the $2n$ first ones.

Under the permanent assumption (6.108), matrix (7.17) is non-singular. In the solution, the inverse of $\mathbf{D}_k$ is also an upper triangular matrix, but with $k+1$ nonsingular super-diagonals.

The considered FODE in the PLE possesses very similar properties to the so-called polynomial matrix description. It is very useful in linear dynamic system connections analysis.

7.1.3 *Response of the linear time-invariant FOS described by the SSE*

In this section the main emphasis will be placed on the solution of the linear time-invariant FODE in the state-space form. The is motivated by the great importance of equations in such a form, when looked at in the context of the dynamic system analysis and synthesis. For practical reasons, the initial time is assumed to be equal to zero $k_0 = 0$. In previous sections, it was proven that in the FO case, for the unique solution evaluation, one needs an infinite number of initial conditions. According to the fundamental property of the function $a^{(\nu)}(k)$, which for $\nu = n \in \mathbb{Z}_+$ equals to zero, for $k > n$ for any particular solution one needs n initial conditions only.

Now, one considers linear state equation (6.41). For simplicity, $k_0 = 0$ will be assumed. A substitution of the FOBD into this equation, with infinite number of initial conditions $\mathbf{x}_{-1}, \mathbf{x}_{-2}, \ldots$, yields

$$\mathbf{x}(k) + \sum_{i=1}^{k-1} a^{(\nu)}(i)\mathbf{x}(k-i) + \sum_{i=1}^{+\infty} a^{(\nu)}(k+i)\mathbf{x}_{-i} = \mathbf{A}\mathbf{x}(k) + \mathbf{B}\mathbf{u}(k). \qquad (7.20)$$

This can be easily transformed to the form

$$(\mathbf{1}_q - \mathbf{A})\,\mathbf{x}(k) = -\sum_{i=1}^{k} a^{(\nu)}(i)\mathbf{x}(k-i) + \mathbf{B}\mathbf{u}(k) - \sum_{i=1}^{+\infty} a^{(\nu)}(k+i)\mathbf{x}_{-i}. \qquad (7.21)$$

Here, one should note that the matrix $\mathbf{1}_q - \mathbf{A}$ invertibility conditions, coincide with those expressed by equation (6.108).

$$\mathbf{x}(1) = (\mathbf{1}_p - \mathbf{A})^{-1}\left[-a^{(\nu)}(1)\mathbf{x}(0) + \mathbf{B}\mathbf{u}(1) - \sum_{i=1}^{+\infty} a^{(\nu)}(i+1)\mathbf{x}_{-i}\right]. \qquad (7.22)$$

From (7.21), for consecutive $k = 0, 1, 2 \ldots$, one obtains

$$(\mathbf{1}_p - \mathbf{A})\,\mathbf{x}(0) = \mathbf{B}\mathbf{u}(0) - \sum_{i=1}^{+\infty} a^{(\nu)}(i)\mathbf{x}_{-i} \qquad (7.23)$$

$$(\mathbf{1}_p - \mathbf{A})\,\mathbf{x}(1) + a^{(\nu)}(1)\mathbf{1}_p\mathbf{x}(0) = \mathbf{B}\mathbf{u}(1) - \sum_{i=1}^{+\infty} a^{(\nu)}(1+i)\mathbf{x}_{-i} \qquad (7.24)$$

$$(\mathbf{1}_p - \mathbf{A})\,\mathbf{x}(2) + a^{(\nu)}(1)\mathbf{1}_p\mathbf{x}(1) + a^{(\nu)}(2)\mathbf{1}_p\mathbf{x}(0) = \mathbf{B}\mathbf{u}(2) - \sum_{i=1}^{+\infty} a^{(\nu)}(2+i)\mathbf{x}_{-i}. \qquad (7.25)$$

The response evaluation procedure can be continued for $k = 3, 4 \ldots$. Collecting all such equations in vector form, one gets

$$
\left[(\mathbf{1}_p - \mathbf{A}) \; a^{(\nu)}(1)\mathbf{1}_p \; a^{(\nu)}(2)\mathbf{1}_p \; \cdots \; a^{(\nu)}(k-1)\mathbf{1}_p \; a^{(\nu)}(k)\mathbf{1}_p \right]
\begin{bmatrix}
\mathbf{x}(k) \\
\mathbf{x}(k-1) \\
\vdots \\
\mathbf{x}(1) \\
\mathbf{x}(0)
\end{bmatrix}
$$

$$
= - \left[a^{(\nu)}(k+1)\mathbf{1}_p \; a^{(\nu)}(k+2)\mathbf{1}_p \; a^{(\nu)}(k+3)\mathbf{1}_p \; \cdots \right]
\begin{bmatrix}
\mathbf{x}_{-1} \\
\mathbf{x}_{-2} \\
\mathbf{x}_{-3} \\
\vdots
\end{bmatrix}
+ \mathbf{B}\mathbf{u}(k). \qquad (7.26)
$$

Collecting all such equations in a matrix-vector form one obtains

$$
\begin{bmatrix}
(\mathbf{1}_p - \mathbf{A}) & a^{(\nu)}(1)\mathbf{1}_p & a^{(\nu)}(2)\mathbf{1}_p & \cdots & a^{(\nu)}(k-1)\mathbf{1}_p & a^{(\nu)}(k)\mathbf{1}_p \\
\mathbf{0}_p & (\mathbf{1}_p - \mathbf{A}) & a^{(\nu)}(1)\mathbf{1}_p & \cdots & a^{(\nu)}(k-2)\mathbf{1}_p & a^{(\nu)}(k-1)\mathbf{1}_p \\
\vdots & \vdots & \vdots & \cdots & \vdots & \vdots \\
\mathbf{0}_p & \mathbf{0}_p & \mathbf{0}_p & \cdots & (\mathbf{1}_p - \mathbf{A}) & a^{(\nu)}(1)\mathbf{1}_p \\
\mathbf{0}_p & \mathbf{0}_p & \mathbf{0}_p & \cdots & \mathbf{0}_p & (\mathbf{1}_p - \mathbf{A})
\end{bmatrix}
\begin{bmatrix}
\mathbf{x}(k) \\
\mathbf{x}(k-1) \\
\vdots \\
\mathbf{x}(1) \\
\mathbf{x}(0)
\end{bmatrix}
$$

$$
= -
\begin{bmatrix}
a^{(\nu)}(k+1)\mathbf{1}_p & a^{(\nu)}(k+2)\mathbf{1}_p & a^{(\nu)}(k+3)\mathbf{1}_p & \cdots \\
a^{(\nu)}(k)\mathbf{1}_p & a^{(\nu)}(k+1)\mathbf{1}_p & a^{(\nu)}(k+2)\mathbf{1}_p & \cdots \\
\vdots & \vdots & \vdots & \\
a^{(\nu)}(2)\mathbf{1}_p & a^{(\nu)}(3)\mathbf{1}_p & a^{(\nu)}(4)\mathbf{1}_p & \cdots \\
a^{(\nu)}(1)\mathbf{1}_p & a^{(\nu)}(2)\mathbf{1}_p & a^{(\nu)}(3)\mathbf{1}_p & \cdots
\end{bmatrix}
\begin{bmatrix}
\mathbf{x}_{-1} \\
\mathbf{x}_{-2} \\
\mathbf{x}_{-3} \\
\vdots
\end{bmatrix}
$$

$$
+
\begin{bmatrix}
\mathbf{B} & \mathbf{0}_{p,q} & \cdots & \mathbf{0}_{p,q} & \mathbf{0}_{p,q} \\
\mathbf{0}_{p,q} & \mathbf{B} & \cdots & \mathbf{0}_{p,q} & \mathbf{0}_{p,q} \\
\vdots & \vdots & \vdots & \vdots \\
\mathbf{0}_{p,q} & \mathbf{0}_{p,q} & \cdots & \mathbf{B} & \mathbf{0}_{p,q} \\
\mathbf{0}_{p,q} & \mathbf{0}_{p,q} & \cdots & \mathbf{0}_{p,q} & \mathbf{B}
\end{bmatrix}
\begin{bmatrix}
\mathbf{u}(k) \\
\mathbf{u}(k-1) \\
\vdots \\
\mathbf{u}(1) \\
\mathbf{u}(0)
\end{bmatrix} .
$$

$$(7.27)$$

Now, investigations will be focused on the matrix placed in the left-hand side of equation (7.27). Using what was introduced in Section 6.7, notation (6.180) and (6.189), the solution of (7.27) has the form

$$
\begin{bmatrix} \mathbf{x}(k) \\ \mathbf{x}(k-1) \\ \vdots \\ \mathbf{x}(1) \\ \mathbf{x}(0) \end{bmatrix}
$$

$$
= \begin{bmatrix} \boldsymbol{\Phi}(0) & \boldsymbol{\Phi}(1) & \cdots & \boldsymbol{\Phi}(k-1) & \boldsymbol{\Phi}(k) \\ \mathbf{0}_p & \boldsymbol{\Phi}(0) & \cdots & \boldsymbol{\Phi}(k-2) & \boldsymbol{\Phi}(k-1) \\ \vdots & \vdots & & \vdots & \vdots \\ \mathbf{0}_p & \mathbf{0}_p & \cdots & \boldsymbol{\Phi}(0) & \boldsymbol{\Phi}(1) \\ \mathbf{0}_p & \mathbf{0}_p & \cdots & \mathbf{0}_p & \boldsymbol{\Phi}(0) \end{bmatrix}
\begin{bmatrix} \mathbf{Bu}(k) \\ \mathbf{Bu}(k-1) \\ \vdots \\ \mathbf{Bu}(1) \\ \mathbf{Bu}(0) \end{bmatrix}
$$

$$
- \begin{bmatrix} \boldsymbol{\Phi}(0) & \boldsymbol{\Phi}(1) & \cdots & \boldsymbol{\Phi}(k-1) & \boldsymbol{\Phi}(k) \\ \mathbf{0}_p & \boldsymbol{\Phi}(0) & \cdots & \boldsymbol{\Phi}(k-2) & \boldsymbol{\Phi}(k-1) \\ \vdots & \vdots & & \vdots & \vdots \\ \mathbf{0}_p & \mathbf{0}_p & \cdots & \boldsymbol{\Phi}(0) & \boldsymbol{\Phi}(1) \\ \mathbf{0}_p & \mathbf{0}_p & \cdots & \mathbf{0}_p & \boldsymbol{\Phi}(0) \end{bmatrix}
\begin{bmatrix} a^{(\nu)}(k+1)\mathbf{1}_p & a^{(\nu)}(k+2)\mathbf{1}_p & \cdots \\ a^{(\nu)}(k)\mathbf{1}_p & a^{(\nu)}(k+1)\mathbf{1}_p & \cdots \\ \vdots & \vdots & \vdots \\ a^{(\nu)}(2)\mathbf{1}_p & a^{(\nu)}(3)\mathbf{1}_p & \cdots \\ a^{(\nu)}(1)\mathbf{1}_p & a^{(\nu)}(2)\mathbf{1}_p & \cdots \end{bmatrix}
\begin{bmatrix} \mathbf{x}_{-1} \\ \mathbf{x}_{-2} \\ \mathbf{x}_{-3} \\ \vdots \end{bmatrix}.
$$

$$\tag{7.28}$$

From the matrix equation (7.28), one gets

$$
\mathbf{x}(k) = \sum_{i=0}^{k} \boldsymbol{\Phi}(i)\mathbf{Bu}(k-i) - \sum_{j=1}^{+\infty}\sum_{i=0}^{k} \boldsymbol{\Phi}(i)a^{(\nu)}(k+j-i)\mathbf{x}_{-j}. \tag{7.29}
$$

The response,

$$
\mathbf{x}(k) = \sum_{i=0}^{k} \boldsymbol{\Phi}(k-i)\mathbf{Bu}(i) \tag{7.30}
$$

is called the forced response whereas

$$
\mathbf{x}(k) = -\sum_{j=1}^{+\infty}\sum_{i=0}^{k} \boldsymbol{\Phi}(i)a^{(\nu)}(k+j-i)\mathbf{x}_{-j} \tag{7.31}
$$

is known as the homogenous one.

The next numerical example shows the states transients of a dynamic FOS being described by the SSE with three states.

Example 7.3. Consider the FODE described by the SSE of the form

$$
\begin{bmatrix} {}^{GL}_{k_0}\Delta_k^{(\nu)}x_1(k) \\ {}^{GL}_{k_0}\Delta_k^{(\nu)}x_2(k) \\ {}^{GL}_{k_0}\Delta_k^{(\nu)}x_3(k) \end{bmatrix} = \begin{bmatrix} 0 & 1 & 0 \\ 0 & 0 & 1 \\ -a_0 & 0 & 0 \end{bmatrix} \begin{bmatrix} x_1(k) \\ x_2(k) \\ x_3(k) \end{bmatrix} + \begin{bmatrix} 0 \\ 0 \\ 1 \end{bmatrix} u(k) \tag{7.32}
$$

where $\nu = 0.75$ and $a_0 = 0.15$. One should find the homogenous solution ($u(k) = 0$ for $k = 0, 1, 2, k_0 = 0, \dots$)

(I) $\mathbf{x}_{-1} = \mathbf{x}_{-2} = \begin{bmatrix} 1 \\ 0 \\ 0 \end{bmatrix}$, $\mathbf{x}_{-i} = \mathbf{0}$ for $i = 3, 4, \dots$

(II) $\mathbf{x}_{-1} = \mathbf{x}_{-2} = \mathbf{x}_{-3} = \mathbf{0}$, $\mathbf{x}_{-4} = \begin{bmatrix} 1 \\ 0 \\ 0 \end{bmatrix}$, $\mathbf{x}_{-i} = \mathbf{0}$ for $i = 5, 6, \dots$

Solution. For initial conditions (I), the simulated three state-variables plots are presented in Fig. 7.5.

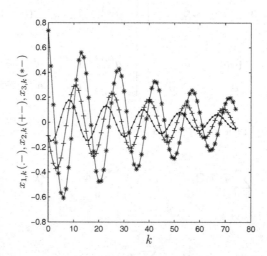

Fig. 7.5   Plots of three state-variables of the FO system (7.32) for initial conditions (I).

The state trajectory in 3D state-space, from two different points of views, is given in Figs. 7.6 and 7.7. An additional point of view at the state-trajectory, enhances a reposing trajectory on the 2D plane in the 3D state-space.

In the next considered case (II), the transient behavior is similar with one exception. The influence of the third initial state $\mathbf{x}_{-3}$ is smaller. When interpreting

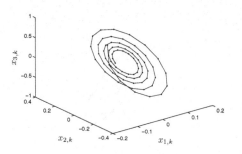

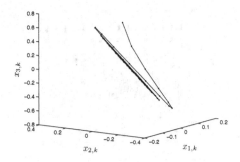

Fig. 7.6   Plot of the state-trajectory.

Fig. 7.7   Plot of the state-trajectory from a different point-of-view.

the initial conditions as elements representing the dynamic system memory, one can conclude that the impact of consecutive initial states $\mathbf{x}_{-i}$ will be increasingly lower. This effect of forgetting past events is described by terms $a^{(\nu)}(k+j)$ in (7.28), which strongly tend towards zero. The plots of state-variables are given in Fig. 7.8 and those of related state trajectory in Fig. 7.9.

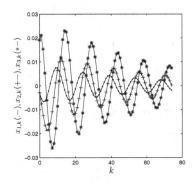

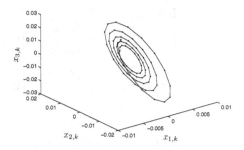

Fig. 7.8   Plots of three state-variables of the FO system (7.32) for initial conditions (II).

Fig. 7.9   Plot of the state-trajectory for initial conditions (II).

For the same initial conditions as posed in case (II), but with a different parameter $a_0 = 0.0926$, the system homogenous response achieves its limit circle. This is revealed in Figs. 7.10 and 7.11, respectively.

The solution of the FODE in the SSE form, expressed by formula (7.27), essentially simplifies for the IO case when $\nu = n = 1$. Then formula (7.27) transforms

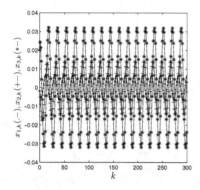

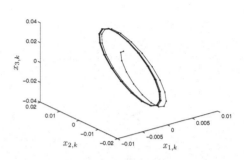

Fig. 7.10 Plots of three state-variables of the FO system (7.32) for initial conditions (II) and $a_0 = 0.0926$.

Fig. 7.11 Plot of the state-trajectory for initial conditions (II) and $a_0 = 0.0926$.

to

$$
\begin{bmatrix}
(\mathbf{1}_p - \mathbf{A}) & -\mathbf{1}_p & \mathbf{0}_p & \cdots & \mathbf{0}_p & \mathbf{0}_p \\
\mathbf{0}_p & (\mathbf{1}_p - \mathbf{A}) & -\mathbf{1}_p & \cdots & \mathbf{0}_p & \mathbf{0}_p \\
\vdots & \vdots & \vdots & \cdots & \vdots & \vdots \\
\mathbf{0}_p & \mathbf{0}_p & \mathbf{0}_p & \cdots & (\mathbf{1}_p - \mathbf{A}) & -\mathbf{1}_p \\
\mathbf{0}_p & \mathbf{0}_p & \mathbf{0}_p & \cdots & \mathbf{0}_p & (\mathbf{1}_p - \mathbf{A})
\end{bmatrix}
\begin{bmatrix}
\mathbf{x}(k) \\
\mathbf{x}(k-1) \\
\vdots \\
\mathbf{x}(1) \\
\mathbf{x}(0)
\end{bmatrix}
$$

$$
= -
\begin{bmatrix}
\mathbf{0}_p & \mathbf{0}_p & \mathbf{0}_p & \cdots \\
\mathbf{0}_p & \mathbf{0}_p & \mathbf{0}_p & \cdots \\
\vdots & \vdots & \vdots \\
\mathbf{0}_p & \mathbf{0}_p & \mathbf{0}_p & \cdots \\
-\mathbf{1}_p & \mathbf{0}_p & \mathbf{0}_p & \cdots
\end{bmatrix}
\begin{bmatrix}
\mathbf{x}_{-1} \\
\mathbf{x}_{-2} \\
\mathbf{x}_{-3} \\
\vdots
\end{bmatrix}
+
\begin{bmatrix}
\mathbf{B} & \mathbf{0}_{p,q} & \cdots & \mathbf{0}_{p,q} & \mathbf{0}_{p,q} \\
\mathbf{0}_{p,q} & \mathbf{B} & \cdots & \mathbf{0}_{p,q} & \mathbf{0}_{p,q} \\
\vdots & \vdots & & \vdots & \vdots \\
\mathbf{0}_{p,q} & \mathbf{0}_{p,q} & \cdots & \mathbf{B} & \mathbf{0}_{p,q} \\
\mathbf{0}_{p,q} & \mathbf{0}_{p,q} & \cdots & \mathbf{0}_{p,q} & \mathbf{B}
\end{bmatrix}
\begin{bmatrix}
\mathbf{u}(k) \\
\mathbf{u}(k-1) \\
\vdots \\
\mathbf{u}(1) \\
\mathbf{u}(0)
\end{bmatrix}. \qquad (7.33)
$$

Recall that in the considered case $a^{(1)}(0) = 1$, $a^{(1)}(1) = -1$ and $a^{(1)}(i) = 0$ for $i = 2, 3, \ldots$. Also, note that in (7.33) the first term in the right-hand side further simplifies to

$$
-
\begin{bmatrix}
\mathbf{0}_p & \mathbf{0}_p & \mathbf{0}_p & \cdots \\
\mathbf{0}_p & \mathbf{0}_p & \mathbf{0}_p & \cdots \\
\vdots & \vdots & \vdots \\
\mathbf{0}_p & \mathbf{0}_p & \mathbf{0}_p & \cdots \\
-\mathbf{1}_p & \mathbf{0}_p & \mathbf{0}_p & \cdots
\end{bmatrix}
\begin{bmatrix}
\mathbf{x}_{-1} \\
\mathbf{x}_{-2} \\
\mathbf{x}_{-3} \\
\vdots
\end{bmatrix}
= -
\begin{bmatrix}
\mathbf{0}_p \\
\mathbf{0}_p \\
\vdots \\
\mathbf{0}_p \\
-\mathbf{1}_p
\end{bmatrix}
\begin{bmatrix}
\mathbf{x}_{-1}
\end{bmatrix}. \qquad (7.34)
$$

Though the matrix $\mathbf{D}_{kp}$, in this special case, is very simple, its inverse $\boldsymbol{\Phi}_{kp}$ is still in the upper triangular block band matrix. Taking into account (7.34), the solution has the form

$$
\begin{bmatrix} \mathbf{x}(k) \\ \mathbf{x}(k-1) \\ \vdots \\ \mathbf{x}(1) \\ \mathbf{x}(0) \end{bmatrix}
$$

$$
= \begin{bmatrix} \boldsymbol{\Phi}(0) & \boldsymbol{\Phi}(1) & \cdots & \boldsymbol{\Phi}(k-1) & \boldsymbol{\Phi}(k) \\ \mathbf{0}_p & \boldsymbol{\Phi}(0) & \cdots & \boldsymbol{\Phi}(k-2) & \boldsymbol{\Phi}(k-1) \\ \vdots & \vdots & & \vdots & \vdots \\ \mathbf{0}_p & \mathbf{0}_p & \cdots & \boldsymbol{\Phi}(0) & \boldsymbol{\Phi}(1) \\ \mathbf{0}_p & \mathbf{0}_p & \cdots & \mathbf{0}_p & \boldsymbol{\Phi}(0) \end{bmatrix} \begin{bmatrix} \mathbf{Bu}(k) \\ \mathbf{Bu}(k-1) \\ \vdots \\ \mathbf{Bu}(1) \\ \mathbf{Bu}(0) \end{bmatrix}
$$

$$
- \begin{bmatrix} \boldsymbol{\Phi}(0) & \boldsymbol{\Phi}(1) & \cdots & \boldsymbol{\Phi}(k-1) & \boldsymbol{\Phi}(k) \\ \mathbf{0}_p & \boldsymbol{\Phi}(0) & \cdots & \boldsymbol{\Phi}(k-2) & \boldsymbol{\Phi}(k-1) \\ \vdots & \vdots & & \vdots & \vdots \\ \mathbf{0}_p & \mathbf{0}_p & \cdots & \boldsymbol{\Phi}(0) & \boldsymbol{\Phi}(1) \\ \mathbf{0}_p & \mathbf{0}_p & \cdots & \mathbf{0}_p & \boldsymbol{\Phi}(0) \end{bmatrix} \begin{bmatrix} \mathbf{0}_p \\ \mathbf{0}_p \\ \vdots \\ \mathbf{0}_p \\ -\mathbf{1}_p \end{bmatrix} \begin{bmatrix} \mathbf{x}_{-1} \end{bmatrix} \qquad (7.35)
$$

from which one derives

$$
\mathbf{x}(k) = \boldsymbol{\Phi}(k)\mathbf{x}_{-1} + \sum_{i=0}^{k} \boldsymbol{\Phi}(k-i)\mathbf{Bu}(i). \qquad (7.36)
$$

Comment. From the first row of equation (7.33) one gets

$$
(\mathbf{1}_p - \mathbf{A})\mathbf{x}(k) - \mathbf{x}(k-1) = \mathbf{Bu}(k) \qquad (7.37)
$$

hence,

$$
\mathbf{x}(k) - \mathbf{x}(k-1) = {}_{k-1}^{GL}\Delta_k^{(1)}\mathbf{x}(k) = \mathbf{Ax}(k) + \mathbf{Bu}(k). \qquad (7.38)
$$

7.1.4 Response of the linear time-invariant FOS described by the FOTF

The response of the dynamic system described by a transfer function is always evaluated for zero initial conditions. Hence, such responses are considered. As was already mentioned for a given FOTF, its Markov parameters define its impulse response (a response to the discrete Dirac pulse). Such a response can be denoted as

$$g(k) = \sum_{i=0}^{+\infty} g_i \delta(k - i) \tag{7.39}$$

Konwing the system impulse response, a response to any signal is calculated as a discrete convolution

$$y(k) = g(k) * u(k) = \sum_{i=0}^{k} g(i)u(k - i) = \sum_{i=0}^{k} g_i u(k - i). \tag{7.40}$$

The mentioned steps will be summarized in the following Algorithm.

Algorithm 7.1.

(I) For a given FOTF performing its numerator by denominator division, evaluate the impulse response $g(k)$,

(II) For a given input signal $u(k)$, calculate the discrete convolution $g(k) * u(k)$.

The steps of Algorithm 7.1 are particularly useful in a situation where there is no straightaway given function describing the input signal $u(k)$, but its measured values collected in a set $u(k) = \{u_0, u_1, \ldots\}$ are available. The application of the considered Algorithm is presented in the following numerical example.

Example 7.4. Given a FO system described by the FOTF

$$G(z) = \frac{b_0}{[(1 - z^{-1})^\nu + a_0]^3}. \tag{7.41}$$

find its response to a measured input signal defined by a set of samples. Processed $k_{max} = 9180$ samples are plotted in Fig. 7.12 as dots connected by continuous intervals.

The signal was measured using the data acquisition system and further processed to get its first-order BD. The noise was so amplified that the shape of the signal was vague and its filtering was necessary.

Solution. Two filter actions are compared. The first filter is a classical third-order digital filter (integer-order filter — IOF), described by the TF

$$G_{IOF} = \left[\frac{b_0}{(1 - z^{-1})^1 + a_0} \right]^3 = \left[\frac{zb_0'}{z - a_0'} \right]^3 \tag{7.42}$$

where $a_0' = \frac{1}{1+a_0}$ and $b_0' = \frac{b_0}{1+a_0}$.

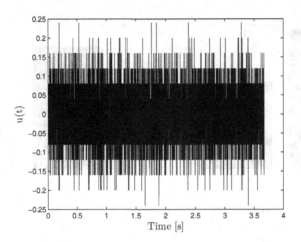

Fig. 7.12   Plot of the measured input signal $u(k)$.

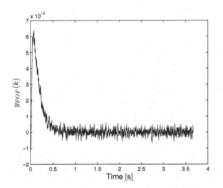

Fig. 7.13   Output signal of the FOF.

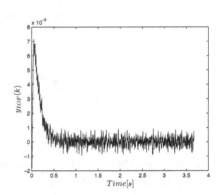

Fig. 7.14   Output signal of the IOF.

In the second FO filter (FOF), the TF is as follows

$$G_{FOF} = \left[\frac{b_0}{(1 - z^{-1})^\nu + a_0} \right]^3 \qquad (7.43)$$

where $\nu \in (0, 2]$. For $a_0 = b_0 = 0.05$, $\nu = 1.2$. The filtering results are given in Figs. 7.13 and 7.14, respectively.

The effect of an ideal filtration should be a continuous horizontal line on zero level for $t > 1[s]$. As the filtration effect criterion a sum of the absolute error (SAE) and a sum of the squared error (SsE) is chosen. In Figs. 7.15 and 7.16, the values of mentioned criteria for FOF and IOF output signals are given.

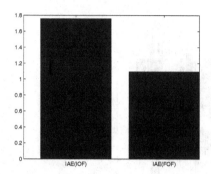

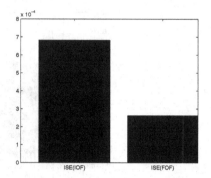

Fig. 7.15   SAE criterion values for the FOI and IOF.

Fig. 7.16   SsE criterion values for the FOI and IOF.

Table 7.1   FOTF parameters.

$w_{r,i}, w_{c,1}$	$v_{r,i}, v_{c,1}$
$w_{r,1} = -0.1000$	$a_5 = 1.0000$
$w_{c,1} = -0.3000 + 0.5180i$	$a_4 = -8.0099e - 1$
$w_{c,2} = -0.1000 + 0.2435i$	$a_3 = 0.5481e - 1$
	$a_2 = -1.1378e - 1$
	$a_1 = 2.4940e - 2$
	$a_0 = -2.4827e - 5$
$v_{r,1} = -2$	$b_1 = -9.9310e - 6$
$v_{c,1} = -1.0000 + 0.5000i$	$b_2 = -3.9724e - 5$
	$b_2 = -5.2137e - 5$
	$b_3 = -2.4827e - 5$

In the IOF, there is only one parameter to select a_0 resulting in the one-variable functions $SAE(a_0)$ and $SsE(a_0)$. Whereas, in the FOF, there are two parameters a_0 and ν causing two-variables functions $IAE(a_0, \nu)$ and $SsE(a_0, \nu)$. An appropriate choice of ν leads to inequalities $SAE(a_0) > SAE(a_0, \nu)$ and $SsE(a_0) > SsE(a_0, \nu)$.

The next numerical example reveals a variety of transient behavior for the FODE subjected to the same input signal.

Example 7.5. Assume the commensurate FODE (6.131) with $q = 3, p = 5, p_r = 1$, $p_c = 2$, $q_r = 1$, $q_c = 1$ with $\nu = 0.8$. All $\bar{r}_i = \bar{s}_i = \underline{r}_i = \underline{s}_i = 1$. Plot the response to the unit step function $\mathbf{1}(k - 1)$ using the convolution sum (7.13).

Solution. In Table 7.1, the FOTF parameters are collected. The unit step response is given in Fig. 7.17. The response tends monotonically to the steady-state value on a level of 1.

In Table 7.2, the FOTF parameters are also collected. Note that in this table, the first column complex pseudo-poles real parts $w_{c,1}$ and $w_{c,2}$ signs were changed. The response achieves its steady-state with oscillations of constant amplitudes. There

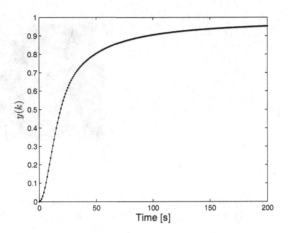

Fig. 7.17   Plot of FOTF (6.132) response $y(k)$ with parameters in Table 7.1.

Table 7.2   FOTF parameters.

$w_{r,i}, w_{c,1}$	$v_{r,i}, v_{c,1}$
$w_{r,1} = -0.1000$	$a_5 = 1.0000$
$w_{c,1} = 0.3000 + 0.5180i$	$a_4 = -0.7000$
$w_{c,2} = 0.1000 + 0.2435i$	$a_3 = 0.4676$
	$a_2 = -0.0585$
	$a_1 = 0.0135$
	$a_0 = 0.0025$
$v_{r,1} = -2$	$b_1 = 0.0010$
$v_{c,1} = -1.0000 + 0.5000i$	$b_2 = 0.0040$
	$b_2 = 0.0052$
	$b_3 = 0.0025$

is more then one constant amplitude. One can state that there is a set of repeating constant amplitudes. The oscillations are around a level 1.

In Table 7.3 the third set of the FOTF parameters are collected. This time the response tends towards minus infinity which means that there is no steady-state solution.

The presented plots reveal a large variety of transient characteristics. One should notice that the FODE is characterized by 5 pseudo-poles, 2 pseudo-zeros and one FO. A small change in them may cause a big change in the shape of the solutions.

An influence of the FO will now be discussed. For the parameters indicated in Table 7.2 and orders $\nu = 0.5$ and $\nu = 0.25$, the responses are presented in Figs. 7.21 and 7.22, respectively.

It may readily be observed from the responses, that the FOBD "discrete differentiation force" decreases due to the decrease of the order. Its fractionality reveals this important property.

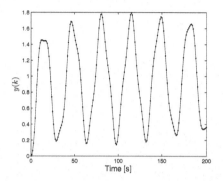

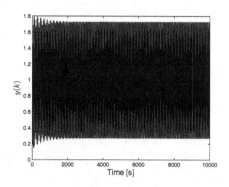

Fig. 7.18 Plot of FOTF (6.132) response $y(k)$ with parameters in Table 7.2 for $k < 200$.

Fig. 7.19 Plot of FOTF (6.132) response $y(k)$ with parameters in Table 7.2 for $k < 10000$.

Table 7.3   FOTF parameters.

$w_{r,i}, w_{c,1}$	$v_{r,i}, v_{c,1}$
$w_{r,1} = 0.0010$	$a_5 = 1.0000$
$w_{c,1} = 0.3000 + 0.5180i$	$a_4 = -0.7000$
$w_{c,2} = 0.1000 + 0.2435i$	$a_3 = 0.4676$
	$a_2 = -0.0585$
	$a_1 = 0.0135$
	$a_0 = 0.0025$
$v_{r,1} = -2$	$b_1 = 0.0010$
$v_{c,1} = -1.0000 + 0.5000i$	$b_2 = 0.0040$
	$b_2 = 0.0052$
	$b_3 = 0.0025$

The convolution sum is an efficient tool in solving the FODE described by the FOTF. The final numerical example shows the FOTF response to a rectangular wave.

Example 7.6. The input signal described by the formula

$$u(k) = \begin{cases} 1 & \text{for } (l-1)i + 1 < k < il \\ 0 & \text{for } li + 1 < k < (l+1)i \end{cases} \qquad (7.44)$$

is plotted in Fig. 7.23. Calculate a response of the FODE with parameters collected in Table 7.2 to the considered input signal.

Solution. Using the convolution sum, one obtains the response of the FODE in version with parameters given in Table 7.2 to input signal (7.17) as plotted in Figs. 7.24 and 7.25 for the two ranges $k < 1000$ and $k < 30000$.

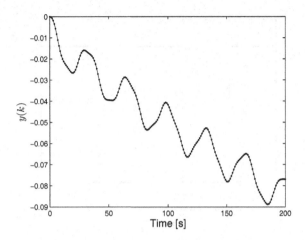

Fig. 7.20   Plot of FOTF (6.132) response $y(k)$ with parameters in Table 7.3.

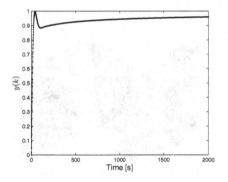

Fig. 7.21   Plot of FOTF (6.132) response $y(k)$ with parameters in Table 7.3 for $\nu = 0.5$.

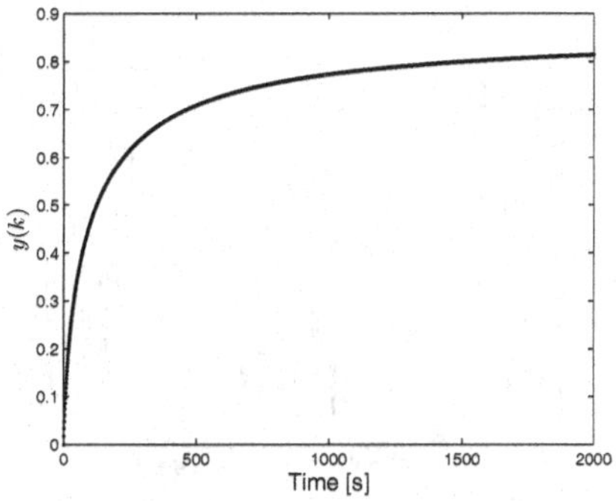

Fig. 7.22   Plot of FOTF (6.132) response $y(k)$ with parameters in Table 7.3 for $\nu = 0.25$.

Now, the FODE with the same parameters, but with two changed FOs $\nu = 0.5$ and $\nu = 0.25$ are considered. Results of computer simulations are presented in Figs. 7.26 and 7.27, respectively.

7.2   Frequency characteristics of the linear time-invariant FO system

The discrete frequency characteristics of the FO systems also carry very important information about system dynamics [Ludovic *et al.*   (1998); Oustaloup *et al.*   (2000); Petras *et al.*   (2000b); Jifeng and Yuankai   (2005); Wang and Li

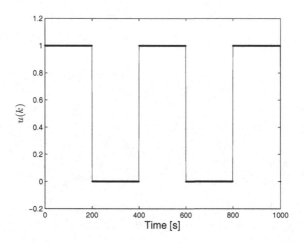

Fig. 7.23   Plot of the rectangular wave described by (7.44).

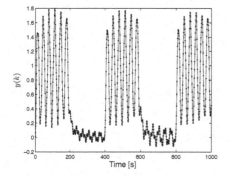

Fig. 7.24   Response $y(k)$ of the FODE with parameters in Table 7.3 for $\nu = 0.8$ to $u(k)$ defined by (7.44) for $k < 1000$.

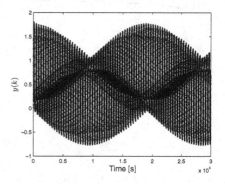

Fig. 7.25   Response $y(k)$ of the FODE with parameters in Table 7.3 for $\nu = 0.8$ to $u(k)$ defined by (7.44) for $k < 30000$.

(2006); Li *et al.* (2011)]. There is a close connection between the transient behavior and the frequency characteristics shape, so to analyze the system dynamics one can chose a more accessible tool. The two opposite ways to proceed are shown in Fig. 7.28.

The frequency characteristics can be readily evaluated from the FOTF. By setting

$$z = e^{j\omega T_s} = e^{j\omega_i} \tag{7.45}$$

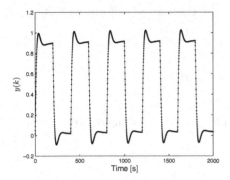

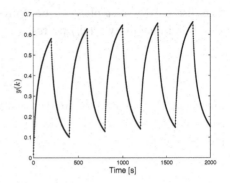

Fig. 7.26 Response $y(k)$ of the FODE with parameters in Table 7.3 for $\nu = 0.5$ to $u(k)$ defined by (7.44).

Fig. 7.27 Response $y(k)$ of the FODE with parameters in Table 7.3 for $\nu = 0.25$ to $u(k)$ defined by (7.44).

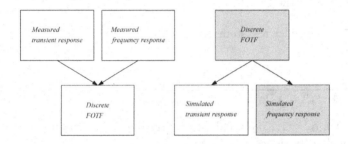

Fig. 7.28 Block diagram of the frequency characteristics identification and simulation.

with $\omega_i = \omega T_s$ as normalized frequency and T_s as a sampling period, into the discrete FOTF, one obtains the discrete Fourier transform of the system described by the FOTF

$$G(z)|_{z=e^{j\omega_i}} = \left.\sum_{i=0}^{+\infty} g(i)z^{-i}\right|_{z=e^{j\omega_i}} = \sum_{i=0}^{+\infty} g(i)z^{-ij\omega_i} = G(e^{j\omega_i}) = G(\omega_i). \quad (7.46)$$

Note that the normalized frequency ω_i range is $[0, \pi]$ and as ω_i goes from 0 to π the value z varies from 1 to -1. The function $G(\omega_i)$ is referred to as the dynamic system frequency response. For $\omega_i \in (0, \pi)$ it is a complex number and for $\omega = 0, \pi$ it is a real number. When exploring complex number forms, one gets different frequency characteristics. The fundamental one is called the discrete Nyquist characteristic (discrete Nyquist plot, discrete polar plot)

$$G(\omega_i) = G\left(e^{\omega_i}\right) = G(z)|_{z=e^{\omega_i}}. \quad (7.47)$$

The discrete Nyquist plot can be represented by the Cartesian form with its real $P(\omega_i)$ and imaginary $Q(\omega_i)$ parts

$$G(\omega_i) = G\left(e^{\omega_i}\right) = P(\omega_i) + jQ(\omega_i). \tag{7.48}$$

Functions $P(\omega_i)$ and $Q(\omega_i)$ are the real and imaginary discrete frequency characteristics of the system described by the FOTS. The polar form of the discrete Nyquist plot is as follows

$$G(\omega_i) = G\left(e^{\omega_i}\right) = |G(\omega_i)|e^{\varphi(\omega_i)}. \tag{7.49}$$

The modulus function $|G(\omega_i)|$ is called a magnitude response whereas the function $\varphi(\omega_i)$ is a phase response. Historically, these characteristics were plotted as $20log_{10}|G(\omega_i)|$ and $\varphi(\omega_i)$ in a logarithmic scale known as the discrete Bode plots.

The following two numerical examples present frequency characteristics of a system described by the FOTF.

Example 7.7. Consider a system described by the FOTF

$$G(z)$$
$$= \frac{b_0}{(1 - z^{-1})^{4\nu} + a_3 (1 - z^{-1})^{3\nu} + a_2 (1 - z^{-1})^{2\nu} + a_1 (1 - z^{-1})^{\nu} + a_0}. \tag{7.50}$$

For $\nu = 0.5$, $a_3 = 1.2$, $a_2 = 0.65$, $a_1 = 0.114$, $a_0 = 0.0126$ and $b_0 = a_0$ simulate its frequency characteristics.

Solution. Simple calculations reveal that (7.50) for assumed coefficients a_i, $i = 0, 1, 2, 3$ can be expressed as

$$G(z) = b_0 \prod_{i=0}^{4} \frac{1}{(1 - z^{-1})^{\nu} - w_i} \tag{7.51}$$

where $w_1 = -0.1 - j\sqrt{0.02}$, $w_2 = w_1^*$, $w_3 = -0.5 - j\sqrt{0.17}$, $w_4 = w_3^*$. The simulated discrete Nyquist plot is given in Fig. 7.29.

In Figs. 7.30 and 7.31, the related magnitude and frequency characteristics are presented, respectively.

Finally the discrete Bode plots are given in Figs. 7.32 and 7.33.

The presented frequency characteristics of the FO system are typical for the classical IO systems. The magnitude characteristics possess the low-pass filtering property. There are no extremes for $\omega_i \in (0, \pi]$. Such a shape of magnitude characteristics suggests a discrete step response without overshoots. This statement confirms the unit step response presented in Fig. 7.34. The response is characterized by a long settling time. This is typical in the FO systems.

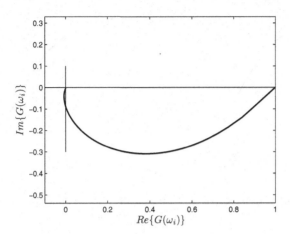

Fig. 7.29   The Nyquist plot of system (7.50).

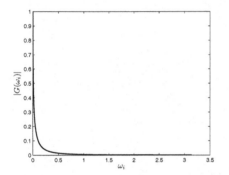

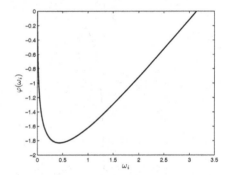

Fig. 7.30 Magnitude characteristic of system (7.50).

Fig. 7.31 Phase characteristic of system (7.50).

In the next numerical example, some rather strange FO system frequency characteristics are analyzed. An analysis of such a system reveals a community of the linear time-invariant IO and FO system dynamic behavior.

Example 7.8. Consider a system from Example 7.7 with new parameters $\nu = 0.5$, $a_3 = -1.2, a_2 = 0.65, a_1 = -0.114, a_0 = 0.0126$ and $b_0 = a_0$. Simulate its frequency characteristics. Following the steps performed in Example 7.7, one gets an equivalent form to (7.51) with $w_1 = 0.1 - j\sqrt{0.02}, w_2 = w_1^*, w_3 = 0.5 - j\sqrt{0.17}$, $w_4 = w_3^*$.

The simulated discrete Nyquist plot in a considered case is given in Fig. 7.35.

In Figs. 7.36 and 7.37 the related magnitude and frequency characteristics are presented, respectively.

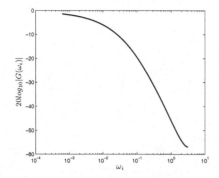
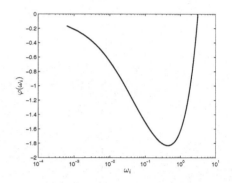

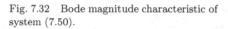

Fig. 7.32 Bode magnitude characteristic of system (7.50).

Fig. 7.33 Bode phase characteristic of system (7.50).

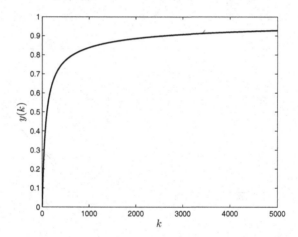

Fig. 7.34   The discrete unit step response of system (7.50).

Finally, the discrete Bode plots are given in Figs. 7.38 and 7.39.

The magnitude plots are characterized by two extremes that follow the classical IO system case, suggesting an oscillatory unit step response. It is given in Fig. 7.40.

The unit step response contains superimposed oscillation frequencies. The first lower frequency is associated with the first peak in the magnitude characteristic. The second peak is responsible for the higher frequency oscillations present in the final plot of the response.

For the discrete-time signal $f(k)$, which is not described by the one-sided $\mathcal{Z}$-Transform one applies the Discrete Fourier Transform (DFT) with discrete

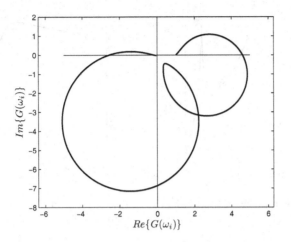

Fig. 7.35   The Nyquist plot of system (7.50) with new parameters.

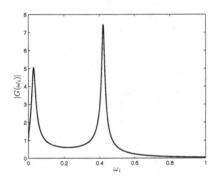

Fig. 7.36   Magnitude characteristic of system (7.50).

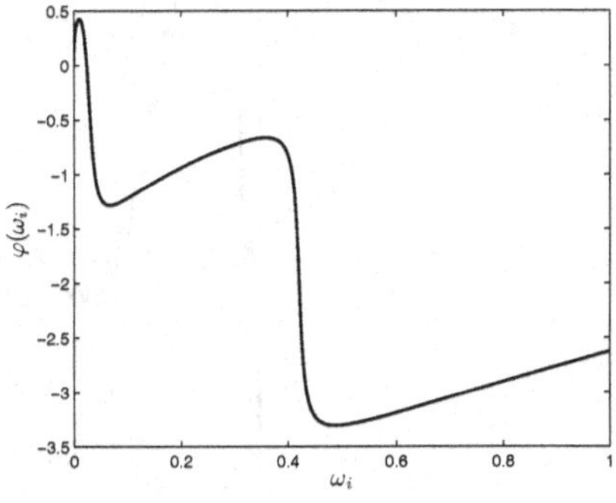

Fig. 7.37   Phase characteristic of system (7.50).

frequencies

$$F(f) = F\left[\omega_i(f)\right] = \beta_D \sum_{k=0}^{N-1} f(k)e^{-jk\omega_i(f)} \tag{7.52}$$

where β_D is a constant coefficient and

$$\omega_i(f) = \frac{2\pi f}{N}, \ f \in \{0, 1, 2, \ldots, N - 1\}. \tag{7.53}$$

An application of the DFT will be presented in a numerical example.

Example 7.9. Given a sequence whose plot is presented in Fig. 7.40. Find its magnitude and phase Bode plots.

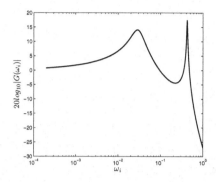

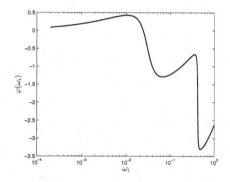

Fig. 7.38   Bode magnitude characteristic of system (7.50).

Fig. 7.39   Bode phase characteristic of system (7.50).

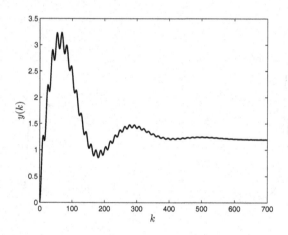

Fig. 7.40   The unit step response of system (7.50) with new parameters.

Solution. Direct application of formula (7.52) for $N = 700$ gives a complex series $F(n)$. Taking the common logarithm of the absolute values of its elements $|F(0)|, |F(1)|, \ldots, |F(N-1)|$ multiplied by 20 gives a Bode magnitude characteristic.

In Fig. 7.42, the Bode phase characteristics are given. Both presented figures reveal a similarity in shapes, but in the DFT some information must be lost due to a discontinuity of the frequencies.

The DFT defined by formula (7.52) is very useful in measured discrete signals which cannot be described as a mathematical formula. In the next numerical example, one evaluates the DFT of a discrete signal obtained from the "Small Dog" image presented in Fig. 1.50. The discrete 1D signal is created by one row of the image where $k_1 = 15 = \text{const}$. It is plotted in Fig. 7.43.

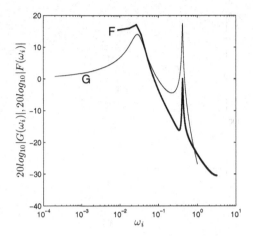

Fig. 7.41 The Bode magnitude plots evaluated according to (7.49) and (7.52).

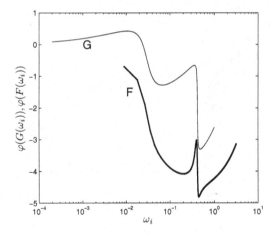

Fig. 7.42 The Bode magnitude plots evaluated according to (7.49) and (7.52).

The magnitude and phase Bode characteristics of $F(15, k_2)$ are presented in Figs. 7.44 and 7.45, respectively.

The DFT should be supplemented by its inverse. The inverse discrete Fourier transform (IDFT) is defined as

$$f(k) = \beta_I \sum_{f=0}^{N-1} F(f)e^{jk\omega_i(f)} \qquad (7.54)$$

where β_I is a coefficient. To preserve the invertibility one should assume $\beta_D\beta_I = 1$.

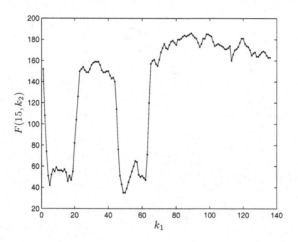

Fig. 7.43   Plot of the discrete signal $F(15, k_2)$.

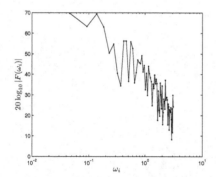 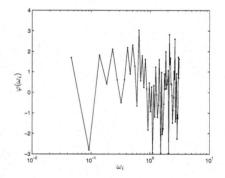

Fig. 7.44   Bode magnitude characteristic of
signal $F(15, k_2)$.

Fig. 7.45   Bode phase characteristic of
signal $F(15, k_2)$.

Then the identity is satisfied

$$f(k) = \beta_I \sum_{f=0}^{N-1} \left[\beta_D \sum_{k=0}^{N-1} f(k) e^{-jk\omega_i(f)} \right] e^{jk\omega_i(f)}$$

$$= \sum_{f=0}^{N-1} \left[\sum_{k=0}^{N-1} f(k) e^{-jk\omega_i(f)} \right] e^{jk\omega_i(f)}. \tag{7.55}$$

To analyse the two discrete-variable functions, one uses 2D discrete Fourier
Transform (2D DFT) . It is a generalization of the DFT defined above.

$$F(f_1, f_2) = F(\omega_i(f_1), \omega_i(f_2)) = \beta_D \sum_{k_1=0}^{N_1-1} \sum_{k_2=0}^{N_2-1} f(k_1, k_2) e^{-jk_1\omega_1(f_1)} e^{-jk_2\omega_1(f_2)}$$

$$(7.56)$$

where β_D is a constant and

$$\omega_i(f_1) = \frac{2\pi f_1}{N_1}, \quad \omega_i(f_2) = \frac{2\pi f_2}{N_2}, \quad f_i \in \{0, 1, 2, \ldots, N_i - 1\}, i = 1, 2. \quad (7.57)$$

Its inverse is very similar to the one-dimensional

$$f(k_1, k_2) = \beta_I \sum_{f_1=0}^{N_1-1} \left[\beta_D \sum_{f_2=0}^{N_2-1} F(f_1, f_2) e^{jk_1\omega_i(f_1)} \right] e^{jk_2\omega_i(f_2)}$$

$$= \sum_{f_1=0}^{N_1-1} \left[\sum_{f_2=0}^{N_2-1} F(f_1, f_2) e^{jk_1\omega_i(f_1)} \right] e^{jk_2\omega_i(f_2)}. \quad (7.58)$$

In general, the 2D DFT $F(f_1, f_2)$ is a complex number which traditionally can be expressed in the Cartesian and polar form, respectively.

$$F(f_1, f_2) = P(f_1, f_2) + jQ(f_1, f_2) \quad (7.59)$$

where

$$P(f_1, f_2) = \text{Re}\{F(f_1, f_2)\}, \quad Q(f_1, f_2) = \text{Im}\{F(f_1, f_2)\} \quad (7.60)$$

are the real and imaginary parts of the 2D DFT and

$$F(f_1, f_2) = |F(f_1, f_2)|e^{j\varphi(f_1, f_2)} \quad (7.61)$$

with

$$|F(f_1, f_2)| = \sqrt{P^2(f_1, f_2) + Q^2(f_1, f_2)}, \quad \varphi(f_1, f_2) = \arctan\left[\frac{Q(f_1, f_2)}{P(f_1, f_2)}\right]. \quad (7.62)$$

are the magnitude and phase characteristics.

In the example, one evaluates numerically the 2D DFT of the image considered in Fig. 1.50.

Example 7.10. Plot the characteristics $P(f_1, f_2)$, $Q(f_1, f_2)$, $|F(f_1, f_2)|$ and $\varphi(f_1, f_2)$.

Solution. The real and imaginary parts of $F(f_1, f_2)$ are given in Figs. 7.46 and 7.47. The magnitude and phase characteristics and its Bode counterparts are presented in Figs. 7.48–7.51, respectively.

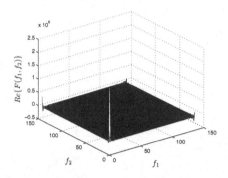

Fig. 7.46   Real part of the 2D DFT of image 1.50.

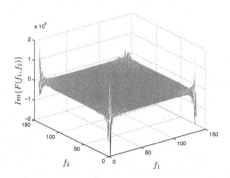

Fig. 7.47   Imaginary part of the 2D DFT of image 1.50.

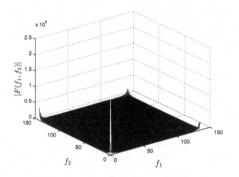

Fig. 7.48   Magnitude plot of the 2D DFT of image 1.50.

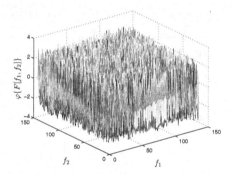

Fig. 7.49   Phase plot of the 2D DFT of image 1.50.

7.3   Linear time-invariant FOS reachability and observability

The considered in this section properites are the most important in dynamical system analysis and synthesis. The FO discrete system reachability and observability is more complicated due to an assumed infinite number of initial states. The long memory of the FO system, represented by an infinite number of initial states with strongly lowering impact due to the distance to the present instance, complicates the state observability notion.

7.3.1   *Linear time-invariant FOS reachability*

One starts with the definition of the reachable state of the FO system described by the SSE. To simplify considerations, one assumes an initial time $k_0 = 0$. This assumption does not influence their generalitisation. The definitions given below,

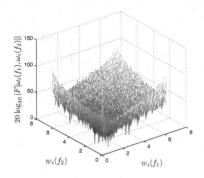
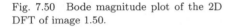

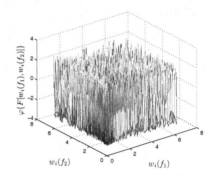

Fig. 7.50 Bode magnitude plot of the 2D DFT of image 1.50.

Fig. 7.51 Bode phase plot of the 2D DFT of image 1.50.

followed by [Kailath (1980); Kaczorek (2011); Klamka (1991); Klamka (2008, 2010)], gives an insight into this problem.

Definition 7.1. A state $\mathbf{x}(L-1)$ is called a reachable one in (given) L steps if there exists a bounded input signal sequence $\{\mathbf{u}(0), \mathbf{u}(1), \ldots, \mathbf{u}(k_{L-1}),\}$ which steers the initial state set containing, in general, an infinite number of zero elements $\mathbf{x}(-1) = \mathbf{x}(-2) = \mathbf{x}(-3) = \cdots = \mathbf{x}(-\infty) = \mathbf{0}$ towards it.

Definition 7.2. If every state $\mathbf{x}(L-1)$ is reachable in L steps by a related boundend input signal sequence $\{\mathbf{u}(0), \mathbf{u}(1), \ldots, \mathbf{u}(k_{L-1}),\}$ which steers the zero initial state set, $\mathbf{x}(-1) = \mathbf{x}(-2) = \mathbf{x}(-3) = \cdots = \mathbf{x}(-\infty) = \mathbf{0}$, than the system is called reachable in L steps.

Definition 7.3. If for every state $\mathbf{x}(L-1)$ there is L for which a system is reachable in L steps by related bounded input signal sequence $\{\mathbf{u}(0), \mathbf{u}(1), \ldots, \mathbf{u}(k_{L-1}),\}$, which steers the zero initial states set $\mathbf{x}(-1) = \mathbf{x}(-2) = \mathbf{x}(-3) = \cdots = \mathbf{x}(-\infty) = \mathbf{0}$ than the system is called reachable one.

The following theorems formulate the linear discrete time-invariant systems described by FO SSE reachability conditions.

Theorem 7.1. *The FOS described by the FOSSEs (6.41) is reachable in L, steps if and only if, a reachability matrix $\mathbf{R}(L-1)$ is of full rank*

$$\mathbf{R}_I(L) = \mathrm{rank}\begin{bmatrix} \mathbf{\Phi}(0)\mathbf{B} & \mathbf{\Phi}(1)\mathbf{B} & \cdots & \mathbf{\Phi}(L-1)\mathbf{B} \end{bmatrix} = p. \tag{7.63}$$

Proof. By the assumption $\mathbf{x}(-1) = \mathbf{x}(-2) = \mathbf{x}(-3) = \cdots = \mathbf{x}(-\infty) = \mathbf{0}$ from

(7.40) there is

$$\mathbf{x}(L-1)=\sum_{i=0}^{L-1}\mathbf{\Phi}(L-1-i)\mathbf{B}\mathbf{u}(i)=\left[\mathbf{\Phi}(L-1)\mathbf{B}\ \mathbf{\Phi}(L-2)\mathbf{B}\ \cdots\ \mathbf{\Phi}(0)\mathbf{B}\right]\begin{bmatrix}\mathbf{u}(0)\\\mathbf{u}(1)\\\vdots\\\mathbf{u}(L-1)\end{bmatrix}$$

$$\left[\mathbf{\Phi}(0)\mathbf{B}\ \mathbf{\Phi}(1)\mathbf{B}\ \cdots\ \mathbf{\Phi}(L-1)\mathbf{B}\right]\begin{bmatrix}\mathbf{u}(L-1)\\\mathbf{u}(L-2)\\\vdots\\\mathbf{u}(0)\end{bmatrix}.$$

$$(7.64)$$

By the Kronecker-Capelly theorem ([Bernstein (2009)]), a solution of the linear matrix algebraic equation with unknown input signal vectors exists for every state $\mathbf{x}(L)$, if and only if, the condition (7.63) is satisfied. $\qquad\square$

The results of theorem 7.1 will be confirmed by a numerical example.

Example 7.11. Consider the FOSSE with matrices

$$\mathbf{A}=\begin{bmatrix}\frac{5}{2}&1&0\\-\frac{13}{3}&-\frac{5}{3}&0\\\frac{26}{3}&\frac{23}{6}&\frac{1}{4}\end{bmatrix},\quad\mathbf{b}=\begin{bmatrix}1\\-1\\3\end{bmatrix}.\tag{7.65}$$

Verify the FOSSE reachability condition.

Solution. The easiest way to find a series of matrices $\mathbf{\Phi}(0),\mathbf{\Phi}(1),\ldots$ is to build a matrix (7.28), then calculate its inverse (7.37) and extract the searched matrices. The steps mentioned above lead for $L=6$ to the following set of matrices

$$\mathbf{\Phi}(0)=\begin{bmatrix}8&3&0\\-13&-\frac{9}{2}&0\\26&\frac{35}{3}&\frac{4}{3}\end{bmatrix},\quad\mathbf{\Phi}(1)=\begin{bmatrix}\frac{25}{2}&\frac{21}{4}&0\\-\frac{91}{4}&-\frac{75}{8}&0\\0&0&\frac{8}{9}\end{bmatrix},\quad\mathbf{\Phi}(2)=\begin{bmatrix}19&\frac{33}{4}&0\\-\frac{143}{4}&-\frac{123}{8}&0\\\frac{143}{2}&\frac{3497}{108}&\frac{22}{27}\end{bmatrix}$$

$$\mathbf{\Phi}(3)=\begin{bmatrix}\frac{893}{32}&\frac{789}{64}&0\\-\frac{3419}{64}&-\frac{3003}{128}&0\\\frac{3419}{32}&\frac{5435}{112}&\frac{65}{81}\end{bmatrix},\quad\mathbf{\Phi}(4)=\begin{bmatrix}\frac{2563}{64}&\frac{2289}{128}&0\\-\frac{9919}{128}&-\frac{8823}{256}&0\\\frac{9919}{64}&\frac{11501}{163}&\frac{763}{937}\end{bmatrix},\quad\mathbf{\Phi}(5)=\begin{bmatrix}\frac{14475}{256}&\frac{7300}{287}&0\\-\frac{8487}{77}&-\frac{6427}{130}&0\\\frac{31964}{145}&\frac{46859}{466}&\frac{1448}{1725}\end{bmatrix}.$$

$$(7.66)$$

Having calculated matrices $\mathbf{\Phi}(k)$ for $k=0,\ldots,5$, one builds the reachability matrix and checks its rank.

$$\mathbf{R}_I(6)=\begin{bmatrix}5&\frac{29}{4}&\frac{43}{4}&\frac{997}{64}&\frac{2837}{128}&\frac{3764}{121}\\-\frac{17}{2}&-\frac{107}{8}&-\frac{163}{8}&-\frac{3835}{128}&-\frac{11015}{256}&-\frac{7537}{124}\\\frac{55}{3}&\frac{995}{36}&\frac{3389}{108}&\frac{9473}{156}&\frac{7297}{84}&\frac{6365}{52}\end{bmatrix}.\tag{7.67}$$

One can easily check that

$$\operatorname{rank}\mathbf{R}_I(6) = 3. \tag{7.68}$$

The considered system is reachable. The reachability matrix can be used to calculate a control signal $\mathbf{u}(k)$, steering the system with zero initial conditions to a desired state $\mathbf{x}(L-1)$. One should only solve an algebraic equation

$$\mathbf{R}_I(6) \begin{bmatrix} u(5) \\ u(4) \\ \vdots \\ u(0) \end{bmatrix} = \mathbf{x}(5). \tag{7.69}$$

To solve the above equation one inserts two involutory matrices (11.19) between the matrix $\mathbf{R}_I(6)$ and the vector $\mathbf{x}(5)$, and then pre-muliplies both sides by elementary row operations matrix $\mathbf{L}_{ep}$

$$\mathbf{L}_{ep} = \begin{bmatrix} -\frac{779}{21} & \frac{913}{42} & \frac{81}{4} \\ \frac{2186}{21} & -\frac{1160}{21} & -54 \\ = \frac{370}{7} & \frac{190}{7} & 27 \end{bmatrix}. \tag{7.70}$$

These operations bring the equation to a more simpler form which enables a clear insight into the solution

$$\begin{bmatrix} 1\,0\,0 & \frac{1}{2} & \frac{19}{12} & \frac{497}{144} \\ 0\,1\,0 & -\frac{121}{48} & -\frac{2017}{288} & -\frac{4169}{294} \\ 0\,0\,1 & \frac{35}{12} & \frac{871}{144} & \frac{293}{27} \end{bmatrix} \begin{bmatrix} u(0) \\ u(1) \\ u(2) \\ u(3) \\ u(4) \\ u(5) \end{bmatrix} = \begin{bmatrix} -\frac{779}{21} & \frac{913}{42} & \frac{81}{4} \\ \frac{2186}{21} & -\frac{1160}{21} & -54 \\ = \frac{370}{7} & \frac{190}{7} & 27 \end{bmatrix} \begin{bmatrix} x_1(5) \\ x_2(5) \\ x_3(5) \end{bmatrix}. \tag{7.71}$$

From the form given above one immediately obtains

$$\begin{bmatrix} u(0) \\ u(1) \\ u(2) \end{bmatrix} = - \begin{bmatrix} \frac{1}{2} & \frac{19}{12} & \frac{497}{144} \\ -\frac{121}{48} & -\frac{2017}{288} & -\frac{4169}{294} \\ \frac{35}{12} & \frac{871}{144} & \frac{293}{27} \end{bmatrix} \begin{bmatrix} u(3) \\ u(4) \\ u(5) \end{bmatrix} + \begin{bmatrix} -\frac{779}{21} & \frac{913}{42} & \frac{81}{4} \\ \frac{2186}{21} & -\frac{1160}{21} & -54 \\ = \frac{370}{7} & \frac{190}{7} & 27 \end{bmatrix} \begin{bmatrix} x_1(5) \\ x_2(5) \\ x_3(5) \end{bmatrix}. \tag{7.72}$$

The simplest solution of equation (7.69) is obtained for $u(3) = u(4) = u(5) = 0$. Then $L = 3$ and a target state is $\mathbf{x}(L-1) = \mathbf{x}(2)$

$$\begin{bmatrix} u(0) \\ u(1) \\ u(2) \end{bmatrix} = \begin{bmatrix} -\frac{779}{21} & \frac{913}{42} & \frac{81}{4} \\ \frac{2186}{21} & -\frac{1160}{21} & -54 \\ = \frac{370}{7} & \frac{190}{7} & 27 \end{bmatrix} \begin{bmatrix} x_1(2) \\ x_2(2) \\ x_3(2) \end{bmatrix}. \tag{7.73}$$

and one obtains a discrete control steering a system zero-state to the desired final one. For assumed

$$\begin{bmatrix} x_1(5) \\ x_2(5) \\ x_3(5) \end{bmatrix} = \begin{bmatrix} 1 \\ 1 \\ 1 \end{bmatrix} \tag{7.74}$$

from (7.73) one obtains

$$\begin{bmatrix} u(0) \\ u(1) \\ u(2) \end{bmatrix} = \begin{bmatrix} \frac{137}{28} \\ -\frac{36}{7} \\ \frac{9}{7} \end{bmatrix}. \tag{7.75}$$

The plot of a controlling signal and the system states short transients are shown in Figs. 7.52 and 7.53.

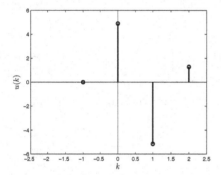

Fig. 7.52 Plot of the controlling signal to the assumed $\mathbf{x}(2)$.

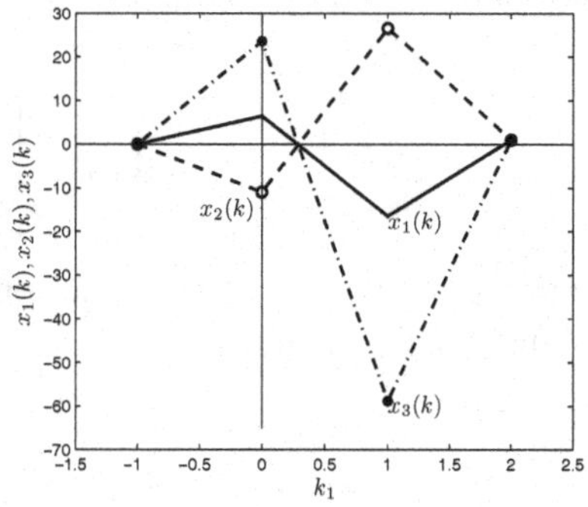

Fig. 7.53 Plot of $x_1(k), x_2(k), x_3(k)$ for zero initial conditions $\mathbf{x}_{-i} = \mathbf{0}$, $i = 1, 2, \ldots$.

These investigations show that in three steps a zero state is steered to the desired final state. The investigations didn't take into account constraints imposed on the values of the controlling signal. This is a very important demand for the practical applications. Usually, lower controlling values can be achieved by an extension of the controlling signal (removing the condition $u(3) = u(4) = u(5) = 0$).

Now, one checks the behavior of a non-reachable system.

Example 7.12. Consider the FOSSE with the state matrix $\mathbf{A}$ defined by (7.65) with a new input matrix

$$\mathbf{b} = \begin{bmatrix} 0 \\ 0 \\ 1 \end{bmatrix}. \tag{7.76}$$

Check its reachability condition.

Solution. Analogous procedure for assumed $L = 6$ leads to the reachability matrix

$$\mathbf{R}_I(6) = \begin{bmatrix} 0 & 0 & 0 & 0 & 0 & 0 \\ 0 & 0 & 0 & 0 & 0 & 0 \\ \frac{4}{3} & \frac{8}{9} & \frac{22}{27} & \frac{65}{81} & \frac{763}{937} & \frac{1448}{1725} \end{bmatrix}. \tag{7.77}$$

Clearly, its rank equals to 1. The system is not reachable.

Now, some inspection of matrix (7.28) and its inverse (7.37) will be performed. This will be helpful in further analysis of the FOSSE reachability. Both matrices are block upper triangular ones and

$$\mathbf{D}_{kp} \mathbf{\Phi}_{kp} = \mathbf{1}_{kp}, \text{ for } k = 0, 1, 2, \ldots. \tag{7.78}$$

Therefore, for $k = 0$

$$(\mathbf{1}_p - \mathbf{A}) \mathbf{\Phi}(0) = \mathbf{1}_p \tag{7.79}$$

and

$$\mathbf{\Phi}(0) = (\mathbf{1}_p - \mathbf{A})^{-1}. \tag{7.80}$$

For $k = 1$

$$\begin{bmatrix} \mathbf{1}_p - \mathbf{A} \, a^{(\nu)}(1)\mathbf{1}_p \\ \mathbf{0}_p \quad \mathbf{1}_p - \mathbf{A} \end{bmatrix} \begin{bmatrix} \mathbf{\Phi}(0) \, \mathbf{\Phi}(1) \\ \mathbf{0}_p \quad \mathbf{\Phi}(0) \end{bmatrix}$$

$$= \begin{bmatrix} \mathbf{1}_p - \mathbf{A} \, a^{(\nu)}(1)\mathbf{1}_p \\ \mathbf{0}_p \quad \mathbf{1}_p - \mathbf{A} \end{bmatrix} \begin{bmatrix} (\mathbf{1}_p - \mathbf{A})^{-1} \quad \mathbf{\Phi}(1) \\ \mathbf{0}_p \quad (\mathbf{1}_p - \mathbf{A})^{-1} \end{bmatrix} = \begin{bmatrix} \mathbf{1}_p \, \mathbf{0}_p \\ \mathbf{0}_p \, \mathbf{1}_p \end{bmatrix}. \tag{7.81}$$

From the equation presented above, one derives

$$(\mathbf{1}_p - \mathbf{A}) \mathbf{\Phi}(1) + a^{(\nu)}(1) (\mathbf{1}_p - \mathbf{A})^{-1} = \mathbf{0}_p \tag{7.82}$$

and determines

$$\mathbf{\Phi}(1) = -a^{(\nu)}(1) (\mathbf{1}_p - \mathbf{A})^{-2}. \tag{7.83}$$

Further investigation leads to

$$\mathbf{\Phi}(2) = -a^{(\nu)}(2) (\mathbf{1}_p - \mathbf{A})^{-2} + \left[a^{(\nu)}(1)\right]^2 (\mathbf{1}_p - \mathbf{A})^{-3}$$

$$\mathbf{\Phi}(3) = -a^{(\nu)}(3) (\mathbf{1}_p - \mathbf{A})^{-2} + 2a^{(\nu)}(1)a^{(\nu)}(2) (\mathbf{1}_p - \mathbf{A})^{-3} - \left[a^{(\nu)}(1)\right]^3 (\mathbf{1}_p - \mathbf{A})^{-4}$$

$$\mathbf{\Phi}(4) = -a^{(\nu)}(4) (\mathbf{1}_p - \mathbf{A})^{-2}$$

$$+ \left\{ a^{(\nu)}(3) + \left[a^{(\nu)}(2)\right]^2 + a^{(\nu)}(1)a^{(\nu)}(3) \right\} (\mathbf{1}_p - \mathbf{A})^{-3}$$

$$-3a^{(\nu)}(2) \left[a^{(\nu)}(1)\right]^2 (\mathbf{1}_p - \mathbf{A})^{-4} + \left[a^{(\nu)}(1)\right]^4 (\mathbf{1}_p - \mathbf{A})^{-5}.$$

$$\tag{7.84}$$

On the basis of the above relationships, one can conclude that

$$\mathbf{\Phi}(k) = \sum_{i=1}^{k+1} \phi^{(\nu)}(i, k+1)\,(\mathbf{1}_p - \mathbf{A})^{-i}$$

$$= \left[(\mathbf{1}_p - \mathbf{A})^{-1}\,(\mathbf{1}_p - \mathbf{A})^{-2}\cdots(\mathbf{1}_p - \mathbf{A})^{-k-1} \right] \begin{bmatrix} \phi^{(\nu)}(1, k+1)\mathbf{1}_p \\ \phi^{(\nu)}(2, k+1)\mathbf{1}_p \\ \vdots \\ \phi^{(\nu)}(k+1, k+1)\mathbf{1}_p \end{bmatrix}. \quad (7.85)$$

The next theorem gives an equivalent FOSSE reachability condition.

Theorem 7.2. *The system described by the FOSSE is reachable, if and only if, the reachability matrix is of full rank*

$$\mathbf{R}_{II}(L) = \mathrm{rank}\left[(\mathbf{1}_p - \mathbf{A})^{-1}\mathbf{B}\,(\mathbf{1}_p - \mathbf{A})^{-2}\mathbf{B}\cdots(\mathbf{1}_p - \mathbf{A})^{-L}\mathbf{B} \right] = p. \quad (7.86)$$

Proof. The equality of ranks of $\mathbf{R}_I$ and $\mathbf{R}_{II}$ will be proved, noting that due to 7.85 there exists a block matrix

$$\left[\mathbf{\Phi}(0)\ \mathbf{\Phi}(1)\ \cdots\ \mathbf{\Phi}(L-1) \right] = \left[(\mathbf{1}_p - \mathbf{A})^{-1}\,(\mathbf{1}_p - \mathbf{A})^{-2}\cdots(\mathbf{1}_p - \mathbf{A})^{-L} \right]$$

$$\times \begin{bmatrix} \phi^{(\nu)}(1,1)\mathbf{1}_p & \phi^{(\nu)}(1,2)\mathbf{1}_p & \phi^{(\nu)}(1,3)\mathbf{1}_p & \cdots & \phi^{(\nu)}(1,L)\mathbf{1}_p \\ \mathbf{0}_p & \phi^{(\nu)}(2,2)\mathbf{1}_p & \phi^{(\nu)}(2,3)\mathbf{1}_p & \cdots & \phi^{(\nu)}(2,L)\mathbf{1}_p \\ \mathbf{0}_p & \mathbf{0}_p & \phi^{(\nu)}(3,3)\mathbf{1}_p & \cdots & \phi^{(\nu)}(3,L)\mathbf{1}_p \\ \vdots & \vdots & \vdots & & \vdots \\ \mathbf{0}_p & \mathbf{0}_p & \mathbf{0}_p & \cdots & \phi^{(\nu)}(L,L)\mathbf{1}_p \end{bmatrix}$$

$$(7.87)$$

and that $\mathbf{R}_I(L)$ can be expressed as

$$\left[\mathbf{\Phi}(0)\mathbf{B}\ \mathbf{\Phi}(1)\mathbf{B}\ \cdots\ \mathbf{\Phi}(L-1)\mathbf{B} \right]$$

$$= \left[\mathbf{\Phi}(0)\ \mathbf{\Phi}(1)\ \cdots\ \mathbf{\Phi}(L-1) \right] \begin{bmatrix} \mathbf{B} & \mathbf{0}_{p,q} & \mathbf{0}_{p,q} & \cdots & \mathbf{0}_{p,q} \\ \mathbf{0}_{p,q} & \mathbf{B} & \mathbf{0}_{p,q} & \cdots & \mathbf{0}_{p,q} \\ \mathbf{0}_{p,q} & \mathbf{0}_{p,q} & \mathbf{B} & \cdots & \mathbf{0}_{p,q} \\ \vdots & \vdots & \vdots & & \vdots \\ \mathbf{0}_{p,q} & \mathbf{0}_{p,q} & \mathbf{0}_{p,q} & \cdots & \mathbf{B} \end{bmatrix}. \quad (7.88)$$

A substitution (7.87) into the right-hand side of (7.88) yields

$$\left[\mathbf{\Phi}(0)\mathbf{B}\ \mathbf{\Phi}(1)\mathbf{B}\ \cdots\ \mathbf{\Phi}(L-1)\mathbf{B}\right] = \left[(\mathbf{1}_p-\mathbf{A})^{-1}\ (\mathbf{1}_p-\mathbf{A})^{-2}\ \cdots\ (\mathbf{1}_p-\mathbf{A})^{-L}\right]$$

$$\times \begin{bmatrix} \phi^{(\nu)}(1,1)\mathbf{1}_p\ \phi^{(\nu)}(1,2)\mathbf{1}_p\ \phi^{(\nu)}(1,3)\mathbf{1}_p\ \cdots\ \phi^{(\nu)}(1,L)\mathbf{1}_p \\ \mathbf{0}_p\ \ \ \ \phi^{(\nu)}(2,2)\mathbf{1}_p\ \phi^{(\nu)}(2,3)\mathbf{1}_p\ \cdots\ \phi^{(\nu)}(2,L)\mathbf{1}_p \\ \mathbf{0}_p\ \ \ \ \mathbf{0}_p\ \ \ \ \phi^{(\nu)}(3,3)\mathbf{1}_p\ \cdots\ \phi^{(\nu)}(3,L)\mathbf{1}_p \\ \vdots\ \ \ \ \vdots\ \ \ \ \vdots\ \ \ \ \vdots \\ \mathbf{0}_p\ \ \ \ \mathbf{0}_p\ \ \ \ \mathbf{0}_p\ \ \ \ \cdots\ \phi^{(\nu)}(L,L)\mathbf{1}_p \end{bmatrix} \begin{bmatrix} \mathbf{B}\ \mathbf{0}_{p,q}\ \mathbf{0}_{p,q}\ \cdots\ \mathbf{0}_{p,q} \\ \mathbf{0}_{p,q}\ \mathbf{B}\ \mathbf{0}_{p,q}\ \cdots\ \mathbf{0}_{p,q} \\ \mathbf{0}_{p,q}\ \mathbf{0}_{p,q}\ \mathbf{B}\ \cdots\ \mathbf{0}_{p,q} \\ \vdots\ \ \vdots\ \ \vdots\ \ \ \ \vdots \\ \mathbf{0}_{p,q}\ \mathbf{0}_{p,q}\ \mathbf{0}_{p,q}\ \cdots\ \mathbf{B} \end{bmatrix}.$$

$$(7.89)$$

Now, one examines a product of matrices

$$\begin{bmatrix} \phi^{(\nu)}(1,1)\mathbf{1}_p\ \phi^{(\nu)}(1,2)\mathbf{1}_p\ \phi^{(\nu)}(1,3)\mathbf{1}_p\ \cdots\ \phi^{(\nu)}(1,L)\mathbf{1}_p \\ \mathbf{0}_p\ \ \ \ \phi^{(\nu)}(2,2)\mathbf{1}_p\ \phi^{(\nu)}(2,3)\mathbf{1}_p\ \cdots\ \phi^{(\nu)}(2,L)\mathbf{1}_p \\ \mathbf{0}_p\ \ \ \ \mathbf{0}_p\ \ \ \ \phi^{(\nu)}(3,3)\mathbf{1}_p\ \cdots\ \phi^{(\nu)}(3,L)\mathbf{1}_p \\ \vdots\ \ \ \ \vdots\ \ \ \ \vdots\ \ \ \ \vdots \\ \mathbf{0}_p\ \ \ \ \mathbf{0}_p\ \ \ \ \mathbf{0}_p\ \ \ \ \cdots\ \phi^{(\nu)}(L,L)\mathbf{1}_p \end{bmatrix} \begin{bmatrix} \mathbf{B}\ \mathbf{0}_{p,q}\ \mathbf{0}_{p,q}\ \cdots\ \mathbf{0}_{p,q} \\ \mathbf{0}_{p,q}\ \mathbf{B}\ \mathbf{0}_{p,q}\ \cdots\ \mathbf{0}_{p,q} \\ \mathbf{0}_{p,q}\ \mathbf{0}_{p,q}\ \mathbf{B}\ \cdots\ \mathbf{0}_{p,q} \\ \vdots\ \ \vdots\ \ \vdots\ \ \ \ \vdots \\ \mathbf{0}_{p,q}\ \mathbf{0}_{p,q}\ \mathbf{0}_{p,q}\ \cdots\ \mathbf{B} \end{bmatrix}$$

$$= \begin{bmatrix} \phi^{(\nu)}(1,1)\mathbf{B}\ \phi^{(\nu)}(1,2)\mathbf{B}\ \phi^{(\nu)}(1,3)\mathbf{B}\ \cdots\ \phi^{(\nu)}(1,L)\mathbf{B} \\ \mathbf{0}_p\ \ \ \ \phi^{(\nu)}(2,2)\mathbf{B}\ \phi^{(\nu)}(2,3)\mathbf{B}\ \cdots\ \phi^{(\nu)}(2,L)\mathbf{B} \\ \mathbf{0}_p\ \ \ \ \mathbf{0}_p\ \ \ \ \phi^{(\nu)}(3,3)\mathbf{B}\ \cdots\ \phi^{(\nu)}(3,L)\mathbf{B} \\ \vdots\ \ \ \ \vdots\ \ \ \ \vdots\ \ \ \ \vdots \\ \mathbf{0}_p\ \ \ \ \mathbf{0}_p\ \ \ \ \mathbf{0}_p\ \ \ \ \cdots\ \phi^{(\nu)}(L,L)\mathbf{B} \end{bmatrix}$$

$$= \begin{bmatrix} \mathbf{B}\ \mathbf{0}_{p,q}\ \mathbf{0}_{p,q}\ \cdots\ \mathbf{0}_{p,q} \\ \mathbf{0}_{p,q}\ \mathbf{B}\ \mathbf{0}_{p,q}\ \cdots\ \mathbf{0}_{p,q} \\ \mathbf{0}_{p,q}\ \mathbf{0}_{p,q}\ \mathbf{B}\ \cdots\ \mathbf{0}_{p,q} \\ \vdots\ \ \vdots\ \ \vdots\ \ \ \ \vdots \\ \mathbf{0}_{p,q}\ \mathbf{0}_{p,q}\ \mathbf{0}_{p,q}\ \cdots\ \mathbf{B} \end{bmatrix} \begin{bmatrix} \phi^{(\nu)}(1,1)\mathbf{1}_q\ \phi^{(\nu)}(1,2)\mathbf{1}_q\ \cdots\ \phi^{(\nu)}(1,L)\mathbf{1}_q \\ \mathbf{0}_q\ \ \ \ \phi^{(\nu)}(2,2)\mathbf{1}_p\ \cdots\ \phi^{(\nu)}(2,L)\mathbf{1}_q \\ \mathbf{0}_q\ \ \ \ \mathbf{0}_q\ \ \ \ \cdots\ \phi^{(\nu)}(3,L)\mathbf{1}_q \\ \vdots\ \ \ \ \vdots\ \ \ \ \vdots \\ \mathbf{0}_q\ \ \ \ \mathbf{0}_q\ \ \ \ \cdots\ \phi^{(\nu)}(L,L)\mathbf{1}_q \end{bmatrix}.$$

$$(7.90)$$

Returning to (7.89) with (7.90) one gets

$$
\left[\boldsymbol{\Phi}(0)\mathbf{B}\ \boldsymbol{\Phi}(1)\mathbf{B}\ \cdots\ \boldsymbol{\Phi}(L-1)\mathbf{B}\right]=\left[(\mathbf{1}_p-\mathbf{A})^{-1}\ (\mathbf{1}_p-\mathbf{A})^{-2}\ \cdots\ (\mathbf{1}_p-\mathbf{A})^{-L}\right]
$$

$$
\times\begin{bmatrix} \mathbf{B} & \mathbf{0}_{p,q} & \mathbf{0}_{p,q} & \cdots & \mathbf{0}_{p,q} \\ \mathbf{0}_{p,q} & \mathbf{B} & \mathbf{0}_{p,q} & \cdots & \mathbf{0}_{p,q} \\ \mathbf{0}_{p,q} & \mathbf{0}_{p,q} & \mathbf{B} & \cdots & \mathbf{0}_{p,q} \\ \vdots & \vdots & \vdots & & \vdots \\ \mathbf{0}_{p,q} & \mathbf{0}_{p,q} & \mathbf{0}_{p,q} & \cdots & \mathbf{B} \end{bmatrix}\begin{bmatrix} \phi^{(\nu)}(1,1)\mathbf{1}_q & \phi^{(\nu)}(1,2)\mathbf{1}_q & \cdots & \phi^{(\nu)}(1,L)\mathbf{1}_q \\ \mathbf{0}_q & \phi^{(\nu)}(2,2)\mathbf{1}_p & \cdots & \phi^{(\nu)}(2,L)\mathbf{1}_q \\ \mathbf{0}_q & \mathbf{0}_q & \cdots & \phi^{(\nu)}(3,L)\mathbf{1}_q \\ \vdots & \vdots & & \vdots \\ \mathbf{0}_q & \mathbf{0}_q & \cdots & \phi^{(\nu)}(L,L)\mathbf{1}_q \end{bmatrix}
$$

$$
=\left[(\mathbf{1}_p-\mathbf{A})^{-1}\mathbf{B}\ \cdots\ (\mathbf{1}_p-\mathbf{A})^{-L}\mathbf{B}\right]\begin{bmatrix} \phi^{(\nu)}(1,1)\mathbf{1}_q & \phi^{(\nu)}(1,2)\mathbf{1}_q & \cdots & \phi^{(\nu)}(1,L)\mathbf{1}_q \\ \mathbf{0}_q & \phi^{(\nu)}(2,2)\mathbf{1}_p & \cdots & \phi^{(\nu)}(2,L)\mathbf{1}_q \\ \mathbf{0}_q & \mathbf{0}_q & \cdots & \phi^{(\nu)}(3,L)\mathbf{1}_q \\ \vdots & \vdots & & \vdots \\ \mathbf{0}_q & \mathbf{0}_q & \cdots & \phi^{(\nu)}(L,L)\mathbf{1}_q \end{bmatrix}.
$$

$$(7.91)$$

Now, realize that for $\nu\in(0,1)$

$$
\phi^{(\nu)}(1,1)=1,\ \phi^{(\nu)}(i,i)=\left[-a^{(\nu)}(i)\right]^i\neq 0,\ \text{for } i=2,3,\ldots,L. \tag{7.92}
$$

This means that

$$
\operatorname{rank}\begin{bmatrix} \phi^{(\nu)}(1,1)\mathbf{1}_q & \phi^{(\nu)}(1,2)\mathbf{1}_q & \cdots & \phi^{(\nu)}(1,L)\mathbf{1}_q \\ \mathbf{0}_q & \phi^{(\nu)}(2,2)\mathbf{1}_p & \cdots & \phi^{(\nu)}(2,L)\mathbf{1}_q \\ \mathbf{0}_q & \mathbf{0}_q & \cdots & \phi^{(\nu)}(3,L)\mathbf{1}_q \\ \vdots & \vdots & & \vdots \\ \mathbf{0}_q & \mathbf{0}_q & \cdots & \phi^{(\nu)}(L,L)\mathbf{1}_q \end{bmatrix}=qL \tag{7.93}
$$

and

$$
\operatorname{rank}\left[\boldsymbol{\Phi}(0)\mathbf{B}\ \boldsymbol{\Phi}(1)\mathbf{B}\ \cdots\ \boldsymbol{\Phi}(L-1)\mathbf{B}\right]
$$

$$
=\operatorname{rank}\left[(\mathbf{1}_p-\mathbf{A})^{-1}\mathbf{B}\ (\mathbf{1}_p-\mathbf{A})^{-2}\mathbf{B}\ \cdots\ (\mathbf{1}_p-\mathbf{A})^{-L}\mathbf{B}\right]. \tag{7.94}
$$

$\square$

7.3.2 *LTI FOS observability*

The definition of the observable state of the FO system described by the SSE is given. To simplify considerations one assumes an initial time $k_0 = 0$ that does not influence its generalities. The definitions given below are similar to those based on the observability notion defined for classical IO systems. But there is a dissimilarity in relation to the observability definition specified for dynamic systems in which there is only one initial condition (initial state). In this sense, the definitions given here are the generalization of the concept of the observability. One considers the FOSSE of the form

$$_0^{GL}\Delta_k^{(\nu)}\mathbf{x}(k) = \mathbf{A}\mathbf{x}(k) + \mathbf{B}\mathbf{u}(k) \tag{7.95}$$

$$\mathbf{y}(k) = \mathbf{C}\mathbf{x}(k) + \mathbf{D}\mathbf{u}(k) \tag{7.96}$$

where $\mathbf{A} \in \mathbb{R}^{p\times s}$, $\mathbf{B} \in \mathbb{R}^{p\times q}$, $\mathbf{C} \in \mathbb{R}^{r\times p}$ and $\mathbf{D} \in \mathbb{R}^{r\times p}$. Vectors $\mathbf{x} \in \mathbb{R}^{p\times 1}$, $\mathbf{u} \in \mathbb{R}^{q\times 1}$ and $\mathbf{y} \in \mathbb{R}^{r\times 1}$ are state, input and output vectors, respectively. The FO belongs to an interval $(0, 1)$. One also assumes that the finite series of input and output signals $\mathbf{u}(0), \mathbf{u}(1), \ldots, \mathbf{u}(L)$ and $\mathbf{u}(0), \mathbf{u}(1), \ldots, \mathbf{u}(L)$ are known. On the basis of the mentioned quantities one should evaluate initial states $\mathbf{x}_{-1}, \mathbf{x}_{-2}, \mathbf{x}_{-3}, \ldots$ also realize that in the considered system, there is an infinite number of initial conditions.

Definition 7.4. An initial state $\mathbf{x}_{-1}$ is called an observable one in (given) L steps if it can be determined from an observed finite output signal sequence $\{\mathbf{y}(0), \mathbf{y}(1), \ldots, \mathbf{y}(k_{L-1})\}$ and known input signal sequence $\{\mathbf{u}(0), \mathbf{u}(1), \ldots, \mathbf{u}(k_{L-1})\}$.

Note, that as an input sequence one may take zero one $\{\mathbf{u}(0) = \mathbf{u}(1) = \ldots = \mathbf{u}(k_{L-1}) = \mathbf{0}\}$. Then, the output signal sequence will be a homogenous system response.

Definition 7.5. A finite set of initial states $\mathbf{x}_{-i}$, for $i = 1, 2, \ldots, l$ is called an observable one in (given) L steps if it can be determined from an observed finite output signal sequence $\{\mathbf{y}(0), \mathbf{y}(1), \ldots, \mathbf{y}(k_{L-1})\}$ and known input sequence $\{\mathbf{u}(0), \mathbf{u}(1), \ldots, \mathbf{u}(k_{L-1})\}$.

The observability of the set of initial conditions $\mathbf{x}_{-i}$, for $i = 1, 2, \ldots$ is related to the block matrix algebraic equation solvability. From solution equations (7.30), (6.42) one gets a matrix equation

$$
\begin{bmatrix} \mathbf{y}(k) \\ \mathbf{y}(k-1) \\ \vdots \\ \mathbf{y}(1) \\ \mathbf{y}(0) \end{bmatrix} = \begin{bmatrix} \mathbf{C} & \mathbf{0} & \cdots & \mathbf{0} \\ \mathbf{0} & \mathbf{C} & \cdots & \mathbf{0} \\ \vdots & \vdots & & \vdots \\ \mathbf{0} & \mathbf{0} & \cdots & \mathbf{C} \end{bmatrix} \begin{bmatrix} \mathbf{x}(k) \\ \mathbf{x}(k-1) \\ \vdots \\ \mathbf{x}(1) \\ \mathbf{x}(0) \end{bmatrix} + \begin{bmatrix} \mathbf{D} & \mathbf{0} & \cdots & \mathbf{0} \\ \mathbf{0} & \mathbf{D} & \cdots & \mathbf{0} \\ \vdots & \vdots & & \vdots \\ \mathbf{0} & \mathbf{0} & \cdots & \mathbf{D} \end{bmatrix} \begin{bmatrix} \mathbf{u}(k) \\ \mathbf{u}(k-1) \\ \vdots \\ \mathbf{u}(1) \\ \mathbf{u}(0) \end{bmatrix}
$$

$$
= \begin{bmatrix} \mathbf{C} & \mathbf{0} & \cdots & \mathbf{0} \\ \mathbf{0} & \mathbf{C} & \cdots & \mathbf{0} \\ \vdots & \vdots & & \vdots \\ \mathbf{0} & \mathbf{0} & \cdots & \mathbf{C} \end{bmatrix} \begin{bmatrix} \boldsymbol{\Phi}(0) & \boldsymbol{\Phi}(1) & \cdots & \boldsymbol{\Phi}(k-1) & \boldsymbol{\Phi}(k) \\ \mathbf{0}_p & \boldsymbol{\Phi}(0) & \cdots & \boldsymbol{\Phi}(k-2) & \boldsymbol{\Phi}(k-1) \\ \vdots & \vdots & & \vdots & \vdots \\ \mathbf{0}_p & \mathbf{0}_p & \cdots & \boldsymbol{\Phi}(0) & \boldsymbol{\Phi}(1) \\ \mathbf{0}_p & \mathbf{0}_p & \cdots & \mathbf{0}_p & \boldsymbol{\Phi}(0) \end{bmatrix} \begin{bmatrix} \mathbf{B} & \mathbf{0} & \cdots & \mathbf{0} \\ \mathbf{0} & \mathbf{B} & \cdots & \mathbf{0} \\ \vdots & \vdots & & \vdots \\ \mathbf{0} & \mathbf{0} & \cdots & \mathbf{B} \end{bmatrix} \begin{bmatrix} \mathbf{u}(k) \\ \mathbf{u}(k-1) \\ \vdots \\ \mathbf{u}(1) \\ \mathbf{u}(0) \end{bmatrix}
$$

$$
- \begin{bmatrix} \mathbf{C} & \mathbf{0} & \cdots & \mathbf{0} \\ \mathbf{0} & \mathbf{C} & \cdots & \mathbf{0} \\ \vdots & \vdots & & \vdots \\ \mathbf{0} & \mathbf{0} & \cdots & \mathbf{C} \end{bmatrix} \begin{bmatrix} \boldsymbol{\Phi}(0) & \boldsymbol{\Phi}(1) & \cdots & \boldsymbol{\Phi}(k-1) & \boldsymbol{\Phi}(k) \\ \mathbf{0}_p & \boldsymbol{\Phi}(0) & \cdots & \boldsymbol{\Phi}(k-2) & \boldsymbol{\Phi}(k-1) \\ \vdots & \vdots & & \vdots & \vdots \\ \mathbf{0}_p & \mathbf{0}_p & \cdots & \boldsymbol{\Phi}(0) & \boldsymbol{\Phi}(1) \\ \mathbf{0}_p & \mathbf{0}_p & \cdots & \mathbf{0}_p & \boldsymbol{\Phi}(0) \end{bmatrix}
$$

$$
\times \begin{bmatrix} a^{(\nu)}(k+1)\mathbf{1}_p & a^{(\nu)}(k+2)\mathbf{1}_p & \cdots \\ a^{(\nu)}(k)\mathbf{1}_p & a^{(\nu)}(k+1)\mathbf{1}_p & \cdots \\ \vdots & \vdots & \vdots \\ a^{(\nu)}(2)\mathbf{1}_p & a^{(\nu)}(3)\mathbf{1}_p & \cdots \\ a^{(\nu)}(1)\mathbf{1}_p & a^{(\nu)}(2)\mathbf{1}_p & \cdots \end{bmatrix} \begin{bmatrix} \mathbf{x}_{-1} \\ \mathbf{x}_{-2} \\ \mathbf{x}_{-3} \\ \vdots \end{bmatrix}.
$$

$$\tag{7.97}$$

The following theorems formulate the linear discrete time-invariant systems described by FOSSE observability conditions.

Theorem 7.3. *The initial states* $\mathbf{x}_{-i}$, *for* $i = 1, 2, \ldots, l$ *of the FO system described by the FOSSE (7.95) and (7.96) are observable in L steps if and only if an observability matrix*

$$
\mathcal{O}_I(L, l) = \begin{bmatrix} \mathbf{O}_t(L-l, l) \\ \mathbf{O}_b(l, l) \end{bmatrix} \tag{7.98}
$$

is of full column rank. Here

$$
\mathbf{O}_t(L-l, l) = \begin{bmatrix} \mathbf{o}_1 & \mathbf{o}_2 & \cdots & \mathbf{o}_l \end{bmatrix} \tag{7.99}
$$

with

$$
\mathbf{o}_i = \begin{bmatrix} \sum_{j=l-i+1}^{L-i} f_{1,i,j} \mathbf{C}\boldsymbol{\Phi}(j) \\ \vdots \\ \sum_{j=l-i+1}^{L-i} f_{L-l,i,j} \mathbf{C}\boldsymbol{\Phi}(j) \end{bmatrix} \quad \text{for } i = 1, 2, \ldots, l \tag{7.100}
$$

$f_{k,i,j}$ — are coefficients and

$$\mathbf{O}_b(l,l) = \begin{bmatrix} \mathbf{C}\boldsymbol{\Phi}(l-1) & \mathbf{C}\boldsymbol{\Phi}(l-2) & \cdots & \mathbf{C}\boldsymbol{\Phi}(1) & \mathbf{C}\boldsymbol{\Phi}(0) \\ \mathbf{C}\boldsymbol{\Phi}(l-2) & \mathbf{C}\boldsymbol{\Phi}(l-3) & \cdots & \mathbf{C}\boldsymbol{\Phi}(0) & \mathbf{0}_p \\ \vdots & \vdots & & \vdots & \vdots \\ \mathbf{C}\boldsymbol{\Phi}(1) & \mathbf{C}\boldsymbol{\Phi}(0) & \cdots & \mathbf{0}_p & \mathbf{0}_p \\ \mathbf{C}\boldsymbol{\Phi}(0) & \mathbf{0}_p & \cdots & \mathbf{0}_p & \mathbf{0}_p \end{bmatrix}. \tag{7.101}$$

Proof. For $k = L - 1$ from (7.97) one gets

$$\begin{bmatrix} \mathbf{y}(L-1) \\ \mathbf{y}(L-2) \\ \vdots \\ \mathbf{y}(1) \\ \mathbf{y}(0) \end{bmatrix} - \begin{bmatrix} \mathbf{D} & \mathbf{0} & \cdots & \mathbf{0} \\ \mathbf{0} & \mathbf{D} & \cdots & \mathbf{0} \\ \vdots & \vdots & & \vdots \\ \mathbf{0} & \mathbf{0} & \cdots & \mathbf{D} \end{bmatrix} \begin{bmatrix} \mathbf{u}(L-1) \\ \mathbf{u}(L-2) \\ \vdots \\ \mathbf{u}(1) \\ \mathbf{u}(0) \end{bmatrix}$$

$$- \begin{bmatrix} \mathbf{C} & \mathbf{0} & \cdots & \mathbf{0} \\ \mathbf{0} & \mathbf{C} & \cdots & \mathbf{0} \\ \vdots & \vdots & & \vdots \\ \mathbf{0} & \mathbf{0} & \cdots & \mathbf{C} \end{bmatrix} \begin{bmatrix} \boldsymbol{\Phi}(0) & \boldsymbol{\Phi}(1) & \cdots & \boldsymbol{\Phi}(L-2) & \boldsymbol{\Phi}(L-1) \\ \mathbf{0}_p & \boldsymbol{\Phi}(0) & \cdots & \boldsymbol{\Phi}(L-3) & \boldsymbol{\Phi}(L-2) \\ \vdots & \vdots & & \vdots & \vdots \\ \mathbf{0}_p & \mathbf{0}_p & \cdots & \boldsymbol{\Phi}(0) & \boldsymbol{\Phi}(1) \\ \mathbf{0}_p & \mathbf{0}_p & \cdots & \mathbf{0}_p & \boldsymbol{\Phi}(0) \end{bmatrix} \begin{bmatrix} \mathbf{B} & \mathbf{0} & \cdots & \mathbf{0} \\ \mathbf{0} & \mathbf{B} & \cdots & \mathbf{0} \\ \vdots & \vdots & & \vdots \\ \mathbf{0} & \mathbf{0} & \cdots & \mathbf{B} \end{bmatrix} \begin{bmatrix} \mathbf{u}(L-1) \\ \mathbf{u}(L-2) \\ \vdots \\ \mathbf{u}(1) \\ \mathbf{u}(0) \end{bmatrix}$$

$$= - \begin{bmatrix} \mathbf{C} & \mathbf{0} & \cdots & \mathbf{0} \\ \mathbf{0} & \mathbf{C} & \cdots & \mathbf{0} \\ \vdots & \vdots & & \vdots \\ \mathbf{0} & \mathbf{0} & \cdots & \mathbf{C} \end{bmatrix} \begin{bmatrix} \boldsymbol{\Phi}(0) & \boldsymbol{\Phi}(1) & \cdots & \boldsymbol{\Phi}(L-2) & \boldsymbol{\Phi}(L-1) \\ \mathbf{0}_p & \boldsymbol{\Phi}(0) & \cdots & \boldsymbol{\Phi}(L-3) & \boldsymbol{\Phi}(L-2) \\ \vdots & \vdots & & \vdots & \vdots \\ \mathbf{0}_p & \mathbf{0}_p & \cdots & \boldsymbol{\Phi}(0) & \boldsymbol{\Phi}(1) \\ \mathbf{0}_p & \mathbf{0}_p & \cdots & \mathbf{0}_p & \boldsymbol{\Phi}(0) \end{bmatrix}$$

$$\times \begin{bmatrix} a^{(\nu)}(L)\mathbf{1}_p & a^{(\nu)}(L+1)\mathbf{1}_p & \cdots & a^{(\nu)}(L+l-1)\mathbf{1}_p \\ a^{(\nu)}(L-1)\mathbf{1}_p & a^{(\nu)}(L)\mathbf{1}_p & \cdots & a^{(\nu)}(L+l-2)\mathbf{1}_p \\ \vdots & \vdots & \vdots & \\ a^{(\nu)}(2)\mathbf{1}_p & a^{(\nu)}(3)\mathbf{1}_p & \cdots & a^{(\nu)}(l+1)\mathbf{1}_p \\ a^{(\nu)}(1)\mathbf{1}_p & a^{(\nu)}(2)\mathbf{1}_p & \cdots & a^{(\nu)}(l)\mathbf{1}_p \end{bmatrix} \begin{bmatrix} \mathbf{x}_{-1} \\ \mathbf{x}_{-2} \\ \vdots \\ \mathbf{x}_{-l} \end{bmatrix}. \tag{7.102}$$

The left hand side of the equation given above is known and will be further denoted as $[\mathbf{y'}^{\mathrm{T}}(L-1),\ \mathbf{y'}^{\mathrm{T}}(L-2),\ \ldots,\ \mathbf{y'}^{\mathrm{T}}(0)]^{\mathrm{T}}$. Hence,

$$
\begin{bmatrix} \mathbf{y}'(L-1) \\ \mathbf{y}'(L-2) \\ \vdots \\ \mathbf{y}'(0) \end{bmatrix} = -
\begin{bmatrix}
\mathbf{C\Phi}(0) & \mathbf{C\Phi}(1) & \cdots & \mathbf{C\Phi}(L-2) & \mathbf{\Phi}(L-1) \\
\mathbf{0}_p & \mathbf{C\Phi}(0) & \cdots & \mathbf{C\Phi}(L-3) & \mathbf{\Phi}(L-2) \\
\vdots & \vdots & & \vdots & \vdots \\
\mathbf{0}_p & \mathbf{0}_p & \cdots & \mathbf{C\Phi}(0) & \mathbf{C\Phi}(1) \\
\mathbf{0}_p & \mathbf{0}_p & \cdots & \mathbf{0}_p & \mathbf{C\Phi}(0)
\end{bmatrix}
$$

$$
\times \begin{bmatrix}
a^{(\nu)}(L)\mathbf{1}_p & a^{(\nu)}(L+1)\mathbf{1}_p & \cdots & a^{(\nu)}(L+l-1)\mathbf{1}_p \\
a^{(\nu)}(L-1)\mathbf{1}_p & a^{(\nu)}(L)\mathbf{1}_p & \cdots & a^{(\nu)}(L+l-2)\mathbf{1}_p \\
\vdots & \vdots & \vdots & \\
a^{(\nu)}(2)\mathbf{1}_p & a^{(\nu)}(3)\mathbf{1}_p & \cdots & a^{(\nu)}(l+1)\mathbf{1}_p \\
a^{(\nu)}(1)\mathbf{1}_p & a^{(\nu)}(2)\mathbf{1}_p & \cdots & a^{(\nu)}(l)\mathbf{1}_p
\end{bmatrix}
\begin{bmatrix} \mathbf{x}_{-1} \\ \mathbf{x}_{-2} \\ \vdots \\ \mathbf{x}_{-l} \end{bmatrix}. \qquad (7.103)
$$

Now, one investigates a second block matrix on the right hand side of (7.102). When performing elementary block matrix operations on block columns, one gets

$$
\begin{bmatrix}
a^{(\nu)}(L)\mathbf{1}_p & a^{(\nu)}(L+1)\mathbf{1}_p & \cdots & a^{(\nu)}(L+l-1)\mathbf{1}_p \\
a^{(\nu)}(L-1)\mathbf{1}_p & a^{(\nu)}(L)\mathbf{1}_p & \cdots & a^{(\nu)}(L+l-2)\mathbf{1}_p \\
\vdots & \vdots & \vdots & \\
a^{(\nu)}(2)\mathbf{1}_p & a^{(\nu)}(3)\mathbf{1}_p & \cdots & a^{(\nu)}(l+1)\mathbf{1}_p \\
a^{(\nu)}(1)\mathbf{1}_p & a^{(\nu)}(2)\mathbf{1}_p & \cdots & a^{(\nu)}(l)\mathbf{1}_p
\end{bmatrix} \mathbf{E}_c
$$

$$
= \begin{bmatrix}
f(L)\mathbf{1}_p & f(L+1)\mathbf{1}_p & \cdots & f(L+l-1)\mathbf{1}_p \\
\vdots & \vdots & & \vdots \\
f(L-1)\mathbf{1}_p & f(L)\mathbf{1}_p & \cdots & f(L+l-2)\mathbf{1}_p \\
\mathbf{0}_p & \mathbf{0}_p & \cdots & \mathbf{1}_p \\
\vdots & \vdots & & \vdots \\
\mathbf{1}_p & \mathbf{0}_p & \cdots & \mathbf{0}_p
\end{bmatrix}. \qquad (7.104)
$$

Taking into account the last equality, the solution (7.102) takes the form

$$
\begin{bmatrix} \mathbf{y}'(L-1) \\ \mathbf{y}'(L-2) \\ \vdots \\ \mathbf{y}'(0) \end{bmatrix} = -
\begin{bmatrix}
\mathbf{C\Phi}(0) & \mathbf{C\Phi}(1) & \cdots & \mathbf{C\Phi}(L-2) & \mathbf{\Phi}(L-1) \\
\mathbf{0}_p & \mathbf{C\Phi}(0) & \cdots & \mathbf{C\Phi}(L-3) & \mathbf{\Phi}(L-2) \\
\vdots & \vdots & & \vdots & \vdots \\
\mathbf{0}_p & \mathbf{0}_p & \cdots & \mathbf{C\Phi}(0) & \mathbf{C\Phi}(1) \\
\mathbf{0}_p & \mathbf{0}_p & \cdots & \mathbf{0}_p & \mathbf{C\Phi}(0)
\end{bmatrix}
$$

$$
\times \begin{bmatrix}
f(L)\mathbf{1}_p & f(L+1)\mathbf{1}_p & \cdots & f(L+l-1)\mathbf{1}_p \\
\vdots & \vdots & & \vdots \\
f(L-1)\mathbf{1}_p & f(L)\mathbf{1}_p & \cdots & f(L+l-2)\mathbf{1}_p \\
\mathbf{0}_p & \mathbf{0}_p & \cdots & \mathbf{1}_p \\
\vdots & \vdots & & \vdots \\
\mathbf{1}_p & \mathbf{0}_p & \cdots & \mathbf{0}_p
\end{bmatrix}
\begin{bmatrix} \mathbf{x}'_{-1} \\ \mathbf{x}'_{-2} \\ \vdots \\ \mathbf{x}'_{-l} \end{bmatrix}. \qquad (7.105)
$$

with

$$
\begin{bmatrix} \mathbf{x}'_{-1} \\ \mathbf{x}'_{-2} \\ \vdots \\ \mathbf{x}'_{-l} \end{bmatrix} = \mathbf{E}_c^{-1} \begin{bmatrix} \mathbf{x}_{-1} \\ \mathbf{x}_{-2} \\ \vdots \\ \mathbf{x}_{-l} \end{bmatrix}. \tag{7.106}
$$

The matrix $\mathbf{E}_c$ represents elementary column operations performed in (7.103). It is always non-singular. Multiplications of both matrices in (7.104) give the observability matrix (7.98)

$$
\begin{bmatrix} \mathbf{y}'(L-1) \\ \mathbf{y}'(L-2) \\ \vdots \\ \mathbf{y}'(0) \end{bmatrix} = \mathcal{O}(L,l) \begin{bmatrix} \mathbf{x}'_{-1} \\ \mathbf{x}'_{-2} \\ \vdots \\ \mathbf{x}'_{-l} \end{bmatrix}. \tag{7.107}
$$

The solvability conditions of (7.104) yield the observability of initial states. □

Comment. For $L = l$ the observability matrix takes the form

$$
\mathcal{O}_I(L,L) = \mathbf{O}_b(L,L). \tag{7.108}
$$

Comment. For $l = 1$ (when only one initial state is to be observed with an assumption that all remaining initial conditions are zero) the observability matrix takes the form

$$
\mathcal{O}_I(L,1) = \begin{bmatrix} \mathbf{O}_t(L-1,1) \\ \mathbf{O}_b(1,1) \end{bmatrix} = \begin{bmatrix} \mathbf{o}_1 \\ \mathbf{C\Phi}(0) \end{bmatrix}. \tag{7.109}
$$

In the considered case equation (7.107) is as follows

$$
\begin{bmatrix} \mathbf{y}'(L-1) \\ \mathbf{y}'(L-2) \\ \vdots \\ \mathbf{y}'(0) \end{bmatrix} = \begin{bmatrix} \mathbf{o}_1 \\ \mathbf{C\Phi}(0) \end{bmatrix} \begin{bmatrix} \mathbf{x}'_{-1} \end{bmatrix}. \tag{7.110}
$$

Now the pre-multiplying of both sides by an elementary row operations matrix $\mathbf{E}_r$ simplifies the considered equation to the form

$$
\begin{bmatrix} \mathbf{y}''(L-1) \\ \mathbf{y}''(L-2) \\ \vdots \\ \mathbf{y}''(0) \end{bmatrix} = \begin{bmatrix} \mathbf{C\Phi}(L-1) \\ \mathbf{C\Phi}(L-2) \\ \vdots \\ \mathbf{C\Phi}(0) \end{bmatrix} \begin{bmatrix} \mathbf{x}'_{-1} \end{bmatrix}. \tag{7.111}
$$

where

$$\begin{bmatrix} \mathbf{y}''(L-1) \\ \mathbf{y}''(L-2) \\ \vdots \\ \mathbf{y}''(0) \end{bmatrix} = \mathbf{E}_r \begin{bmatrix} \mathbf{y}'(L-1) \\ \mathbf{y}'(L-2) \\ \vdots \\ \mathbf{y}'(0) \end{bmatrix}, \quad \begin{bmatrix} \mathbf{C}\boldsymbol{\Phi}(L-1) \\ \mathbf{C}\boldsymbol{\Phi}(L-2) \\ \vdots \\ \mathbf{C}\boldsymbol{\Phi}(0) \end{bmatrix} = \mathbf{E}_r \begin{bmatrix} \mathbf{o}_1 \\ \mathbf{C}\boldsymbol{\Phi}(0) \end{bmatrix} \qquad (7.112)$$

and the initial condition state is observable if and only if the inner block matrix is of full rank.

For the full rank observability matrix with $L > l$ one can apply the pseudo-inverse to calculate a finite elements initial conditions set.

$$\begin{bmatrix} \mathbf{x}'_{-1} \\ \mathbf{x}'_{-2} \\ \vdots \\ \mathbf{x}'_{-l} \end{bmatrix} = \left[\mathcal{O}_I^T(L,l)\mathcal{O}_I(L,l) \right]^{-1} \mathcal{O}_I^T(L,l) \begin{bmatrix} \mathbf{y}'(L-1) \\ \mathbf{y}'(L-2) \\ \vdots \\ \mathbf{y}'(0) \end{bmatrix}. \qquad (7.113)$$

Comment. As it was mentioned earlier, the impact of the chosen initial condition $\mathbf{x}_{-l}$ on the system response strongly diminishes due to the distance between the presence (represented by discrete time instant $k_0 = 0$) and the past measured in a reverse direction $-l$. This is caused by a shape of a function $a^{(\nu)}(k)$. Assuming that all initial states are equal, one can estimate their impact on the system response. The highest will be for the initial state $\mathbf{x}_{-1}$ which is closest to $k_0 = 0$. Considering its weight as 100%, one can plot the percentage rate of remaining consecutive initial conditions. Such plots for different FOs are presented in Fig. 7.54. The plots reveal the number of states giving greater then assumed constant impact to the response (for instance 10%). Such a relation is presented in Fig. 7.55. Evidently, for $\nu = 1$ there is only one initial state needed for a unique homogenous response.

Considerations concerning the observability will be supported by a numerical example.

Example 7.13. Consider the FOSSE with $\nu = 0.75$ and matrices

$$\mathbf{A} = \begin{bmatrix} 1 & 2 & 3 \\ 0 & -\frac{3}{2} & 2 \\ 0 & 0 & -2 \end{bmatrix}, \quad \mathbf{B} = \begin{bmatrix} 12 \\ -10 \\ 3 \end{bmatrix}, \quad \mathbf{C} = \begin{bmatrix} -1 & 1 & -1 \end{bmatrix}, \quad \mathbf{D} = [0]. \qquad (7.114)$$

Having measured the system homogenous response $\mathbf{y}(k)$ for $L = 50$ it is to check the system observability condition and to find three latest initial conditions $\mathbf{x}_{-1}, \mathbf{x}_{-2}$ and $\mathbf{x}_{-3}$.

Solution. To perform the initial state observation one simulates the homogenous system response. For

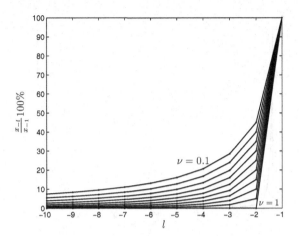

Fig. 7.54   Percentage impact of equal initial states on the FOS response.

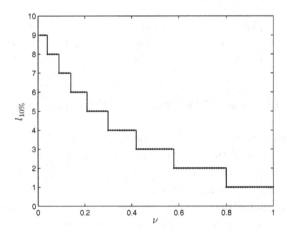

Fig. 7.55   Number of initial conditions of higher impact then 10% vs. FO.

$$\mathbf{x}_{-1} = \begin{bmatrix} 1 \\ 0 \\ 0 \end{bmatrix}, \ \mathbf{x}_{-2} = \begin{bmatrix} 1 \\ 1 \\ 0 \end{bmatrix}, \ \mathbf{x}_{-3} = \begin{bmatrix} 1 \\ 1 \\ 1 \end{bmatrix}, \tag{7.115}$$

one gets the homogenous system response given in Fig. 7.56.

In the considered case $\mathbf{y}'(k) = \mathbf{y}(k)$ for $k = 0, 1, \ldots$. For $L = 9$ the observability matrix has rank 9. The system is observable. Yet for $L = 273$ from (7.113) and (7.106) one gets a satisfactory solution

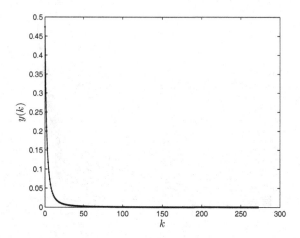

Fig. 7.56   Homogenous system response.

$$\mathbf{x}_{-1} = \begin{bmatrix} 1.0004 \\ 0.0003 \\ 0.0012 \end{bmatrix}, \ \mathbf{x}_{-2} = \begin{bmatrix} 0.9986 \\ 1.0001 \\ 0.0003 \end{bmatrix}, \ \mathbf{x}_{-3} = \begin{bmatrix} 1.0026 \\ 1.0019 \\ 1.0002 \end{bmatrix}. \qquad (7.116)$$

7.4   Linear time-invariant FO system stability analysis

The stability of the dynamic system is the most important notion in system analysis. The stability conditions of the discrete linear time-invariant IO systems are well defined [Ogata (1987); Kaczorek (1992)]. The IO system stability region bounded by a unit circle on a complex z-plane is presented in Fig. 7.57. Following this stability region for FO continuous-time system [Matignon (1996, 1998)] in a natural way one tries to evaluate similar regions for the FO systems [Bonnet and Partington (2000); Chen and Moore (2002a); Hofreiter and Zitek (2003); Wang and Li (2006); Ahmed and Hala (2006); Dzieliński and Sierociuk (2006b); Deng (2007); El-Salam and El-Sayed (2007); Busłowicz (2008); Busłowicz and Kaczorek (2009); Tavazoei *at al.* (2008); Petráš (2009b); Guermah *et al.* (2010); Ostalczyk (2010a); Stanisławski (2013)].

Now consider the FODE of the form

$$_{0}^{GL}\Delta_{k}^{(p\nu)}y(k) + a_{p-1}y(k)_{0}^{GL}\Delta_{k}^{(p-1)\nu)}y(k) + \cdots$$
$$+ a_{1}y(k)_{0}^{GL}\Delta_{k}^{(\nu)}y(k) + a_{0}y(k) = b_{0}u(k). \qquad (7.117)$$

Assuming zero initial conditions $0 = y_{-1} = y_{-2} = \cdots$ one applies the one-sided $\mathcal{Z}$-transform

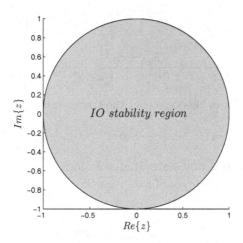

Fig. 7.57   Stability region for linear time-invariant IO system.

$$\left[\left(1-z^{-1}\right)^{p\nu} + a_{p-1}\left(1-z^{-1}\right)^{(p-1)\nu} Y + \cdots + a_0\right] Y(z) = b_0 U(z). \qquad (7.118)$$

where $Y(z), U(z)$ are appropriate transforms. In this algebraic equation $a_i, b_0 \in \mathbb{R}$ for $i = p-1, p-2, \ldots, 0$. Related FOTF has the form

$$
\begin{aligned}
G(z) = \frac{Y(z)}{U(z)} &= \frac{b_0}{\left(1-z^{-1}\right)^{p\nu} + a_{p-1}\left(1-z^{-1}\right)^{(p-1)\nu} + \cdots + a_1\left(1-z^{-1}\right)^{\nu} + a_0} \\
&= \frac{b_0}{\sum_{i=0}^{+\infty} a^{(p\nu)}(i)z^{-i} + a_{p-1}\sum_{i=0}^{+\infty} a^{((p-1)\nu)}(i)z^{-i} + \cdots + a_1\sum_{i=0}^{+\infty} a^{(\nu)}(i)z^{-i} + a_0} \\
&= \lim_{L\to+\infty} \frac{b_0}{\sum_{i=0}^{L} a^{(p\nu)}(i)z^{-i} + \cdots + a_1\sum_{i=0}^{L} a^{(\nu)}(i)z^{-i} + a_0} \\
&= \lim_{L\to+\infty} \frac{b_0 z^{L}}{\sum_{i=0}^{L} a^{(p\nu)}(i)z^{L-i} + \cdots + a_1\sum_{i=0}^{L} a^{(\nu)}(i)z^{L-i} + a_0 z^{L}} \\
&= \lim_{L\to+\infty} \frac{b_0' z^{L}}{\prod_{i=1}^{L}(z - z_i)}.
\end{aligned}
$$

$$(7.119)$$

where $b_0' = \frac{b_0}{1+a_{p-1}+\cdots+a_1+a_0}$ and z_i are the transfer function poles. The considered system is asymptotically stable, if and only if, all solution's absolute values satisfy $|z_i| < 1$ for $i = 1, 2, \ldots$ [Kaczorek (2011)], [Stanisławski (2013)]. This means that all solutions should be within a so-called unit circle in the complex plane (a circle centered at the origin with a radius 1). To verify the system stability one should check a configuration of the solutions. There are plenty of stability criteria [Ogata (1987)] for LDTS. Following the IO procedures on transforms (7.118) to

$$G(z) = \frac{Y(z)}{U(z)}$$

$$= \frac{b_0 z^{p\nu}}{(z-1)^{p\nu} + a_{p-1} z^{\nu} (z-1)^{(p-1)\nu} + \cdots + a_1 z^{(p-1)\nu} (z-1)^{\nu} + z^{p\nu} a_0}. \quad (7.120)$$

A characteristic equation in this case is of the transcendental form

$$(z-1)^{p\nu} + a_{p-1} z^{\nu} (z-1)^{(p-1)\nu} + \cdots + a_1 z^{(p-1)\nu} (z-1)^{\nu} + z^{p\nu} a_0 = 0. \quad (7.121)$$

The FOTF is asymptotically stable if all solutions of (7.121) lie inside the unit circle. The following non-linear coordinate transformation maps the unit circle into another stability contour

$$u = f(z) = \left(1 - z^{-1}\right)^{\nu} - 1, \text{ for } \nu \in (0,1]. \quad (7.122)$$

Under the above transformation the FOTF (7.118) takes a form of the IOTF with a complex variable u

$$G(u) = G\left[u(z)\right] = \frac{Y(u)}{U(u)}$$

$$= \frac{b_0}{(u+1)^p + a_{p-1}(u+1)^{(p-1)} + \cdots + a_1(u+1) + a_0}$$

$$= \frac{b_0}{u^p + \bar{a}_{p-1} u^{p-1} + \cdots + \bar{a}_1 u + \bar{a}_0}$$

$$= \frac{b_0}{\prod_{i=0}^{p} (u - u_i)}. \quad (7.123)$$

where $\bar{a}_i$, $i = 0,1,\ldots,p-1$ are coefficients related to p and a_i. For instantce, $\bar{a}_{p-1} = p + a_{p-1}$ and $\bar{a}_0 = a_0 + a_1 + \cdots + a_{p-1} + 1$ and u_i denotes a single or multiple real or complex pole of $G(u)$. Note that in a case of a presence of the complex u_i there exists its complex conjugate counterpart. The number of poles p by assumption is finite. This fact is fundamental in the FOS stability analysis. Now a short discussion concerning the transformation (7.122) will be presented.

The unit circle in the z-domain which bounds the stability region is described as $z = e^{j\phi}$ where $0 \leq \phi < 2\pi$. Hence,

$$u(\phi) = \left(1 - e^{-j\phi}\right)^{\nu} - 1, \text{ for } 0 \leq \phi < 2\pi. \quad (7.124)$$

The exterior of the unit circle defined by a closed positively oriented contour C defined as

$$C = \bigcup_{i=1}^{4} C_i \quad (7.125)$$

where

$$\mathcal{C}_1 = \left\{ z : z = e^{j\omega_i} \text{ for } \omega_i \in \left[\pi + \arcsin\left(\frac{\delta}{1}\right), -\pi - \arcsin\left(\frac{\delta}{1}\right)\right) \right\}$$

$$\mathcal{C}_2 = \left\{ z : z = \omega_i - j\delta \text{ for } \omega_i \in \left[-\sqrt{1 - \delta^2}, -\sqrt{R^2 - \delta^2} \right\} \right.$$

$$\mathcal{C}_3 = \left\{ z : z = Re^{j\omega_i} \text{ for } \omega_i \in \left[\pi - \arcsin\left(\frac{\delta}{R}\right), -\pi + \arcsin\left(\frac{\delta}{R}\right)\right) \right\}$$

$$\mathcal{C}_4 = \left\{ z : z = \omega_i + j\delta \text{ for } \omega_i \in \left[-\sqrt{R^2 - \delta^2}, -\sqrt{1 - \delta^2} \right\}. \right. \tag{7.126}$$

is presented in Fig. 7.58. The transformed domain $u(\mathcal{D})$ is given in Fig. 7.59.

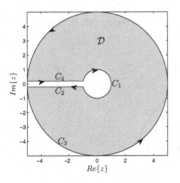

Fig. 7.58   Exterior of the unit circle.

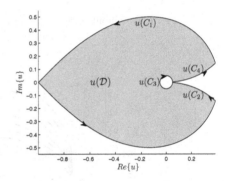

Fig. 7.59 Unit circle transformed by (7.121).

For $\delta \to 0$ and $R \to +\infty$ the FOS, stability region reshapes to that presented in Fig. 7.60 in gray color. This means that the interior of the unit circle is transformed to the gray region in question. The FOS stability regions edges for different orders from a set $\{0.1, 0.2, \ldots, 1.0\}$ are plotted in Fig. 7.61. Note that for $\nu \to 1$ the edge tends to the unit circle. This fact confirms the unity of the FOSs with the IOSs one.

Theorem 7.4. *The LTI FOS is asymptotically stable if and only if all IOTF (7.123) poles are outside the white region in Fig. 7.60.*

Proof. The transformation of the unit circle elements $z \in (C)_1$ by (7.122) can be expressed as

$$z = e^{j\omega_i} = \cos(\omega_i) + j\sin(\omega_i) \text{ for } \omega_i \in [\pi, -\pi). \tag{7.127}$$

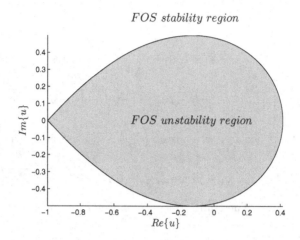

Fig. 7.60   Stability region for linear time-invariant FO system with $\nu = 0.5$.

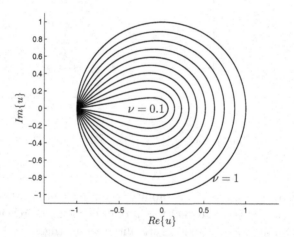

Fig. 7.61   Stability regions edges for different orders for $\{0.1, 0.2, \ldots, 1.0\}$.

Then,

$$
\begin{aligned}
u(\omega_i) = u_R(\omega_i) + j u(\omega_i)_I &= [1 - \cos(\omega_i) + j\sin(\omega_i)]^{\nu} - 1 \\
&= \left[2\cos\left(\frac{\omega_i}{2}\right)\right]^{\nu} e^{j\nu \arctan\left(\tan\frac{\omega_i}{2}\right)} - 1 \\
&= \left[2\cos\left(\frac{\omega_i}{2}\right)\right]^{\nu} \cos\left(\frac{\nu(\pi - \omega_i)}{2}\right) - 1 \\
&\quad + j\left[2\cos\left(\frac{\omega_i}{2}\right)\right]^{\nu} \left[\sin\left(\frac{\nu(\pi - \omega_i)}{2}\right)\right].
\end{aligned} \tag{7.128}
$$

One can easily verify that for $\omega_i \in [\pi, -\pi)$

$$u(\pi) = -1$$

$$u\left(\frac{\pi}{2}\right) = 2^{\frac{\nu}{2}} \cos \frac{\nu\pi}{4} - 1 + j2^{\frac{\nu}{2}} \sin \frac{\nu\pi}{4}$$

$$u(0) = 2^{\nu} \cos \frac{\nu\pi}{2} - 1 + j2^{\nu} \sin \frac{\nu\pi}{2}$$

$$u\left(-\frac{\pi}{2}\right) = 2^{\frac{\nu}{2}} \cos \frac{\nu 3\pi}{4} - 1 + j2^{\frac{\nu}{2}} \sin \frac{\nu 3\pi}{4}. \tag{7.129}$$

Poles z_i of the FOTF (7.119) are transformed to p poles of IOTF (7.123). Conversely

$$z = \frac{1}{1 - (u+1)^{\frac{1}{\nu}}}. \tag{7.130}$$

For any $u = u_R + ju_I$ one gets

$$z(u) = z_R(u) + jz_I(u) = \frac{1 + A\cos\alpha}{1 - 2A\cos\alpha + A^2} - j\frac{A\sin\alpha}{1 - 2A\cos\alpha + A^2} \tag{7.131}$$

where

$$A = \left[(u_R - 1)^2 + u_I^2\right]^{\frac{1}{2\nu}}, \quad \alpha = \frac{\arctan \frac{u_I}{u_R - 1}}{\nu} \tag{7.132}$$

and for the assumed range of ω_i, this map is a bijection.   □

For a better insight into the coordinate transformation (7.122), selected curves in the z-plane are transformed into a u-plane for two FOs. The unit circle in the z-plane and its transformation will be denoted as $\mathcal{U}$ and $u(\mathcal{U})$, respectively. One defines an interval

$$\mathcal{C} = \{z : z = z_R, z_R \in (0,1]\}. \tag{7.133}$$

The transformed intervals $\mathcal{C}$ for $\nu = 1/2$ and $\nu = 1/3$ are presented in Figs. 7.62 and 7.63, respectively.

The intervals $\mathcal{C} = \{z : z = z_R, z_R \in [1,2]\}$ and appropriate transformations for $\nu = 1/2$ and $\nu = 1/3$ are given in Figs. 7.64 and 7.65, respectively. The transform of intervals

$$\mathcal{C} = \{z : z = jz_I, z_I \in (z_b, z_e]\}, \tag{7.134}$$

for $z_b = (0,1]$ and $z_e = [1,2]$ are shown in Figs. 7.66–7.69, respectively.

Finally, transformation (7.122) will be applied to some regions in the z-plane (in gray). The results (in black) are presented in Figs. 7.70–7.75, respectively.

A stable region (inside the unit circle) is transformed to a closed region outside the region enclosed by a curve $u(\mathcal{U})$. The region crossing the unit circle is transformed to the black area belonging to the stability and instability area.

Theorem 7.5. *If the IOS is asymptotically stable then the FOS with the same parameters is also asymptotically stable.*

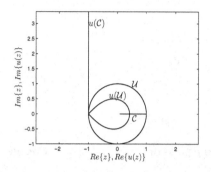

Fig. 7.62   Graphical representations of sets: $\mathcal{U}, u(\mathcal{U}), \mathcal{C}, u(\mathcal{C})$ for $\nu = 1/2$.

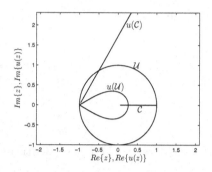

Fig. 7.63   Graphical representations of sets: $\mathcal{U}, u(\mathcal{U}), \mathcal{C}, u(\mathcal{C})$ for $\nu = 1/3$.

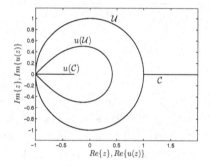

Fig. 7.64   Graphical representations of sets: $\mathcal{U}, u(\mathcal{U}), \mathcal{C}, u(\mathcal{C})$ for $\nu = 1/2$.

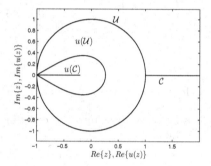

Fig. 7.65   Graphical representations of sets: $\mathcal{U}, u(\mathcal{U}), \mathcal{C}, u(\mathcal{C})$ for $\nu = 1/3$.

Proof. The theorem is a consequence of the fact that the instability regions are included in each other $u\left[\mathcal{U}(\nu_1)\right] \in u\left[\mathcal{U}(\nu_2)\right]$ for $\nu_1 < \nu_2$. The biggest is evaluated for $\nu = 1$. Hence, an asymptotic stability of the IOS guarantees an asymptotic stability of the FOS. $\square$

Theorem 7.6. *If all u_i of the IOS are real satisfying inequality*

$$u_i \in (-\infty, -1) \cup (2^\nu - 1, +\infty) \tag{7.135}$$

then the related FOS is also asymptotically stable.

Proof. For $z = z_R = -1$ one gets $u(z_R) = u(-1) = 2^\nu$-1. The system is asymptotically stable if $u_i > 2^\nu - 1$. On the negative real axis the stability is guaranteed for $u_1 < -1$. $\square$

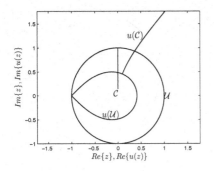

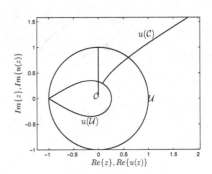

Fig. 7.66 Graphical representations of sets: $\mathcal{U}, u(\mathcal{U}), \mathcal{C}, u(\mathcal{C})$ for $\nu = 1/2$.

Fig. 7.67 Graphical representations of sets: $\mathcal{U}, u(\mathcal{U}), \mathcal{C}, u(\mathcal{C})$ for $\nu = 1/3$.

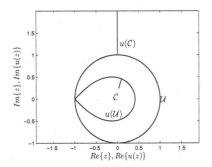

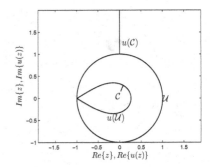

Fig. 7.68 Graphical representations of sets: $\mathcal{U}, u(\mathcal{U}), \mathcal{C}, u(\mathcal{C})$ for $\nu = 1/2$.

Fig. 7.69 Graphical representations of sets: $\mathcal{U}, u(\mathcal{U}), \mathcal{C}, u(\mathcal{C})$ for $\nu = 1/3$.

The results of the above theorems are confirmed by numerical examples.

Example 7.14. Check the stabilty of the FOS described by the FOTF

$$G(z) = \frac{b_1 \left(1 - z^{-1}\right)^{2\nu} + b_0}{\left(1 - z^{-1}\right)^{3\nu} + a_2 \left(1 - z^{-1}\right)^{2\nu} + a_1 \left(1 - z^{-1}\right)^{\nu} + a_0}. \tag{7.136}$$

Verify its stability conditions.

Solution. An application of the coordinate transformation (7.122) to (7.136) leads to the form

$$
\begin{aligned}
G(u) = G\left(u(z)\right) &= \frac{b_1(u+1) + b_0}{(u+1)^3 + a_2(u+1)^2 + a_1(u+1) + a_0} \\
&= \frac{b_1\left(u - u_{z,1}\right)}{\prod_{i=1}^{3}\left(u - u_{p,i}\right)}
\end{aligned}
\tag{7.137}
$$

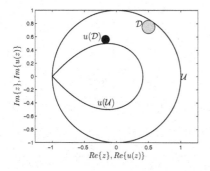

Fig. 7.70   Graphical representations of sets: $\mathcal{U}, u(\mathcal{U}), \mathcal{D}, u(\mathcal{D})$ for $\nu = 1/2$.

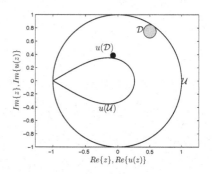

Fig. 7.71   Graphical representations of sets: $\mathcal{U}, u(\mathcal{U}), \mathcal{D}, u(\mathcal{D})$ for $\nu = 1/3$.

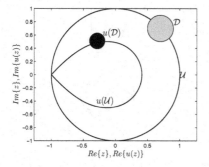

Fig. 7.72   Graphical representations of sets: $\mathcal{U}, u(\mathcal{U}), \mathcal{D}, u(\mathcal{D})$ for $\nu = 1/2$.

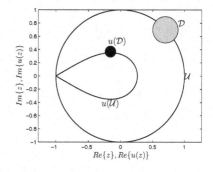

Fig. 7.73   Graphical representations of sets: $\mathcal{U}, u(\mathcal{U}), \mathcal{D}, u(\mathcal{D})$ for $\nu = 1/3$.

where

$$u_{z,1} = -\frac{b_1 + b_0}{b_1}$$

$$u_{p,1} + u_{p,2} + u_{p,3} = 3 + a_2$$

$$u_{p,1}u_{p,2} + u_{p,1}u_{p,3} + u_{p,2}u_{p,3} = 3 + 2a_2 + a_1$$

$$u_{p,1}u_{p,2}u_{p,3} = -1 - a_2 - a_1 - a_0. \tag{7.138}$$

Solving the above equations for $u_{p,1}, u_{p,2}, u_{p,3}$ one checks its configuration relative to the area bounded by $u(\mathcal{U})$. For all $u_{p,1}, u_{p,2}, u_{p,3}$ outside $u(\mathcal{U})$, the considered system is asymptotically stable. For at least one $u_{p,i}$ on the closed curve $u(\mathcal{U})$ that remains stable, the system is on the stability limit. Below, the selected pole configurations and related unit step responses are plotted.

For $b_1 = 1$, $b_0 = -3.3052$ and $a_2 = -4.2955$, $a_1 = 6.3483$, $a_0 = -3.3052$ one gets poles $u_{p,1} = -0.5$, $u_{p,2} = -0.2 + j0.7$, $u_{p,3} = -0.2 - j0.7$. The pole

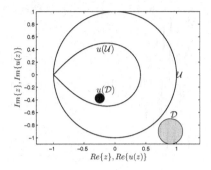

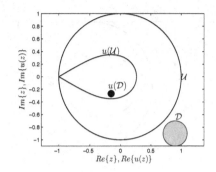

Fig. 7.74 Graphical representations of sets: $\mathcal{U}, u(\mathcal{U}), \mathcal{D}, u(\mathcal{D})$ for $\nu = 1/2$.

Fig. 7.75 Graphical representations of sets: $\mathcal{U}, u(\mathcal{U}), \mathcal{D}, u(\mathcal{D})$ for $\nu = 1/3$.

configuration on a background of the curve $u(\mathcal{U})$ and the unit circle is plotted in Fig. 7.82. A related step response is given in Fig. 7.77.

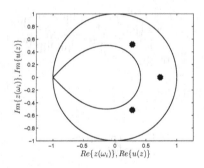

Fig. 7.76 Pole configuration for $a_2 = -4.2955$, $a_1 = 6.3483$, $a_0 = -3.3052$ on the background of $\mathcal{U}$ and $u(\mathcal{U})$ for $\nu = 1/2$.

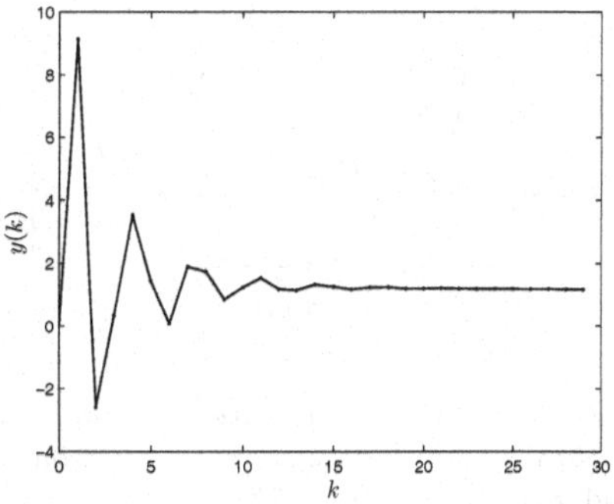

Fig. 7.77 Unit step response related to the configuration on the left.

For $b_1 = 1$, $b_0 = 0.4745$ and $a_2 = -2.886$, $a_1 = 1.7555$, $a_0 = 0.4745$ one gets poles $u_{p,1} = 0.6330$, $u_{p,2} = 0.4530$, $u_{p,3} = -1.2$. The pole configuration in the beckground of the curve $u(\mathcal{U})$ and the unit circle is plotted in Fig. 7.78. A related step response is given in Fig. 7.79.

For $b_1 = 1$, $b_0 = -0.8073$ and $a_2 = -2.5735$, $a_1 = 2.2103$, $a_0 = -0.8073$ one gets poles $u_{p,1} = 0.4142$, $u_{p,2} = -0.4204 + j0.4846$, $u_{p,3} = -0.4204 - j0.4846$. The pole configuration in the beckground of the curve $u(\mathcal{U})$ and the unit circle is plotted in Fig. 7.80. Related step response is given in Fig. 7.81.

For $b_1 = 1$, $b_0 = -2.3966$ and $a_2 = -3.9164$, $a_1 = 5.2333$, $a_0 = -2.3966$ one gets poles $u_{p,1} = -1$, $u_{p,2} = -0.4359 + j0.9$, $u_{p,3} = -0.4359 - j0.9$. The pole

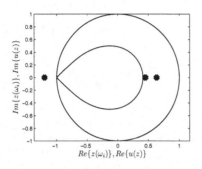

Fig. 7.78 Pole configuration for $a_2 = -2.886$, $a_1 = 1.7555$, $a_0 = 0.4745$ on the background of $\mathcal{U}$ and $u(\mathcal{U})$ for $\nu = 1/2$.

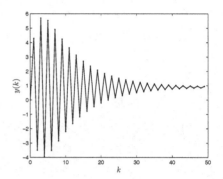

Fig. 7.79 Unit step response related to the configuration on the left.

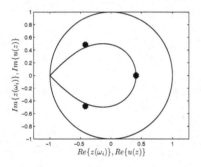

Fig. 7.80 Pole configuration for $a_2 = -2.5735$, $a_1 = 2.2103$, $a_0 = -0.8073$ on $\mathcal{U}$ and $u(\mathcal{U})$ for $\nu = 1/2$.

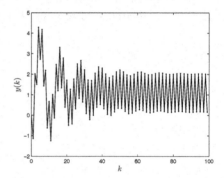

Fig. 7.81 Unit step response related to the configuration on the left.

configuration in the beckground of the curve $u(\mathcal{U})$ and the unit circle is plotted in Fig. 7.82. Related step response is given in Fig. 7.83.

For $b_1 = 1$, $b_0 = -0.9102$ and $a_2 = -2.7289$, $a_1 = 2.4924$, $a_0 = -0.9102$ one gets poles $u_{p,1} = 4327$, $u_{p,2} = -0.3519 + j0.4639$, $u_{p,3} = -0.3519 - j0; 4639$. The pole configuration on in the beckground of the curve $u(\mathcal{U})$ and the unit circle is plotted in Fig. 7.84. A related unstable step response is given in Fig. 7.85.

Now, one states the result known in the stability theorem as the Mikhailov stability criterion [Busłowicz (2008)]. In the criterion, one examines a configuration of the roots of a polynomial with real coefficients. The Mikhailov's criterion is one of the frequency criteria for the stability of the LCTIS. Though initially invented to the LDTIS, it can be generalized to the FO LDTIS. Consider a simple closed curve in a complex u-plane traversed once in an anti-clockwise direction. As such a curve

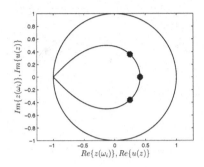

Fig. 7.82 Pole configuration for $a_2 = -3.9164$, $a_1 = 5.2333$, $a_0 = -2.3966$ on the back-ground of $\mathcal{U}$ and $u(\mathcal{U})$ for $\nu = 1/2$.

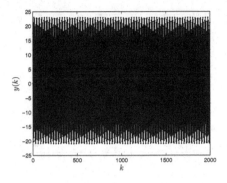

Fig. 7.83 Unit step response related to the configuration on the left.

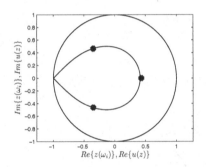

Fig. 7.84 Pole configuration with $u_{p,1} = 4327$, $u_{p,2} = -0.3519 + j0.4639$, $u_{p,3} = -0.3519 - j0;4639$ on the background of $\mathcal{U}$ and $u(\mathcal{U})$ for $\nu = 1/2$.

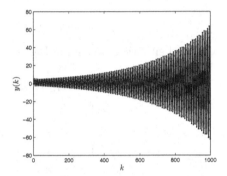

Fig. 7.85 Unit step response related to the configuration on the left.

one can take (7.124). Then, if u_i lies in a stability region presented in Fig. 7.60 then

$$\angle_{0\leq\phi\leq 2\pi}\{u(\phi) - u_i\} = 0 \text{ for } u_i \in \mathcal{D}. \tag{7.139}$$

For u_i inside an instability region then

$$\angle_{0\leq\phi\leq 2\pi}\{u(\phi) - u_i\} = 2\pi \text{ for } u_i \in \mathcal{CD}. \tag{7.140}$$

The discussed cases are presented in Figs. 7.86 and 7.87, respectively.

To emphasize that in considered systems for a complex root there exists its complex conjugate counterpart in both figures, additional roots are indicated. A case when u_i lies sharply on the curve is slightly more complicated. A vector indicating an argument increase $u(\omega_i) - u_i$ instantly changes its direction when passing through a position of u_i. This causes the lack of increae of π. Hence in a

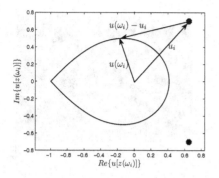

Fig. 7.86 Argument increase for stable pole.

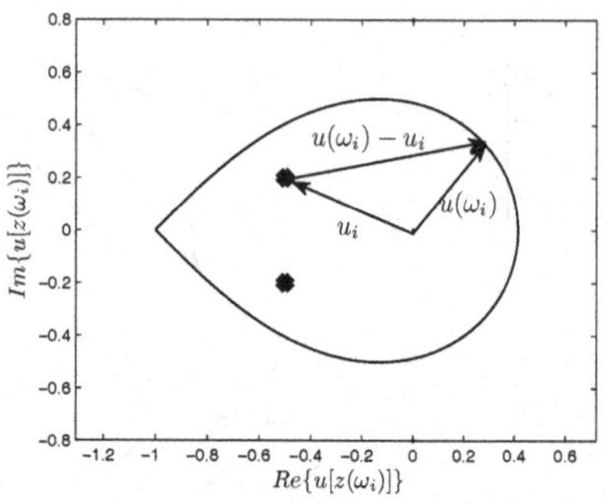

Fig. 7.87 Argument increase for unstable pole.

considered case

$$\angle_{0 \leq \phi \leq 2\pi} \{u(\phi) - u_i\} = \pi \text{ for } u_i \in u(\mathcal{U}). \tag{7.141}$$

A related situation is given in Fig. 7.88.

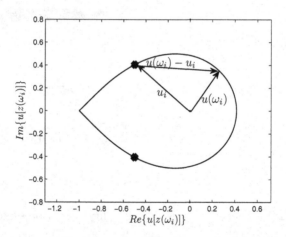

Fig. 7.88 Argument increase for u_i on the edge of stability region.

Denoting now a denominator of the transformed FOTF (7.123) as

$$D(u) = \prod_{i=1}^{p} (u - u_i) \tag{7.142}$$

it is obvious that

$$\angle_{0 \leq \phi \leq 2\pi} D\left[u(\phi)\right] = \angle_{0 \leq \phi \leq 2\pi} \left[\prod_{i=1}^{p} (u - u_i) \right] = 0 p_s + 2\pi p_{us} + \pi p_{sl} \tag{7.143}$$

where

$$p = p_s + p_{us} + p_{sl} \qquad (7.144)$$

and

p – total number of poles

p_s – number of stable poles

p_{us} – number of unstable poles

p_{sl} – number of poles on a stability limit.

In light of the above analysis, one can formulate a following stability criterion.

Theorem 7.7. *The denominator polynomial $D(u)$ is asymptotically stable, if and only if,*

$$\angle_{0 \le \phi \le 2\pi} D\left[u(\phi)\right] = 0. \qquad (7.145)$$

The complex valued function $D\left[u(\phi)\right]$ is known as the Mikhailov hodograph. An increase of 2π is equivalent to one encirclement in a anticlockwise direction of the complex plane origin by the Mikhailov hodograph. Hence, theorem 7.7 can be reformulated in the form.

Theorem 7.8. *The denominator polynomial $D(u)$ is asymptotically stable if and only if the Mikhailov hodograph does not encircle the complex plane origin.*

In the numerical example, one checks the stability of the FOS using the theorems given above.

Example 7.15. Consider a FOTF of the form

$$G(z) = \frac{a_0}{(1 - z^{-1})^{2\nu} + a_1(1 - z^{-1})^{\nu} + a_0}$$

$$= \frac{a_0}{u^2 + (2 + a_1)u + (1 + a_1 + a_0)} = \frac{a_0}{(u - u_1)(u - u_2)}. \qquad (7.146)$$

Check the system stability due to u_1, u_2 configuration.

Solution. For selected u_1, u_2 configurations, the related Mikhailov hodograph will be given. On the base of its shape, the system stability will be stated. Finally, the considered system unit step response will be presented. By a small cross $+$ the complex plane origin is denoted. The poles are represented by x. All considered cases concern the FO $\nu = 0.5$.

For

$$u_1 = -0.1 + j0.48586826605, \ u_2 = -0.1 - j0.48586826605$$

$$a_1 = -1.8, \ a_0 = 1.046067971954434. \qquad (7.147)$$

in Figs. 7.89 and 7.90 the u_1, u_2 configuration and the Mikhailov hodograph are presented, respectively.

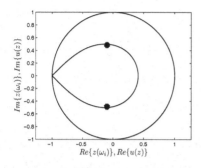

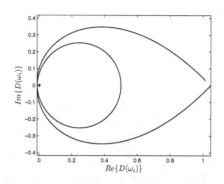

Fig. 7.89   Poles u_1, u_2 configuration.

Fig. 7.90   The Mikhailov hodograph for 7.89.

The Mikhailov hodograph encircles the origin twice. The system is unstable. This statement confirms the unit step responses presented in Fig. 7.91.

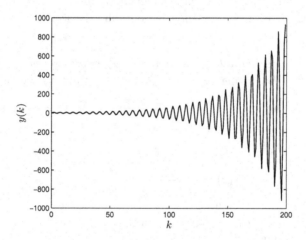

Fig. 7.91   Unit step response of the FOTF with 7.89.

For

$$u_1 = 0.6, \quad u_2 = -0.9$$
$$a_1 = -1.7, \quad a_0 = 1.6. \tag{7.148}$$

in Figs. 7.92 and 7.93 the u_1, u_2 configuration and the Mikhailov hodograph are presented, respectively.

The Mikhailov hodograph encircles the origin ones. The system is still unstable. This statement confirms the unit step responses presented in Fig. 7.94.

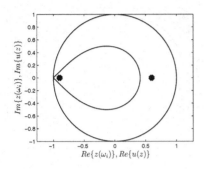

Fig. 7.92   Poles u_1, u_2 configuration.

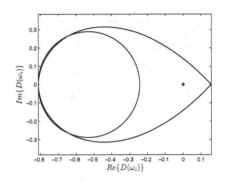

Fig. 7.93   The Mikhailov hodograph for 7.92.

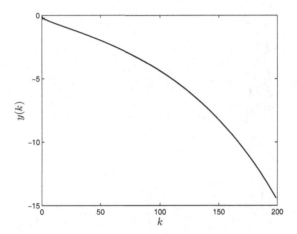

Fig. 7.94   Unit step response of the FOTF with 7.92.

For

$$u_1 = 0.5 + j0.2, \ u_2 = 0.95 - j0.2$$
$$a_1 = -3, \ a_0 = 2.29. \tag{7.149}$$

in Figs. 7.95 and 7.96 the u_1, u_2 configuration and the Mikhailov hodograph are presented, respectively.

The Mikhailov hodograph doesn't encircle the origin. The system is asymptotically stable. This statement confirms the unit step responses presented in Fig. 7.97.

For

$$u_1 = 0.55, \ u_2 = 0.65$$
$$a_1 = -3.2, \ a_0 = 2.5575 \tag{7.150}$$

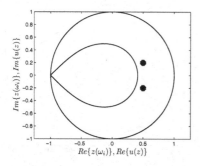

Fig. 7.95   Poles u_1, u_2 configuration.

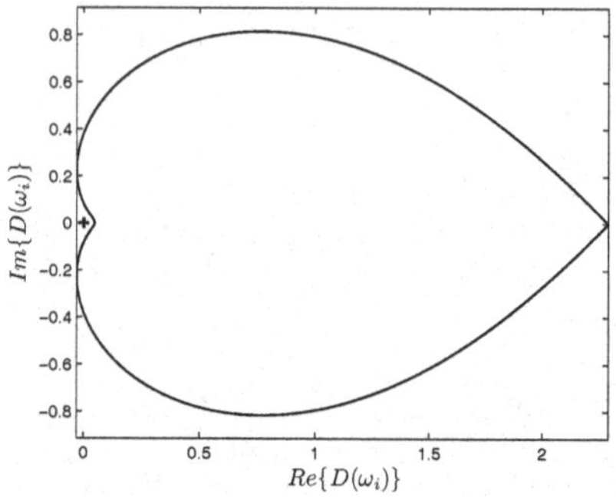

Fig. 7.96   The Mikhailov hodograph for 7.95.

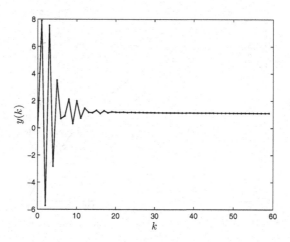

Fig. 7.97   Unit step response of the FOTF with 7.95.

in Figs. 7.98 and 7.99 the u_1, u_2 configuration and the Mikhailov hodograph are presented, respectively.

The Mikhailov hodograph doesn't encircle the origin. The system is asymptotically stable. This statement confirms the unit step responses presented in Fig. 7.100.

For

$$u_1 = -1.3, \ u_2 = 0.55$$
$$a_1 = -1.25, \ a_0 = -0.465 \tag{7.151}$$

in Figs. 7.101 and 7.102 the u_1, u_2 configuration and the Mikhailov hodograph are presented, respectively.

The Mikhailov hodograph doesn't encircle the origin. The system is asymptotically stable. This statement confirms the unit step responses presented in Fig. 7.103.

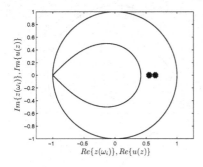

Fig. 7.98   Poles u_1, u_2 configuration.

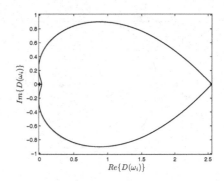

Fig. 7.99   The Mikhailov hodograph for 7.98.

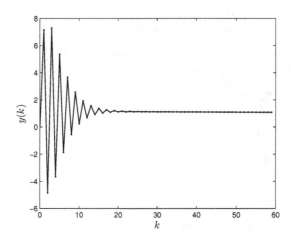

Fig. 7.100   Unit step response of the FOTF with 7.98.

Finally, one presents a system being on the stability limits. Such a FOSs transient behaviour is very interesting because it reveals possible oscillations of the FO generators. Hence, for

$$u_1 = u_2 = \sqrt{(2)} - 1$$
$$a_1 = -2\sqrt{2}, \ a_0 = 2 \tag{7.152}$$

in Figs. 7.104 and 7.105 the u_1, u_2 configuration and the Mikhailov hodograph are presented, respectively.

The Mikhailov hodograph crosses the origin twice. The system is on the stability limit. This statement confirms the unit step responses presented in Figs. 7.106 and 7.107 for two discrete time ranges.

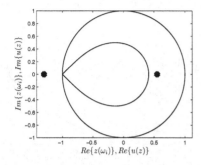

Fig. 7.101   Poles u_1, u_2 configuration.

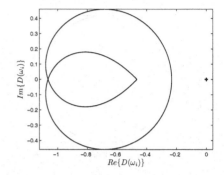

Fig. 7.102   The Mikhailov hodograph for 7.101.

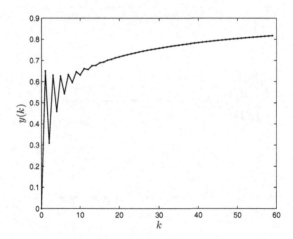

Fig. 7.103   Unit step response of the FOTF with 7.101.

For

$$u_1 = j0.485868245, \quad u_2 = -j0.485868245$$

$$a_1 = -2, \quad a_0 = 1.236067951499380 \tag{7.153}$$

in Figs. 7.108 and 7.109 the u_1, u_2 configuration and the Mikhailov hodograph are presented, respectively.

The Mikhailov hodograph crosses the origin twice. The system is on the stability limit. This statement confirms the unit step responses presented in Figs. 7.110 and 7.111 for two discrete time ranges.

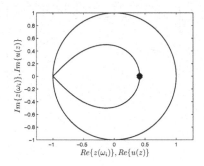

Fig. 7.104　Poles u_1, u_2 configuration.

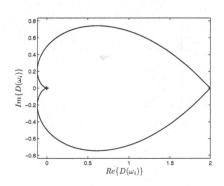

Fig. 7.105　The Mikhailov hodograph for 7.104.

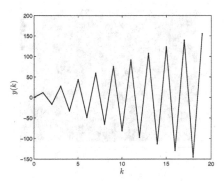

Fig. 7.106　Unit step response of the FOTF with 7.108 for $k = 0, \ldots, 20$.

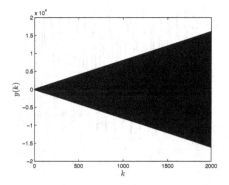

Fig. 7.107　Unit step response of the FOTF with 7.108 for $k = 0, \ldots, 2000$.

For

$$u_1 = -1, \ u_2 = \sqrt{2} - 1$$

$$a_1 = -\sqrt{(2)}, \ a_0 = 0, \ B_0 = 1 \tag{7.154}$$

in Figs. 7.112 and 7.113 the u_1, u_2 configuration and the Mikhailov hodograph are presented, respectively.

The Mikhailov hodograph crosses the origin twice. The system is on the stability limit. This statement confirms the unit step responses presented in Figs. 7.114 and 7.115 for two discrete time ranges.

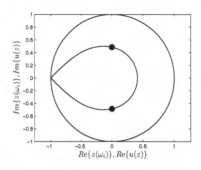

Fig. 7.108   Poles u_1, u_2 configuration.

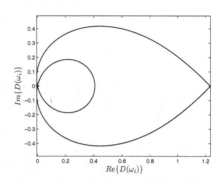

Fig. 7.109   The Mikhailov hodograph for 7.108.

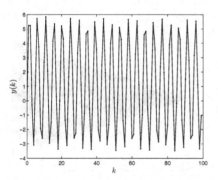

Fig. 7.110   Unit step response of the FOTF with 7.108 for $k = 0, \ldots, 20$.

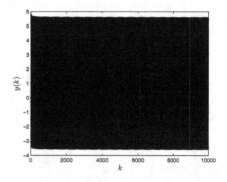

Fig. 7.111   Unit step response of the FOTF with 7.108 for $k = 0, \ldots, 2000$.

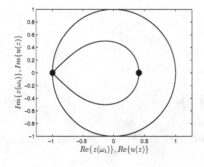

Fig. 7.112   Poles u_1, u_2 configuration.

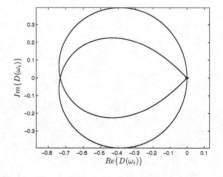

Fig. 7.113   The Mikhailov hodograph for 7.112.

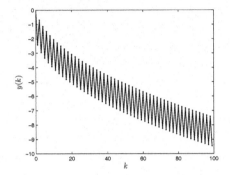 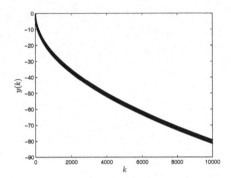

Fig. 7.114   Unit step response of the FOTF with 7.112 for $k = 0, \ldots, 100$.

Fig. 7.115   Unit step response of the FOTF with 7.112 for $k = 0, \ldots, 20000$.

Chapter 8

The linear FO discrete-time fundamental elements

This Chapter presents selected dynamic properties of typical elements [Steiglitz (1974); Ogata (1987); Halanay and Samuel (1997)] in the FO version. Such linear elements are described by the simplest FODEs. To following mentioned items belong in the FO: discrete ideal integrator, ideal discrete differentiator (element realizing the difference), discrete lag element and oscillation element. Their dynamic properties reveal continuity with the IO system properties for FOs tending to integer values.

8.1 The FO ideal summator (FO discrete integrator)

In the case of continuous-time dynamical system analysis the ideal integrator is the most important linear element [Oustaloup *et al.* (2000); Chen and Moore (2002a,b); Ostalczyk (2003); Chen *et al.* Vinagre and Podlubny (2004b); Vinagre *et al.* Monje and Tejado (2007)]. This element serves as a "brick" in building more general dynamic structures when considering the discrete-time systems case. The FO ideal summator (FOS) realizing discrete integration, is formally described by the equation of the form

$$\substack{GL \\ 0} \Delta_k^{(\nu)} y(k) = b_0 u(k) \tag{8.1}$$

or equivalently by its discrete FO transfer function

$$G_I(z^{-1}) = \frac{b_0}{(1 - z^{-1})^{\nu}}. \tag{8.2}$$

The transient characteristics were partially considered in previous Chapters. One should mention that the response for $k_0 = 0$ to any input signal (assuming zero initial conditions $0 = y_{-1} = y_{-2} = \cdots$) may be also expressed in a matrix vector form

$$\mathbf{A}_k^{(\nu)} \mathbf{y}(k) = b_0 \mathbf{u}(k) \tag{8.3}$$

which can also be expressed as

$$\mathbf{D}_{I,k}\mathbf{y}(k) = \mathbf{N}_{I,k}\mathbf{u}(k) \tag{8.4}$$

with

$$\mathbf{D}_{I,k} = \mathbf{A}_k^{(\nu)}, \ \mathbf{N}_{I,k} = \mathbf{I}_k. \tag{8.5}$$

The FO ideal discrete summator (discrete adder) response equals

$$\mathbf{y}(k) = [\mathbf{D}_{I,k}]^{-1}\mathbf{N}_{I,k}\mathbf{u}(k). \tag{8.6}$$

Now investigations will concentrate on the FO frequency characteristics. A substitution of $z = e^{j\omega_s} = \cos\omega_s + j\omega_s$ where ω_s is a so called sampling frequency [Ogata (1987)] with $\omega_s \in [0, \pi]$. Then

$$G_I(z^{-1})\big|_{z=e^{j\omega_s}} = G_I(e^{-j\omega_s}) = \frac{b_0}{(1 - e^{-j\omega_s})^{\nu}}. \tag{8.7}$$

It can be represented in a polar form (also known as the discrete Nyquist plot)

$$G_I(z^{-1})\big|_{z=e^{j\omega_s}} = \left|G_I(e^{-j\omega_s})\right| e^{\phi(\omega_s)} \tag{8.8}$$

where $\left|G_I(e^{-j\omega_s})\right|$ is a magnitude of $G_I(e^{-j\omega_s})$

$$\left|G_I(e^{-j\omega_s})\right| = \frac{1}{\left[2\sin\left(\frac{\omega_s}{2}\right)\right]^{\nu}}, \tag{8.9}$$

and $\phi(\omega_s)$ is the phase shift imparted on the input signal by the FO summator at the frequency ω_s

$$\phi(\omega_s) = \frac{\nu}{2}(\omega_s - \pi). \tag{8.10}$$

In Fig. 8.1, the discrete Nyquist plots of the considered ideal FO summator are shown.

In Figs. 8.2 and 8.3 the discrete magnitude and phase characteristics of the considered ideal FO summator are presented.

The operation of the FO ideal integrator is presented in the numerical example below.

Example 8.1. Using an A/D converter, there was a measured voltage signal that was treated as an output of an electrical device. In Figs. 8.6 and 8.7 the signal and its 1DP image are presented.

One can see that the analyzed signal contains a noise. The noise is caused by a number of well-known factors mentioned here: an A/D converter quantization noise, coefficients quantization errors, roundoff errors from quantizing results of arithmetic operations and arithmetic overflow [Ifeachor and Jervis (1993); Proakis (2007)]. The signal is treated as an input of the considered FO ideal integrator.

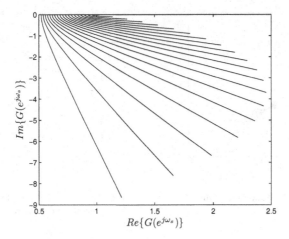

Fig. 8.1 Discrete Nyquist plots of the FO ideal summator for $\nu \in [0.10.2\cdots0.9]$.

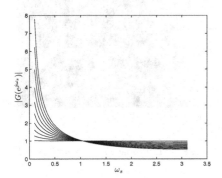

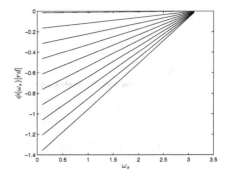

Fig. 8.2 Plot of the magnitude function (8.9) for $\nu \in [0.10.2\cdots0.9]$.

Fig. 8.3 Plot of the phase angle function (8.10) for $\nu \in [0.10.2\cdots0.9]$.

Its outputs and its 1DP images are due to three FOs $\nu \in \{0.3, 0.6, 0.8\}$ and one IO $\nu = 1$ which are presented in Figs. 8.8 and 8.9, respectively.

The plots in Figs. 8.8, 8.10, 8.12 reveal, that for fractional orders from a range $(0, 1)$, the output signals contain decreasing noise due to increasing order. This indicates that the FO ideal summator possesses the noise filtering property. This property is stronger and higher in accordance to the summation order. Also, the plots related to FOs have one maximum, whereas for the integer order 1, the output is monotonically increasing. This property may be used in the edge detection of an image.

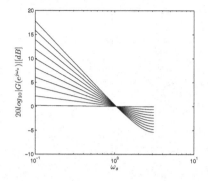

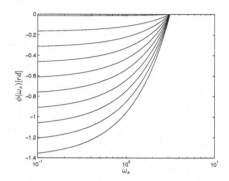

Fig. 8.4   The discrete Bode plot of function (8.9) for $\nu \in [0.10.2 \cdots 0.9]$.

Fig. 8.5   The phase angle of function (8.10) in a logarithmic scale for $\nu \in [0.10.2 \cdots 0.9]$.

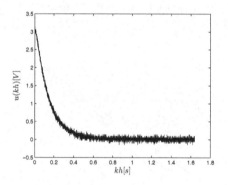

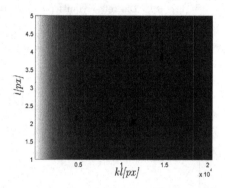

Fig. 8.6   Measured signal.

Fig. 8.7   1DP image of measured signal.

8.2   Discrete FO ideal differentiator

In the case of continuous-time dynamical system analysis, the ideal differentiator is not used in practice despite its great theoretical importance [Al-Alaoui (1993); Chen and Moore (2002b)]. This is caused by the necessity of an input signal pre-filtering. It is formally described by the equation of the form

$$y(k) = b_0 {}_0^{GL}\Delta_k^{(\nu)} y(k) \tag{8.11}$$

or equivalently by its discrete FO transfer function

$$G_D(z^{-1}) = b_0 \left(1 - z^{-1}\right)^{\nu}. \tag{8.12}$$

The transient characteristics were already partially considered in previous chapters. Here, investigations will concentrate on the FO discrete differentiator frequency

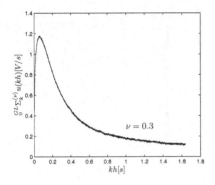

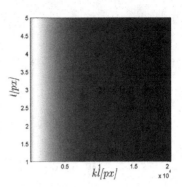

Fig. 8.8 FO ideal integrator output due to the measured signal.

Fig. 8.9 1DP image of the FO ideal integrator output due to the measured signal.

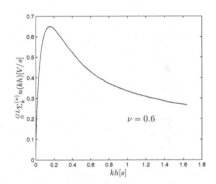

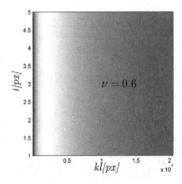

Fig. 8.10 FO $\nu = 0.6$ ideal integrator output due to the measured signal.

Fig. 8.11 1DP image of the FO $\nu = 0.6$ ideal integrator output due to the measured signal.

characteristics. Therefore,

$$G_D(z^{-1})\big|_{z=e^{j\omega_s}} = G_D(e^{-j\omega_s}) = b_0 \left(1 - e^{-j\omega_s}\right)^{\nu} \tag{8.13}$$

which can be transformed to the polar form

$$G_D(z^{-1})\big|_{z=e^{j\omega_s}} = \left|G_D(e^{-j\omega_s})\right| e^{\phi(\omega_s)} \tag{8.14}$$

where $\left|G_D(e^{-j\omega_s})\right|$ is a magnitude of $G_D(e^{-j\omega_s})$

$$\left|G_D(e^{-j\omega_s})\right| = \left[2\sin\left(\frac{\omega_s}{2}\right)\right]^{\nu} \tag{8.15}$$

and $\phi_D(\omega_s)$ is the phase shift imparted on the input signal by the FO discrete differentiator at the frequency ω_s

$$\phi(\omega_s) = \frac{\nu}{2}(\pi - \omega_s). \tag{8.16}$$

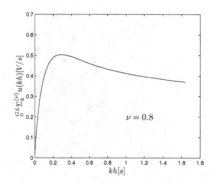

Fig. 8.12   FO $\nu = 0.8$ ideal integrator output due to the measured signal.

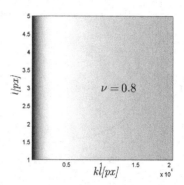

Fig. 8.13   1DP image of the FO $\nu = 0.8$ ideal integrator output due to the measured signal.

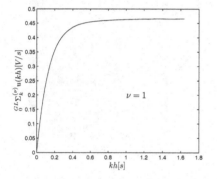

Fig. 8.14   FO $\nu = 1$ ideal integrator output due to the measured signal.

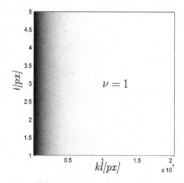

Fig. 8.15   1DP image of the FO $\nu = 1$ ideal integrator output due to the measured signal.

In Fig. 8.16 the discrete Nyquist plots of the considered ideal FO discrete differentiator are shown.

In Figs. 8.17 and 8.18 the discrete magnitude and phase characteristics of the considered ideal FO differentiator are presented.

Finally, in Figs. 8.19 and 8.20 the Bode characteristics are presented.

The operation of the FO ideal discrete differentiator is presented in the numerical example below.

Example 8.2. Now, the same measured signal as considered in Example 8.1 will be subjected to FO discrete differentiation. The outputs of the FO discrete differentiator and related 1DP images are given in Figs. 8.21, 8.23, 8.25, 8.27 and 8.22, 8.24, 8.26, 8.28, respectively.

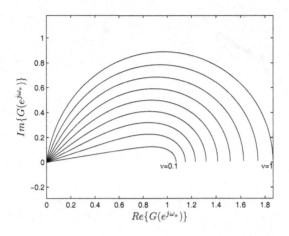

Fig. 8.16   Discrete Nyquist plots of the FO ideal differentiator for $\nu \in [0.10.2 \cdots 0.9]$.

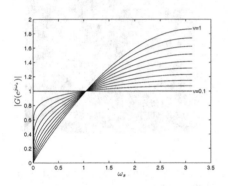

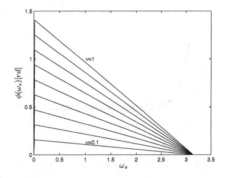

Fig. 8.17   Plot of the magnitude function (8.15) for $\nu \in [0.10.2 \cdots 0.9]$.

Fig. 8.18   Plot of the phase angle function (8.16) for $\nu \in [0.10.2 \cdots 0.9]$.

The plots in Figs. 8.21, 8.23 and 8.25 reveal that for fractional orders from a range $(0,1)$ the output signals contain increasing noise due to increasing order. This indicates that the FO ideal discrete differentiator strongly amplifies noise. The highest noise is obtained for a classical discrete first-order differentiator given in Ref. 8.27.

In the next numerical example the FO discrete differentiation of a special function is analyzed.

Example 8.3. Consider a spline function defined by the formula

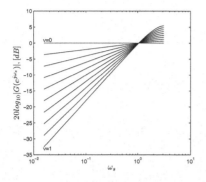

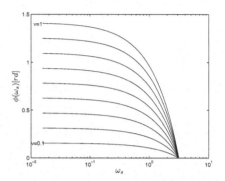

Fig. 8.19 The discrete Bode plot of function (8.12) for $\nu \in [0.10.2 \cdots 0.9]$.

Fig. 8.20 The phase angle of function (8.12) in a logarithmic scale for $\nu \in [0.10.2 \cdots 0.9]$.

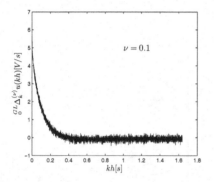

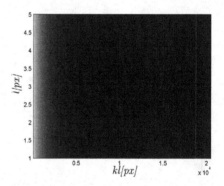

Fig. 8.21 FO discrete differentiator output for $\nu = 0.1$.

Fig. 8.22 1DP image of the FO ideal differentiator output.

$$u(k) = \begin{cases} 0 & \text{for } 0 \leqslant k_a \\ a_1 k^2 + b_1 k + c_1 & \text{for } k_a < k \leqslant k_b \\ a_2 k^2 + b_2 k + c_2 & \text{for } k_b < k \leqslant k_c \\ u_{k_c} & \text{for } k_c < k \end{cases} \tag{8.17}$$

where

$$\begin{bmatrix} a_1 \\ b_1 \\ c_1 \end{bmatrix} = \begin{bmatrix} k_a^2 & k_a & 1 \\ 2k_a & 1 & 0 \\ k_b^2 & k_b & 1 \end{bmatrix}^{-1} \begin{bmatrix} 0 \\ 0 \\ u(k_b) \end{bmatrix}, \begin{bmatrix} a_2 \\ b_2 \\ c_2 \end{bmatrix} = \begin{bmatrix} k_b^2 & k_b & 1 \\ k_c^2 & k_c & 1 \\ 2k_c & 1 & 0 \end{bmatrix}^{-1} \begin{bmatrix} u(k_b) \\ u(k_c) \\ 0 \end{bmatrix} \tag{8.18}$$

and $k_a = 250, k_b = 500, k_c = 750, u(k_b) = 0.5, u(k_c) = 1$.

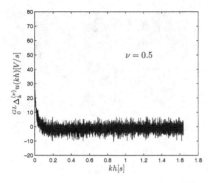

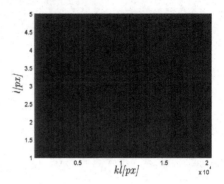

Fig. 8.23   FO discrete differentiator output for $\nu = 0.5$.

Fig. 8.24   1DP image of the FO ideal differentiator output.

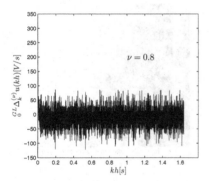

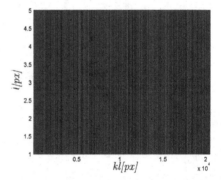

Fig. 8.25   FO discrete differentiator output for $\nu = 0.8$.

Fig. 8.26   1DP image of the FO ideal differentiator output.

The plot and related 1DP image of function (8.12) is shown in Figs. 8.29 and 8.30, respectively.

The next figures contain the outputs of the FO discrete ideal differentiator when subjected to the input signal described by spline function (8.12). The 1DP images of the output signals are attached to the appropriate plots.

The calculated FO ideal discrete differentiator outputs reveal wider opportunities for signal processing in relation to the classical (1st order) difference. All output signals processed that are related to FOs are characterized by one maximum with shape asymmetry on their opposite sides. This special property can be used in 2D image edge detection. Only 1st (integer) order discrete differentiator output is symmetrical with respect to its maximum.

The transient characteristics analysis of the FO ideal discrete differentiator does not need to be limited to the orders from an interval $(0, 1]$. One can also consider

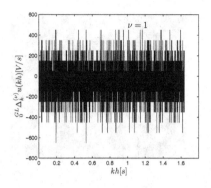

Fig. 8.27 FO discrete differentiator output for $\nu = 1$.

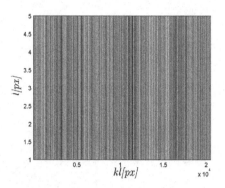

Fig. 8.28 1DP image of the FO ideal differentiator output.

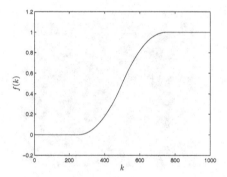

Fig. 8.29 Spline function (8.17) plot.

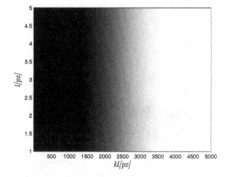

Fig. 8.30 1DP image of the spline function (8.17).

higher discrete differentiating orders the most important of which form an image processing point of view, with a range $[1, 2)$. Hence, in Figs. 8.39, 8.41, 8.43 and 8.45 the FO ideal discrete differentiator outputs for the selected orders from the mentioned range are presented. Figs. 8.40, 8.42, 8.44, 8.46 contain related images.

8.3 Discrete FO lag element

This element is a counterpart to continuous first-order lag element, which is fundamental in dynamic system analysis. The FO element is described by the FODE

$$\substack{GL \\ k_0}\Delta_k^{(\nu)}y(k) + a_0 y(k) = b_0 u(k) \tag{8.19}$$

where ν is the FO, and a_0, b_0 are constant parameters. Regularly, the FODE is

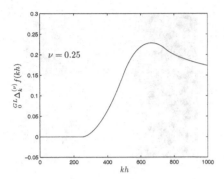

Fig. 8.31 Output of the FO ideal differentiator subjected to the input signal described by spline function (8.17) plot for $\nu = 0.25$.

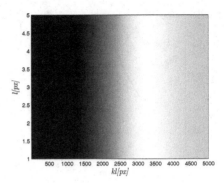

Fig. 8.32 1DP image of the output of the FO ideal differentiator subjected to the input signal described by spline function (8.17).

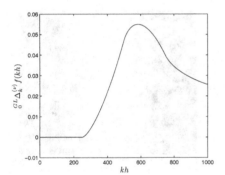

Fig. 8.33 Output of the FO ideal differentiator subjected to the input signal described by spline function (8.17) plot for $\nu = 0.5$.

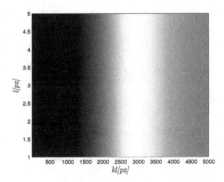

Fig. 8.34 1DP image of the output of the FO ideal differentiator subjected to the input signal described by spline function (8.17).

written in the form

$$\tau {}_{k_0}^{GL}\Delta_k^{(\nu)} y(k) + y(k) = b_0' u(k) \tag{8.20}$$

where τ is the so called time constant that characterizes the transient properties of the element. The discrete transfer function has the form

$$G_L(z) = \frac{Y(z)}{U(z)} = \frac{b_0}{(1 - z^{-1})^\nu + a_0} = \frac{b_0'}{\tau (1 - z^{-1})^\nu + 1}. \tag{8.21}$$

The FO discrete lag element can be obtained from the ideal FO discrete integrator being contoured by a negative feedback. The block diagram given in Fig. 8.47 presents this.

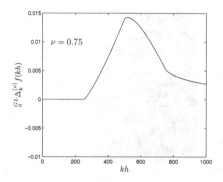

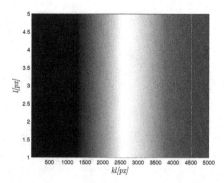

Fig. 8.35 Output of the FO ideal differentiator subjected to the input signal described by spline function (8.17) plot for $\nu = 0.75$.

Fig. 8.36 1DP image of the output of the FO ideal differentiator subjected to the input signal described by spline function (8.17).

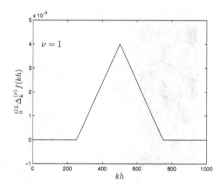

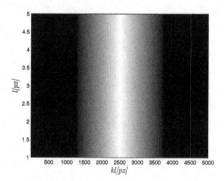

Fig. 8.37 Output of the FO ideal differentiator subjected to the input signal described by spline function (8.17) plot for $\nu = 1$.

Fig. 8.38 1DP image of the output of the FO ideal differentiator subjected to the input signal described by spline function (8.17).

The impulse response of the considered element is of the form

$$g_L(k) = \mathcal{Z}^{-1}\left\{G_l(z)\right\} = \mathcal{Z}^{-1}\left\{\sum_{i=0}^{+\infty} g_l(i)z^{-i}\right\} \qquad (8.22)$$

where $g_L(0), g_L(1), \ldots$ are Markov parameters. The simplest way to obtain the aforementioned Markov parameters is to perform a numerator and denominator polynomials long-term division. Having the function $g_L(k)$ one can immediately evaluate a response to any input signal using the discrete convolution formula. Hence,

$$y(k) = \mathcal{Z}^{-1}\left\{G_L(z)U(z)\right\} = g_L(k) * u(k). \qquad (8.23)$$

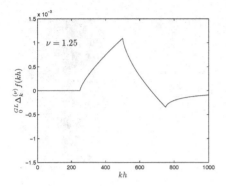

Fig. 8.39 Output of the FO ideal differentiator subjected to the input signal described by spline function (8.17) plot for $\nu = 1.25$.

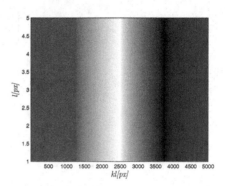

Fig. 8.40 1DP image of the output of the FO ideal differentiator subjected to the input signal described by spline function (8.17).

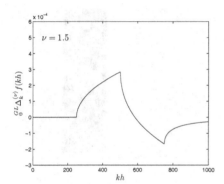

Fig. 8.41 Output of the FO ideal differentiator subjected to the input signal described by spline function (8.17) plot for $\nu = 1.5$.

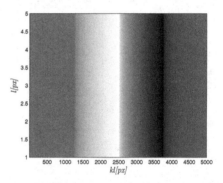

Fig. 8.42 1DP image of the output of the FO ideal differentiator subjected to the input signal described by spline function (8.17).

Expanding the FOBD in FODE (8.14) one gets

$$\begin{bmatrix} 1 \ a^{(\nu)}(1) \ a^{(\nu)}(2) \ \cdots \ a^{(\nu)}(k-1) \ a^{(\nu)}(k) \end{bmatrix} \begin{bmatrix} y(k) \\ y(k-1) \\ y(k-2) \\ \vdots \\ y(1) \\ y(0) \end{bmatrix} + a_0 y(k) = b_0 u(k). \quad (8.24)$$

Simple rearrangements of the above equation lead to the form

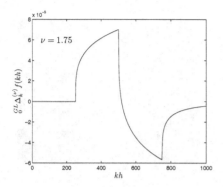

Fig. 8.43 Output of the FO ideal differentiator subjected to the input signal described by spline function (8.17) plot for $\nu = 1.75$.

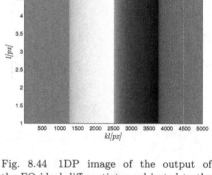

Fig. 8.44 1DP image of the output of the FO ideal differentiator subjected to the input signal described by spline function (8.17).

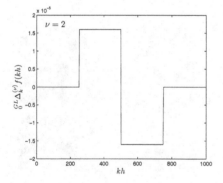

Fig. 8.45 Output of the FO ideal differentiator subjected to the input signal described by spline function (8.17) plot for $\nu = 2$.

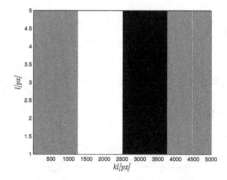

Fig. 8.46 1DP image of the output of the FO ideal differentiator subjected to the input signal described by spline function (8.17).

$$\left[1 + a_0 \; a^{(\nu)}(1) \; a^{(\nu)}(2) \; \cdots \; a^{(\nu)}(k-1) \; a^{(\nu)}(k)\right] \begin{bmatrix} y(k) \\ y(k-1) \\ y(k-2) \\ \vdots \\ y(1) \\ y(0) \end{bmatrix} = b_0 u(k). \qquad (8.25)$$

which is valid for all $k, k-1, k-2, \ldots, 1, 0$. Here, zero initial conditions are assumed: $0 = y_{-1} = y_{-2} = \cdots$. Repeating all such equation collection procedures in one matrix-vector form results in

$$\mathbf{D}_{L,k}\mathbf{f}(k) = \mathbf{N}_{L,k}\mathbf{u}(k) \qquad (8.26)$$

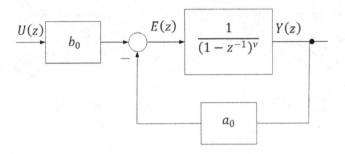

Fig. 8.47 Block diagram of the FO lag element (8.16)

where

$$\mathbf{D}_{L,k} = \mathbf{D}_{I,k} + a_0 \mathbf{1}_k$$
$$\mathbf{N}_{L,k} = \mathbf{N}_{I,k}. \qquad (8.27)$$

Assuming that $a_0 + 1 \neq 0$ the FO lag element solution to any input signal is of the form

$$\mathbf{f}(k) = [\mathbf{D}_{L,k}]^{-1} \mathbf{N}_{L,k} \mathbf{u}(k). \qquad (8.28)$$

In Figs. 8.48 and 8.49 the FO lag element impulse responses are presented for two ranges of orders $\nu \in [0.1, 0.2, \ldots, 0.9]$ and $\nu \in [1.0, 1.1, \ldots, 1.9]$, respectively.

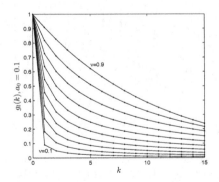

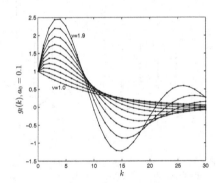

Fig. 8.48 FO lag element impulse response for $\nu \in [0.1, 0.2, \ldots, 0.9]$ and $b_0 = a_0 + 1$.

Fig. 8.49 FO lag element impulse response for $\nu \in [1.0, 1.1, \ldots, 1.9]$ and $b_0 = a_0 + 1$.

In Figs. 8.50 and 8.51, the FO lag element unit step responses are presented for two ranges of orders: $\nu \in [0.1, 0.2, \ldots, 0.9]$ and $\nu \in [1.0, 1.1, \ldots, 1.9]$, respectively.

To show the long memory property of the FO system, a response to the series of discrete Dirac pulses is defined as

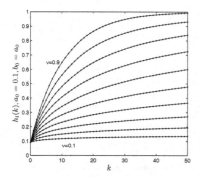

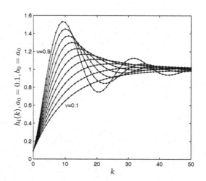

Fig. 8.50   FO lag element step response for $\nu \in [0.1, 0.2, \ldots, 0.9]$ and $b_0 = a_0 + 1$.

Fig. 8.51   FO lag element step response for $\nu \in [1.0, 1.1, \ldots, 1.9]$ and $b_0 = a_0 + 1$.

$$u(k) = \sum_{i=0}^{+\infty} \delta(k - iH) \tag{8.29}$$

where L denotes a distance between consecutive impulses is presented in Fig. 8.52. In Fig. 8.53, the first-order lag element response is given. One realises that in the FO case, the transient state is longer then in the IO case.

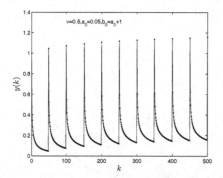

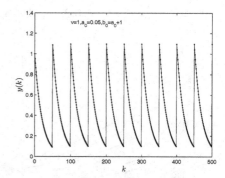

Fig. 8.52   FO lag element response to (8.19) for $\nu \in [0.1, 0.2, \ldots, 0.9]$ and $b_0 = a_0 + 1$.

Fig. 8.53   FO lag element response to (8.19) for $\nu \in [1.0, 1.1, \ldots, 1.9]$ and $b_0 = a_0 + 1$.

The frequency characteristics are obtained by the substitution $z = e^{j\omega_s}$

$$G_L(z)\big|_{z=e^{j\omega_s}} = \frac{b_0}{(1 - e^{-j\omega_s})^{\nu} + a_0}. \tag{8.30}$$

Simple calculations reveal that the amplitude and phase frequency characteristics have the forms

$$\left|G_L(e^{j\omega_s})\right| = \frac{b_0}{\sqrt{\left\{\left(2\sin\frac{\omega_s}{2}\right)^\nu \cos\left[\frac{\nu}{2}(\pi - \omega_s)\right] + a_0\right\}^2 + \left\{\left(2\sin\frac{\omega_s}{2}\right)^\nu \sin\left[\frac{\nu}{2}(\pi - \omega_s)\right]\right\}^2}} \tag{8.31}$$

$$\varphi_L(\omega_s) = \frac{\left(2\sin\frac{\omega_s}{2}\right)^\nu \sin\left[\frac{\nu}{2}(\pi - \omega_s)\right]}{\left(2\sin\frac{\omega_s}{2}\right)^\nu \cos\left[\frac{\nu}{2}(\pi - \omega_s)\right] + a_0}. \tag{8.32}$$

Note that for $a_0 = 0$ one gets $G_L(z) = G_I(z)$. In Figs. 8.54 and 8.55 the Nyquist plots of the FO lag elements for orders $\nu \in [0.1, 0.2, \ldots, 0.9]$ and $\nu \in [1.0, 1.1, \ldots, 1.9]$ are plotted, respectively.

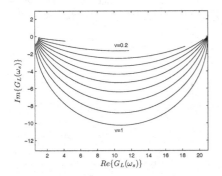

Fig. 8.54   FO lag element Nyquist plots for $\nu \in [0.1, 0.2, \ldots, 0.9]$ and $b_0 = a_0 + 1$.

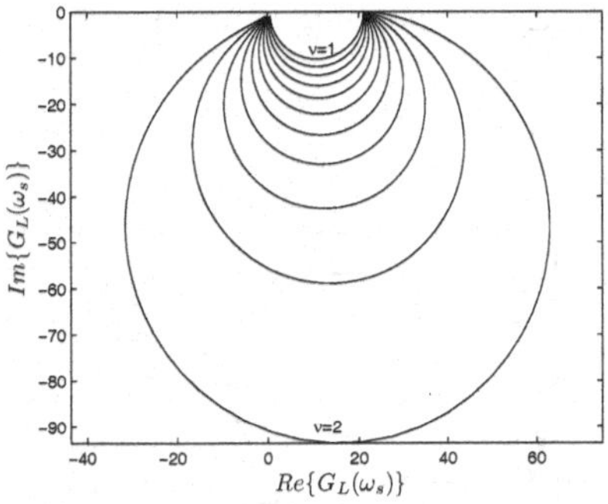

Fig. 8.55   FO lag element Nyquist plots for $\nu \in [1.0, 1.1, \ldots, 1.9]$ and $b_0 = a_0 + 1$.

In Figs. 8.56 and 8.57 the magnitude frequency characteristics of the FO lag elements for orders $\nu \in [0.1, 0.2, \ldots, 0.9]$ and $\nu \in [1.0, 1.1, \ldots, 1.9]$ are plotted, respectively. Related phase plots are given in Figs. 8.58 and 8.59.

In digital signal processing and discrete dynamical system analysis, it is useful to consider the magnitude frequency characteristics $20\log_{10}|G_L(\omega_s)|$ when expressed in decibels (dB) along the frequency axis in a logarithmic scale. Such characteristics are known as the Bode plots.

The Bode plots given in Fig. 8.60 reveal the filtering property of the FO lag element for all orders from a range $\nu \in (0, 1]$. For higher orders, $\nu \in (1, 2]$ all the Bode plots have one maximum. The slope of the descending parts of the graphs are equal to $-20\nu[\frac{dB}{dec}]$. This property is indicated in Figs. 8.62 and 8.63, where a line of a related slope is added.

Until now, the analysis concentrated on the FO's influence on the transient and frequency properties of the lag element. Evidently, two additional coefficients, b_0 and a_0 in (8.14) also contribute to the dynamics of the FO lag element. This will be analyzed now. A ratio between b_0 and a_0 represents the element gain, so one

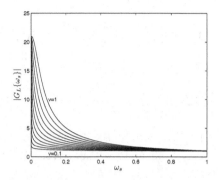

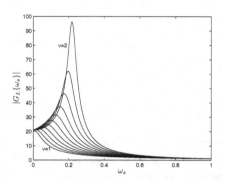

Fig. 8.56 FO lag element the magnitude frequency characteristics for $\nu \in [0.1, 0.2, \ldots, 0.9]$ and $b_0 = a_0 + 1$.

Fig. 8.57 FO lag element the magnitude frequency characteristics for $\nu \in [1.0, 1.1, \ldots, 1.9]$ and $b_0 = a_0 + 1$.

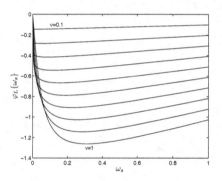

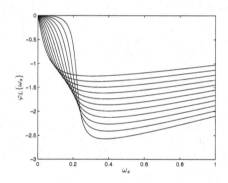

Fig. 8.58 FO lag element the phase frequency characteristics for $\nu \in [0.1, 0.2, \ldots, 0.9]$ and $a_0 = 0.05$, $b_0 = a_0 + 1$.

Fig. 8.59 FO lag element the phase frequency characteristics for $\nu \in [1.0, 1.1, \ldots, 1.9]$ and $a_0 = 0.05$, $b_0 = a_0 + 1$.

assumes $a_0 = b_0$ when wishing to see the influence of a_0 on the element dynamics. In Figs. 8.64 and 8.65 the FO lag element unit step responses for constant FOs $\nu = 0.5$ and $\nu = 1.5$, as well as the different coefficient a_0 values from the set $\{0.1, 0.2, \ldots 1.0\}$ are presented.

The last figures reveal that the settling time of the response rises due to the decrease of the coefficient a_0. For $a_0 \in \mathbb{R}_+$ the FO lag element is stable, whereas for $a_0 \in \mathbb{R}_-$ is not. It is a commonly known fact that oscillations in the step responses are related to the peaks in the magnitude frequency characterisitc. This is also satisfied in the FO lag element case. This statement confirms Figs. 8.51, 8.57 and 8.61.

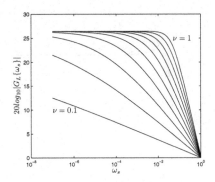

Fig. 8.60 FO lag element the Bode plots for $\nu \in [0.1, 0.2, \ldots, 0.9]$ and $a_0 = 0.05$, $b_0 = a_0 + 1$.

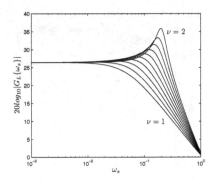

Fig. 8.61 FO lag element the Bode plots for $\nu \in [1.0, 1.1, \ldots, 1.9]$ and $a_0 = 0.05$, $b_0 = a_0 + 1$.

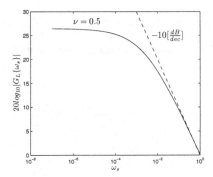

Fig. 8.62 FO lag element the Bode plot for $\nu = 0.5$ and $a_0 = 0.05$, $b_0 = a_0 + 1$ with a line indicated a slope $-20\nu[dB/dec]$.

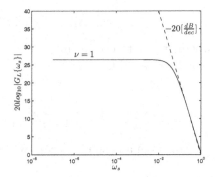

Fig. 8.63 FO lag element the Bode plots for $\nu = 1]$ and $a_0 = 0.05$, $b_0 = a_0 + 1$ with a line indicated a slope $-20\nu[dB/dec]$.

8.4 FO oscillation element

An element name refers to IO dynamical linear element name having oscillatory properties. Under specified conditions imposed on its parameters, it can be a second-order inertial or integration with lag. The common features are two parameters a_1 and a_0 ($b_0 = a_0$). The FODE describing it has the form

$$_{k_0}^{GL}\Delta_k^{(2\nu)}y(k) + a_1 {}_{k_0}^{GL}\Delta_k^{(\nu)}y(k) + a_0 y(k) = b_0 u(k-1). \tag{8.33}$$

Here, one considers the commensurate FODE. Following the notation used in the classical discrete linear systems description [Ogata (1987)] it can be equivalently described as

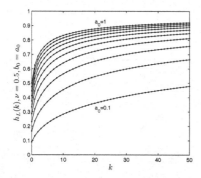
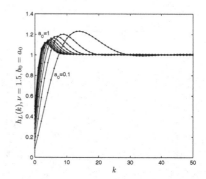

Fig. 8.64 FO lag element step responses for $\nu = 0.5$, $b_0 = a_0$ and $a_0 \in \{0.1, 0.2, \ldots, 1.0\}$.

Fig. 8.65 FO lag element step responses for $\nu = 1.5$, $b_0 = a_0$ and $a_0 \in \{0.1, 0.2, \ldots, 1.0\}$.

$$\tau^2 {}_{k_0}^{GL}\Delta_k^{(2\nu)}y(k) + 2\tau\zeta {}_{k_0}^{GL}\Delta_k^{(\nu)}y(k) + y(k) = \kappa u(k), \qquad (8.34)$$

where: τ — a time constant, ζ — a damping factor, and κ — a gain of the element. The discrete TF has a form

$$G_O(z) = \frac{Y(z)}{U(z)} = \frac{b_0}{\left(1 - z^{-1}\right)^{2\nu} + a_1\left(1 - z^{-1}\right)^{\nu} + a_0}. \qquad (8.35)$$

Now from a denominator of (8.35) one defines a polynomial

$$w^2 + a_1 w + a_0 = (w - w_1)(w - w_2) \qquad (8.36)$$

by a substitution

$$w = (1 - z^{-1})^{\nu} \qquad (8.37)$$

and considers three cases of (8.35) depending on its roots. In Table 8.1 all possible mentioned cases are indicated.

In this section, the most important is a scenario when $\Delta < 0$, because this condition is not covered by ideal FO discrete integrator and lag element. In the numerical example, selected plots of the unit step responses of the FO oscillation element are analyzed.

Example 8.4. Plot numerically evaluated unit step responses of the FO oscillation elements related to cases c1.1-c1.3.

Solution. In Figs. 8.66 and 8.67 the step responses of the FO element (8.35) with $a_1 = -0.2, a_0 = 0.0244$ and $a_1 = 0, a_0 = 0.0144$ are given.

Table 8.1 Possible cases of w_1, w_2 roots.

	w_1	w_2		case number
$\Delta < 0$	w_1	w_1^*	$Re\{w_1\} < 0$	c1.1
			$Re\{w_1\} = 0$	c1.2
			$Re\{w_1\} > 0$	c1.3
$\Delta = 0$	w_1	w_1	$w_1 < 0$	c2.1
			$w_1 = 0$	c2.2
			$w_1 > 0$	c2.3
$\Delta > 0$	w_1	w_2	$w_1 < 0, w_2 < 0$	c3.1
			$w_1 < 0, w_2 > 0$	c3.2
			$w_1 > 0, w_2 > 0$	c3.3
			$w_1 = 0, w_2 < 0$	c3.4
			$w_1 = 0, w_2 > 0$	c3.5

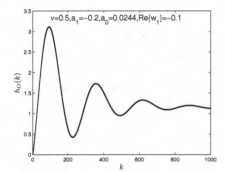

Fig. 8.66 FO oscillation element step response for $\nu = 0.5$, $b_0 = a_0$ and $a_1 = -0.2, a_0 = 0.0244$.

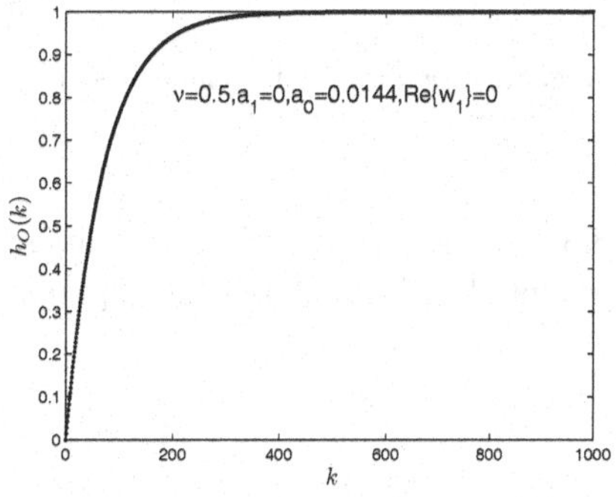

Fig. 8.67 FO oscillation element step response for $\nu = 0.5$, $b_0 = a_0$ and $a_1 = 0, a_0 = 0.0144$.

The last of the considered elements is particularly interesting because for $\nu = 0.5$ and $a_1 = 0$ the equation (8.33) simplifies to an equation

$$\underset{k_0}{\overset{GL}{}}\Delta_k^{(1)} y(k) + a_0 y(k) = b_0 u(k) \tag{8.38}$$

which describes the classical first-order difference equation. The plot in Fig. 8.67 confirms the response character. Finally, one presents a response marked as a c1.3 case. For $\nu = 0.5$, $b_0 = a_0$ and $a_1 = 0, a_0 = 0.0144$ one obtains a plot given in Fig. 8.68.

8.4.1 *Doubled FO lag and discrete integrator element*

Now one considers two special cases of the FO oscillation element. They are marked as c2.2 and c2.3 in Table 8.1, respectively. In the first considered case, equation

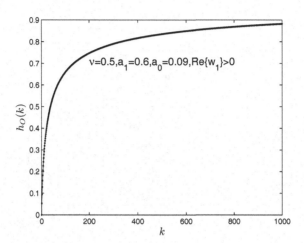

Fig. 8.68 FO oscillation element step response for $\nu = 0.5$, $b_0 = a_0$ and $a_1 = 0$, $a_0 = 0.0144$.

(8.33) reduces to two FO ideal integrators connected in series. Such a connection is presented in Fig. 8.69. This explains the name — doubled FO discrete integrator.

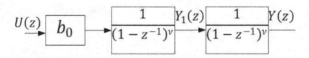

Fig. 8.69 Block diagram of doubled discrete integrator

The unit step responses are similar to the FOBS of a constant function (unit step function). Here, one should mention that the doubled FO discrete integrator related to order $\nu = 0.5$ is equal to the first-order classical discrete integrator. Moreover, a connection in a series of three FO ideal integrators of orders $\nu = 1/3$, results in the same first-order classical integrator and so on.

The special case marked as c2.3 is a doubled FO lag element. Its block diagram is presented in Fig. 8.70.

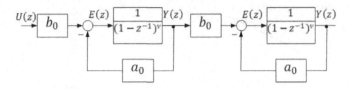

Fig. 8.70   Block diagram of doubled FO discrete lag element.

The elements marked by c.3.1 and c3.2 appear to be non-stable. This is because they are out of interest in the dynamic system analysis. The unit step responses of the doubled lag element marked by c3.3 and calculated numerically for $\nu = 1/2$, $\nu = 1/3$ and $a_1 = 1, a_0 = $ are plotted in Figs. 8.71 and 8.72, respectively.

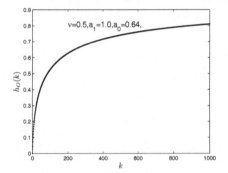

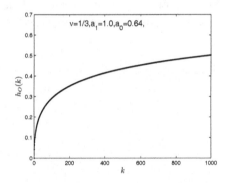

Fig. 8.71   FO oscillation element step response for $\nu = 1/2$, $b_0 = a_0$ and $a_1 = 1.0, a_0 = 0.64$.

Fig. 8.72   FO oscillation element step response for $\nu = 1/3$, $b_0 = a_0$ and $a_1 = 1.0, a_0 = 0.64$.

Comparing the plots given in Figs. 8.71 and 8.72 and computed for the same parameters a_1, a_0 and different orders $\nu = 1/2$ and $\nu = 1/3$, one deduces that the differentiation "force" increases due to the order value. The next case c3.4 represents an unstable element and will not be considered here. However, the last element c3.5, named as the FO discrete integrator with lag can be treated as a connection in series of the FO discrete integrator and the FO lag element acting as pre-filter. In Figs. 8.73 and 8.74, the unit step responses of two exemplary elements defined by the two orders $\nu = 0.5$ and $\nu = 1.5$ and the same coefficients $a_1 = 1, a_0 = 0, b_0 = 1$ are plotted.

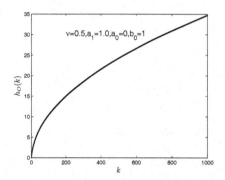
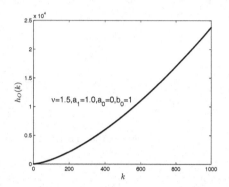

Fig. 8.73 FO discrete integrator with lag
for $\nu = 0.5$, $b_0 = 1$ and $a_1 = 1$, $a_0 = 0$.

Fig. 8.74 FO discrete integrator with lag
for $\nu = 1.5$, $b_0 = 1$ and $a_1 = 1$, $a_0 = 0$.

In both cases, the unit step responses tend to infinity. One can say that the responses
related to $\nu = 1$ are placed "between" the plots presented in Figs. 8.73 and 8.74.
This reveals a great potential in dynamic transient characteristics shaping. It is
successfully used in the dynamic system identification.

Chapter 9

FO discrete-time system structures

The main advantage of the SS, TF and PML description of the linear systems lies in the possibility to build more complicated structures by using three types of connections: serial, parallel and a closed-loop (with usually negative feedback). Among the dynamic blocks representing the dynamic system, one assumes two additional elements: an adder and a branch point [Steiglitz (1974); Ogata (1987); Halanay and Samuel (1997); Franklin *et al.* (1998); Dorf and Bishop (2005); Bhattacharya (2008)]. The adder is a block with multi-inputs one output block that can be described by an algebraic equation representing a sum of all input signals with appropriate signs. The branch point has one-input with multi-outputs signals that are all the same as the input (it spreads out signals without any change). Non-linear blocks performing signals multiplication or division will not be used in these considerations. The most important connection in the control system analysis and synthesis is a closed-loop connection [Kaczorek (1992); Franklin *et al.* (2002)].

9.1 FO system serial connection

One considers two single-input single-output (SISO) systems described by the FOTSs and connection in series. The block diagram of them is given in Fig. 9.1. Here, it is very important to note that one assumes that the connection of the second system does not influence the parameters of the first one. This means that the second dynamic system does not play a role on the load of its predecessor. Although there are no contraindications to connect non-commensurate subsystems, a preliminary transformation to one commensurate form is recommended. From Fig. 9.1 one gets five algebraic equations

$$U(z) = U_1(z)$$
$$Y_1(z) = G_1(z)U_1(z)$$
$$Y_1(z) = U_2(z)$$
$$Y_2(z) = G_2(z)U_2(z)$$
$$Y(z) = Y_2(z). \tag{9.1}$$

Eliminating in a set of equations (9.1) variables $U_1(z), Y_1(z), U_2(z), Y_2(z), U_2(z)$ one gets an equivalent relation. The block diagram in Fig. 9.1 depicts it. However, from a mathematical point of view when considering the case of the SISO system, it is

$$G(z) = G_2(z)G_1(z) = G_1(z)G_2(z). \tag{9.2}$$

So one can change an order but in practice, for $G_1(z)G_2(z)$, the inner signal $U_1(z)$ may be different while the output signal $Y(z)$ will be the same. This may essentially influence the total system response due to the inner signal $Y_1(z)$ constraints. In the multi-input multi-output (MIMO) systems serial connection where FOTFs are replaced by the FOTMs this operation is forbidden because of non-commutativity of matrix multiplication.

The numerical example related to the FO and IO SISO systems serial connection clarifies the presented procedure.

Example 9.1. Consider a serial connection of two subsystems consecutively described by

$$G_1(z) = \frac{b_{1,0}}{(1 - z^{-1})^\nu + a_{1,0}}, \text{ with } \nu = \frac{1}{3}$$

$$G_2(z) = \frac{b_{2,1}(1 - z^{-1}) + b_{2,0}}{(1 - z^{-1})^2 + a_{2,1}(1 - z^{-1})^1 + a_{2,0}}. \tag{9.3}$$

The TF $G_2(z)$ describes the classical second order discrete system. Then according to (9.2) one has

$$G(z) = G_2(z)G_1(z) = \frac{b_{2,1}(1 - z^{-1}) + b_{2,0}}{(1 - z^{-1})^2 + a_{2,1}(1 - z^{-1})^1 + a_{2,0}} \frac{b_{1,0}}{(1 - z^{-1})^\nu + a_{1,0}}$$

$$\frac{b_{1,0}b_{2,1}(1 - z^{-1}) + b_{1,0}b_{2,0}}{(1 - z^{-1})^{\frac{7}{3}} + a_{1,0}(1 - z^{-1})^{\frac{6}{3}} + a_{2,1}(1 - z^{-1})^{\frac{4}{3}} + a_{2,1}(1 - z^{-1})^{\frac{3}{3}} + a_{2,0}(1 - z^{-1})^{\frac{1}{3}} + a_{1,0}}. \tag{9.4}$$

The serial connection of two subsystems described by the FOSSEs is given in Fig. 9.3.

In the case when commensurate FOSSEs describe two subsystems, there are seven equations

$$u(k) = u_1(k),$$
$${}^{GL}_0\Delta_k^{(\nu)}\mathbf{x}_i(k) = \mathbf{A}_i\mathbf{x}_i(k) + \mathbf{b}_i u_i(k), \quad \text{for } i = 1, 2$$
$$y(k) = \mathbf{c}_i\mathbf{x}_2(k) + d_i u_2(k), \quad \text{for } i = 1, 2$$
$$y_1(k) = u_2(k)$$
$$y(k) = y_2(k). \tag{9.5}$$

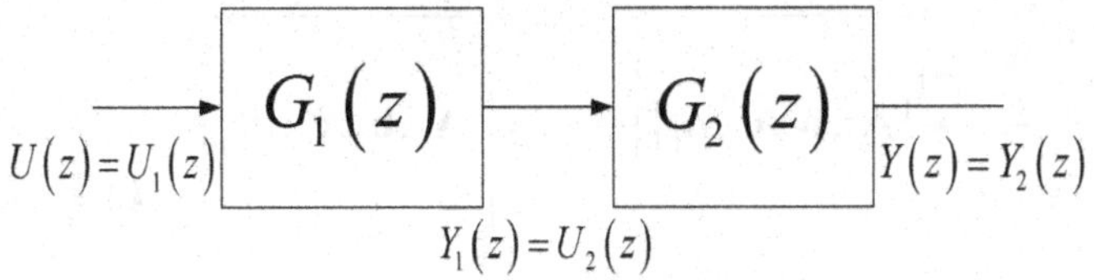

Fig. 9.1   Serial connection of two FOTFs.

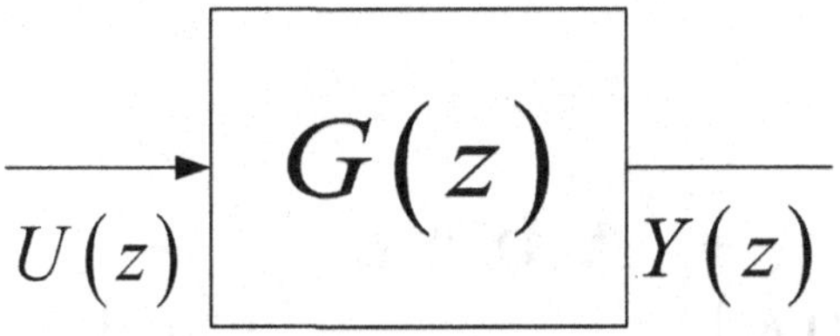

Fig. 9.2   Equivalent block of the FOTF.

Now one defines a new state vector

$$\mathbf{x}(k) = \begin{bmatrix} \mathbf{x}_1(k) \\ \mathbf{x}_2(k) \end{bmatrix}. \tag{9.6}$$

Then from (9.5) with (9.6) the FOSSE is of the form

$$\begin{bmatrix} {}^{GL}_{0}\Delta^{(\nu)}_{k}\mathbf{x}_1(k) \\ {}^{GL}_{0}\Delta^{(\nu)}_{k}\mathbf{x}_2(k) \end{bmatrix} = \begin{bmatrix} \mathbf{A}_1 & \mathbf{0} \\ \mathbf{b}_2\mathbf{c}_1 & \mathbf{A}_2 \end{bmatrix} \begin{bmatrix} \mathbf{x}_1(k) \\ \mathbf{x}_2(k) \end{bmatrix} + \begin{bmatrix} \mathbf{b}_1 \\ d_1\mathbf{b}_2 \end{bmatrix} u(k) \tag{9.7}$$

$$y(k) = \begin{bmatrix} \mathbf{0} & \mathbf{c}_2 \end{bmatrix} \begin{bmatrix} \mathbf{x}_1(k) \\ \mathbf{x}_2(k) \end{bmatrix} + [d_1 d_2]u(k). \tag{9.8}$$

Hence the block diagram in Fig. 9.3 is equivalent to Fig. 9.4 with

$$\mathbf{A} = \begin{bmatrix} \mathbf{A}_1 & \mathbf{0} \\ \mathbf{b}_2\mathbf{c}_1 & \mathbf{A}_2 \end{bmatrix}, \quad \mathbf{b} = \begin{bmatrix} \mathbf{b}_1 \\ d_1\mathbf{b}_2 \end{bmatrix}, \quad \mathbf{c} = \begin{bmatrix} \mathbf{0} & \mathbf{c}_2 \end{bmatrix}, \quad d = [d_1 d_2]. \tag{9.9}$$

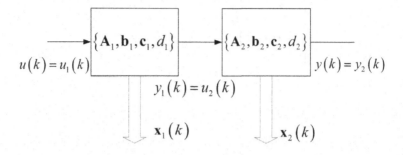

Fig. 9.3 Serial connection of two FOSSs.

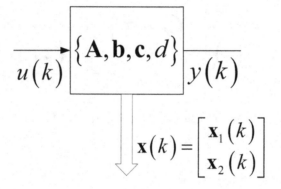

Fig. 9.4 Equivalent block of the FOSS.

Finally, one considers a serial connection of two subsystems described by the FOPLMs. The connection is shown in Fig. 9.5. From this figure one derives

$$\mathbf{u}_1(k) = \mathbf{u}(k)$$
$$\mathbf{y}_i(k) = \mathbf{D}_{i,k}^{-1}\mathbf{N}_{i,k}\mathbf{u}_i(k), \quad \text{for } i = 1, 2$$
$$\mathbf{y}_1(k) = \mathbf{u}_2(k)$$
$$\mathbf{y}_2(k) = \mathbf{y}(k). \tag{9.10}$$

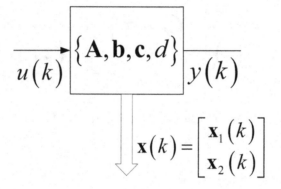

Fig. 9.5 Serial connection of elements described by the PLM.

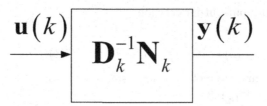

Fig. 9.6 Equivalent block described by the PLM.

Combining the above equations to eliminate variables $\mathbf{u}_1(k), \mathbf{u}_2(k), \mathbf{y}_1(k), \mathbf{y}_2(k)$, one obtains the equivalent description

$$\mathbf{y}(k) = \mathbf{D}_{2,k}^{-1}\mathbf{N}_{2,k}\mathbf{D}_{1,k}^{-1}\mathbf{N}_{1,k}. \tag{9.11}$$

In most real dynamic systems, the matrices $\mathbf{N}_{i,k}$ are singular. This property is related to the zero value d_i in the FOSSEs and a condition $q_i < p_i$ in the FOTFs. Hence, one gets the equivalent block presented in Fig. 9.6 with

$$\mathbf{D} = \mathbf{D}_2, \ \mathbf{N} = \mathbf{N}_{2,k}\mathbf{D}_{1,k}^{-1}\mathbf{N}_{1,k}. \tag{9.12}$$

9.2 FO system parallel connection

The parallel connection of two subsystems described by the FOTSs is presented in Fig. 9.7.

From this figure one derives three equations

$$Y_i(z) = G_i(z)U(z), \ \text{for } i = 1, 2$$
$$Y(z) = Y_1(z) + Y_2(z). \tag{9.13}$$

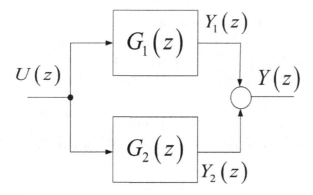

Fig. 9.7 Parallel connection of elements described by the FOTF.

They give the final input-output relation

$$Y(z) = \left[G_1(z) + G_{(s)} \right] U(z) = G(z)U(z). \tag{9.14}$$

The parallel connection of two subsystems described by the commensurate FOSSEs is shown in Fig. 9.8.

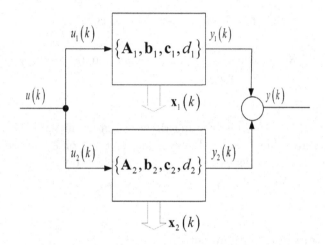

Fig. 9.8   Parallel connection of elements described by the FOSSEs.

Note that now $u_1(k) = u_2(k) = u(k)$. Following a procedure related to a case of serial connections, one creates a new state vector (9.6).

$$\begin{aligned}
{}^{GL}_0\Delta^{(\nu)}_k \mathbf{x}_i(k) &= \mathbf{A}_i\mathbf{x}_i(k) + \mathbf{b}_i u(k), \text{ for } i = 1, 2 \\
y_i(k) &= \mathbf{c}_i\mathbf{x}_i(k) + d_i u(k), \text{ for } i = 1, 2 \\
y(k) &= y_1(k) + y_2(k).
\end{aligned} \tag{9.15}$$

Then the equivalent system description is

$$\begin{bmatrix} {}^{GL}_0\Delta^{(\nu)}_k \mathbf{x}_1(k) \\ {}^{GL}_0\Delta^{(\nu)}_k \mathbf{x}_2(k) \end{bmatrix} = \begin{bmatrix} \mathbf{A}_1 & \mathbf{0} \\ \mathbf{0} & \mathbf{A}_2 \end{bmatrix} \begin{bmatrix} \mathbf{x}_1(k) \\ \mathbf{x}_2(k) \end{bmatrix} + \begin{bmatrix} \mathbf{b}_1 \\ \mathbf{b}_2 \end{bmatrix} u(k) \tag{9.16}$$

$$y(k) = \begin{bmatrix} \mathbf{c}_1 & \mathbf{c}_2 \end{bmatrix} \begin{bmatrix} \mathbf{x}_1(k) \\ \mathbf{x}_2(k) \end{bmatrix} + [d_1 + d_2]u(k). \tag{9.17}$$

This means that in the equivalent system given in Fig. 9.4, the appropriate matrices are as follows

$$\mathbf{A} = \begin{bmatrix} \mathbf{A}_1 & \mathbf{0} \\ \mathbf{0} & \mathbf{A}_2 \end{bmatrix}, \ \mathbf{b} = \begin{bmatrix} \mathbf{b}_1 \\ \mathbf{b}_2 \end{bmatrix}, \ \mathbf{c} = \begin{bmatrix} \mathbf{c}_1 & \mathbf{c}_2 \end{bmatrix}, \ d = [d_1 + d_2]. \tag{9.18}$$

Parallel system connection analysis can be supported by a numerical example.

Example 9.2. Consider two FO systems described by the FOSSEs

$$\begin{matrix}GL\\0\end{matrix}\Delta_k^{(\nu)}[x_1](k) = \left[-\frac{1}{2}\right][x_1(k)] + [1]u(k) \tag{9.19}$$

$$y_1(k) = [2]x_1(k) \tag{9.20}$$

$$\begin{matrix}GL\\0\end{matrix}\Delta_k^{(\nu)}\begin{bmatrix}x_2(k)\\x_3(k)\end{bmatrix} = \begin{bmatrix}-\frac{1}{2} & 1\\0 & -\frac{1}{3}\end{bmatrix}\begin{bmatrix}x_2(k)\\x_3(k)\end{bmatrix} + \begin{bmatrix}2\\2\end{bmatrix}u(k) \tag{9.21}$$

$$y_2(k) = \begin{bmatrix}1 & -1\end{bmatrix}\begin{bmatrix}x_2(k)\\x_3(k)\end{bmatrix}. \tag{9.22}$$

Using (9.16) and (9.17) one immediately obtains

$$\begin{bmatrix}GL\\0\end{matrix}\Delta_k^{(\nu)}\mathbf{x}_1(k)\\ \begin{matrix}GL\\0\end{matrix}\Delta_k^{(\nu)}\mathbf{x}_2(k)\\ \begin{matrix}GL\\0\end{matrix}\Delta_k^{(\nu)}\mathbf{x}_3(k)\end{bmatrix} = \begin{bmatrix}-\frac{1}{2} & 0 & 0\\0 & -\frac{1}{2} & 1\\0 & 0 & \frac{1}{3}\end{bmatrix}\begin{bmatrix}x_1(k)\\x_2(k)\\x_3(k)\end{bmatrix} + \begin{bmatrix}1\\2\\2\end{bmatrix}u(k) \tag{9.23}$$

$$y(k) = \begin{bmatrix}2 & 1 & -1\end{bmatrix}\begin{bmatrix}x_1(k)\\x_2(k)\\x_3(k)\end{bmatrix}. \tag{9.24}$$

To check the calculation correctness one first evaluates the FOTFs of both subsystems

$$G_1(z) = \mathbf{c}_1\left[(1-z^{-1})^\nu\mathbf{1} - \mathbf{A}_1\right]^{-1}\mathbf{b}_1$$

$$= [2]\left[(1-z^{-1})^\nu + \frac{1}{2}\right]^{-1}[1] = \frac{2}{(1-z^{-1})^\nu + \frac{1}{2}} \tag{9.25}$$

$$G_2(z) = \mathbf{c}_2\left[(1-z^{-1})^\nu\mathbf{1} - \mathbf{A}_2\right]^{-1}\mathbf{b}_2$$

$$= \begin{bmatrix}1 & -1\end{bmatrix}\begin{bmatrix}(1-z^{-1})^\nu + \frac{1}{2} & -1\\0 & (1-z^{-1})^\nu + \frac{1}{3}\end{bmatrix}^{-1}\begin{bmatrix}2\\2\end{bmatrix}$$

$$= -\frac{\frac{7}{3}}{\left[(1-z^{-1})^\nu + \frac{1}{2}\right]\left[(1-z^{-1})^\nu + \frac{1}{3}\right]}. \tag{9.26}$$

Then

$$G(z) = G_1(z) + G_2(z)$$

$$= \frac{2}{(1-z^{-1})^\nu + \frac{1}{2}} - \frac{\frac{7}{3}}{\left[(1-z^{-1})^\nu + \frac{1}{2}\right]\left[(1-z^{-1})^\nu + \frac{1}{3}\right]}$$

$$= \frac{2(1-z^{-1})^\nu - \frac{5}{3}}{\left[(1-z^{-1})^\nu + \frac{1}{2}\right]\left[(1-z^{-1})^\nu + \frac{1}{3}\right]} = \frac{2(1-z^{-1})^\nu - \frac{5}{3}}{(1-z^{-1})^{2\nu} + \frac{5}{6}(1-z^{-1})^\nu + \frac{1}{6}}. \tag{9.27}$$

Direct calculation of $G(z)$ using (9.23)–(9.24) leads to the same result

$$G(z) = \mathbf{c}\left[(1-z^{-1})^{\nu}\mathbf{1} - \mathbf{A}\right]^{-1}\mathbf{b}$$

$$= \begin{bmatrix} 2 & 1 & -1 \end{bmatrix} \begin{bmatrix} (1-z^{-1})^{\nu}+\frac{1}{2} & 0 & 0 \\ 0 & (1-z^{-1})^{\nu}+\frac{1}{2} & 1 \\ 0 & 0 & (1-z^{-1})^{\nu}+\frac{1}{3} \end{bmatrix}^{-1} \begin{bmatrix} 1 \\ 2 \\ 2 \end{bmatrix}$$

$$= \frac{2(1-z^{-1})^{\nu} - \frac{5}{3}}{\left[(1-z^{-1})^{\nu}+\frac{1}{2}\right]\left[(1-z^{-1})^{\nu}+\frac{1}{3}\right]}. \quad (9.28)$$

The parallel connection of two subsystems described by the PLMEs is very simple because from Fig. 9.9 one obtains

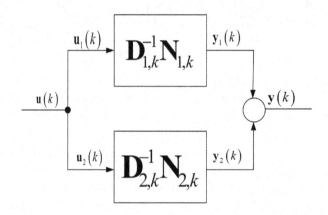

Fig. 9.9   Parallel connection of elements described by the FOPLM.

$$\mathbf{y}_i(k) = \mathbf{D}_{i,k}^{-1}\mathbf{N}_{i,k} \text{ for } i = 1, 2$$

$$\mathbf{y}(k) = \mathbf{y}_1(k) + \mathbf{y}_2(k) \quad (9.29)$$

from which one gets

$$\mathbf{y}(k) = \left(\mathbf{D}_{1,k}^{-1}\mathbf{N}_{1,k} + \mathbf{D}_{2,k}^{-1}\mathbf{N}_{2,k}\right)\mathbf{u}(k)$$

$$= \mathbf{D}_{1,k}^{-1}\left(\mathbf{N}_{1,k} + \mathbf{D}_{1,k}\mathbf{D}_{2,k}^{-1}\mathbf{N}_{2,k}\right)\mathbf{u}(k) = \mathbf{D}_{2,k}^{-1}\left(\mathbf{D}_{2,k}\mathbf{D}_{1,k}^{-1}\mathbf{N}_{1,k} + \mathbf{N}_{2,k}\right)\mathbf{u}(k) \quad (9.30)$$

and

$$\mathbf{D}_k = \mathbf{D}_{1,k}, \ \mathbf{N}_k = \mathbf{N}_{1,k} + \mathbf{D}_{1,k}\mathbf{D}_{2,k}^{-1}\mathbf{N}_{2,k}$$
$$\mathbf{D}_k = \mathbf{D}_{2,k}, \ \mathbf{N}_k = \mathbf{D}_{2,k}\mathbf{D}_{1,k}^{-1}\mathbf{N}_{1,k} + \mathbf{N}_{2,k}. \tag{9.31}$$

As was mentioned earlier matrices $\mathbf{N}_{i,k}$ are usually singular. Several properties of the upper triangular band matrices are given in a form of the Lemma 9.1.

Lemma 9.1. *Consider singular $\mathbf{N}_{i,k}$ and non-singular $\mathbf{D}_{i,k}$ ($i = 1, 2$) upper triangular band matrices. Then*

(I) $\mathbf{N}_{1,k} + \mathbf{N}_{2,k}$ *– singular*
(II) $\mathbf{N}_{1,k}\mathbf{N}_{2,k}$ *– singular*
(III) $\mathbf{N}_{1,k} + \mathbf{D}_{1,k}$ *– non-singular*
(IV) $\mathbf{N}_{1,k}\mathbf{D}_{1,k}$ *– singular*
(V) $\mathbf{D}_{1,k}\mathbf{D}_{2,k}$ *– non-singular.*

Proof. The product and sum of two upper triangular matrices are also upper triangular band ones. In such matrices elements on the main diagonals are $n_{i,11} = 0$ and $d_{i,11} \neq 0$. Then $n_{1,11} + n_{2,11} = 0$, $n_{1,11}n_{2,11} = 0$, $n_{1,11} + d_{1,11} \neq 0$, $n_{1,11}d_{1,11} = 0$ and $d_{1,11}d_{2,11} \neq 0$. □

9.3   FO closed-loop system

The configuration of the dynamic systems with a negative feedback is the most important of the previously considered systems. The block diagram of systems described by the FOTSs is presented in Fig. 9.10.

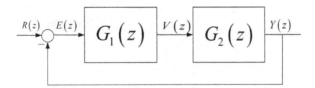

Fig. 9.10   FOTF of the closed-loop system with negative unity feedback.

Note that there are two dynamic systems connected in series with a negative unity feedback. There are three equations describing this connection

$$V(z) = G_1(z)E(z)$$
$$Y(z) = G_2(z)V(z)$$
$$E(z) = R(z) - Y(z). \tag{9.32}$$

where

> $R(z)$ – reference signal
> $Y(z)$ – closed-loop system output signa l(the plant output)
> $V(z)$ – controlling signal (the plant input)
> $E(z)$ – closed-loop system error signal
> $G_1(z)$ – controller FOTF
> $G_2(z)$ – plant FOTF.

When combining equations (9.32) to eliminate signals $E(z).V(z)$ one obtains the IO FOTF of the closed-loop system

$$G_{YR}(z) = \frac{Y(z)}{R(z)} = \frac{G_2(z)G_1(z)}{1 + G_2(z)G_1(z)}. \tag{9.33}$$

The same equations allow one to also determine the so-called sensitivity TF (error TF)

$$G_{ER}(z) = \frac{E(z)}{R(z)} = \frac{1}{1 + G_2(z)G_1(z)}. \tag{9.34}$$

The closed-loop system described by the FOSSEs and presented in Fig. 9.11 consists of seven equations

$$\begin{aligned}
{}^{GL}_{0}\Delta^{(\nu)}_k \mathbf{x}_i(k) &= \mathbf{A}_i\mathbf{x}_i(k) + \mathbf{b}_i u_i(k), \quad \text{for } i = 1,2 \\
y_i(k) &= \mathbf{c}_i\mathbf{x}_i(k) + d_i u_i(k), \quad \text{for } i = 1,2 \\
u_1(k) &= e(k) \\
y_1(k) &= u_2(k) \\
y_2(k) &= y(k) \\
e(k) &= r(k) - y(k).
\end{aligned} \tag{9.35}$$

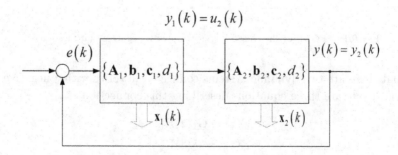

Fig. 9.11 FOSSE of the closed-loop system with negative unity output feedback.

From these equations, one gets

$$\begin{bmatrix} {}_{0}^{GL}\Delta_{k}^{(\nu)}\mathbf{x}_1(k) \\ {}_{0}^{GL}\Delta_{k}^{(\nu)}\mathbf{x}_2(k) \end{bmatrix}$$

$$= \begin{bmatrix} \mathbf{A}_1 - \frac{d_2}{1+d_1d_2}\mathbf{b}_1\mathbf{c}_1 & -\frac{1}{1+d_1d_2}\mathbf{b}_1\mathbf{c}_2 \\ \frac{1}{1+d_1d_2}\mathbf{b}_2\mathbf{c}_1 & \mathbf{A}_2 - \frac{d_1}{1+d_1d_2}\mathbf{b}_1\mathbf{c}_1 \end{bmatrix} \begin{bmatrix} \mathbf{x}_1(k) \\ \mathbf{x}_2(k) \end{bmatrix} + \begin{bmatrix} \frac{1}{1+d_1d_2}\mathbf{b}_1 \\ \frac{d_1}{1+d_1d_2}\mathbf{b}_2 \end{bmatrix} r(k) \quad (9.36)$$

$$y(k) = \begin{bmatrix} \frac{d_2}{1+d_1d_2}\mathbf{c}_1 & \frac{1}{1+d_1d_2}\mathbf{c}_2 \end{bmatrix} \begin{bmatrix} \mathbf{x}_1(k) \\ \mathbf{x}_2(k) \end{bmatrix} + \begin{bmatrix} \frac{d_1d_2}{1+d_1d_2} \end{bmatrix} r(k). \quad (9.37)$$

These equations are valid for $1 + d_1d_2 \neq 0$. In practice, this condition is always satisfied. For $d_2 = 0$, (9.36) and (9.37) essentially simplify

$$\begin{bmatrix} {}_{0}^{GL}\Delta_{k}^{(\nu)}\mathbf{x}_1(k) \\ {}_{0}^{GL}\Delta_{k}^{(\nu)}\mathbf{x}_2(k) \end{bmatrix}$$

$$= \begin{bmatrix} \mathbf{A}_1 & -\mathbf{b}_1\mathbf{c}_2 \\ \mathbf{b}_2\mathbf{c}_1 & \mathbf{A}_2 - d_1\mathbf{b}_1\mathbf{c}_1 \end{bmatrix} \begin{bmatrix} \mathbf{x}_1(k) \\ \mathbf{x}_2(k) \end{bmatrix} + \begin{bmatrix} \mathbf{b}_1 \\ d_1\mathbf{b}_2 \end{bmatrix} r(k) \quad (9.38)$$

$$y(k) = \begin{bmatrix} \mathbf{0} & \mathbf{c}_2 \end{bmatrix} \begin{bmatrix} \mathbf{x}_1(k) \\ \mathbf{x}_2(k) \end{bmatrix} + [0]\, r(k). \quad (9.39)$$

and additionally, for $d_1 = 0$

$$\begin{bmatrix} {}_{0}^{GL}\Delta_{k}^{(\nu)}\mathbf{x}_1(k) \\ {}_{0}^{GL}\Delta_{k}^{(\nu)}\mathbf{x}_2(k) \end{bmatrix} = \begin{bmatrix} \mathbf{A}_1 & -\mathbf{b}_1\mathbf{c}_2 \\ \mathbf{b}_2\mathbf{c}_1 & \mathbf{A}_2 \end{bmatrix} \begin{bmatrix} \mathbf{x}_1(k) \\ \mathbf{x}_2(k) \end{bmatrix} + \begin{bmatrix} \mathbf{b}_1 \\ \mathbf{0} \end{bmatrix} r(k) \quad (9.40)$$

$$y(k) = \begin{bmatrix} \mathbf{0} & \mathbf{c}_2 \end{bmatrix} \begin{bmatrix} \mathbf{x}_1(k) \\ \mathbf{x}_2(k) \end{bmatrix} + [0]\, r(k). \quad (9.41)$$

The same result can be obtained from a system described by (9.7) and (9.7) with negative unity feedback. The state-space description provides an additional advantage over the TF one, because one can make a feedback from state vectors as is presented in Fig. 9.12.

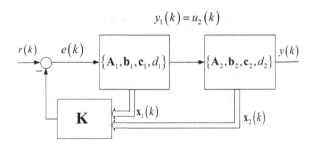

Fig. 9.12 FOSSE of the closed-loop system with negative state feedback.

Here, $\mathbf{K} = [\mathbf{k}_1 \ \mathbf{k}_2]$ is a block matrix and the last equation in a set (9.35) must be replaced by

$$e(k) = r(k) - \begin{bmatrix} \mathbf{k}_1 & \mathbf{k}_2 \end{bmatrix} \begin{bmatrix} \mathbf{x}_1(k) \\ \mathbf{x}_2(k) \end{bmatrix}. \tag{9.42}$$

Then the closed-loop system FOSSE is as follows

$$\begin{bmatrix} {}^{GL}_0\Delta_k^{(\nu)}\mathbf{x}_1(k) \\ {}^{GL}_0\Delta_k^{(\nu)}\mathbf{x}_2(k) \end{bmatrix} = \begin{bmatrix} \mathbf{A}_1 - \mathbf{b}_1\mathbf{k}_1 & -\mathbf{b}_1\mathbf{k}_2 \\ \mathbf{b}_2(\mathbf{c}_1 - \mathbf{k}_1) & \mathbf{A}_2 - d_1\mathbf{b}_2\mathbf{k}_2 \end{bmatrix} \begin{bmatrix} \mathbf{x}_1(k) \\ \mathbf{x}_2(k) \end{bmatrix} + \begin{bmatrix} \mathbf{b}_1 \\ d_1\mathbf{b}_2 \end{bmatrix} r(k) \tag{9.43}$$

$$y(k) = \begin{bmatrix} d_2(\mathbf{c}_1 - \mathbf{k}_1) & \mathbf{c}_2 - d_1\mathbf{k}_2 \end{bmatrix} \begin{bmatrix} \mathbf{x}_1(k) \\ \mathbf{x}_2(k) \end{bmatrix} + [d_1 d_2] \, r(k). \tag{9.44}$$

In practice, a feedback from the controller is not used. This means that $\mathbf{k}_1 = \mathbf{0}$. Keeping the previously entered assumption $d_2 = 0$, a simplified description takes the form

$$\begin{bmatrix} {}^{GL}_0\Delta_k^{(\nu)}\mathbf{x}_1(k) \\ {}^{GL}_0\Delta_k^{(\nu)}\mathbf{x}_2(k) \end{bmatrix} = \begin{bmatrix} \mathbf{A}_1 & -\mathbf{b}_1\mathbf{k}_2 \\ \mathbf{b}_2\mathbf{c}_1 & \mathbf{A}_2 - d_1\mathbf{b}_2\mathbf{k}_2 \end{bmatrix} \begin{bmatrix} \mathbf{x}_1(k) \\ \mathbf{x}_2(k) \end{bmatrix} + \begin{bmatrix} \mathbf{b}_1 \\ d_1\mathbf{b}_2 \end{bmatrix} r(k) \tag{9.45}$$

$$y(k) = \begin{bmatrix} \mathbf{0} & \mathbf{c}_2 - d_1\mathbf{k}_2 \end{bmatrix} \begin{bmatrix} \mathbf{x}_1(k) \\ \mathbf{x}_2(k) \end{bmatrix} + [0] \, r(k). \tag{9.46}$$

Finally, a description of a closed-loop system described by the PMLEs will be analyzed. It is given in Fig. 9.13.

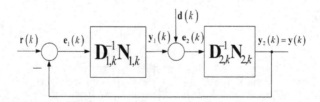

Fig. 9.13   PML of the closed-loop system with negative state feedback.

An additional disturbance signal $\mathbf{d}(k)$ was introduced. In practice, such a signal is beyond any control. From Fig. 9.13, one derives five equations

$$e_1(k) = \mathbf{r}(k) - \mathbf{y}_2(k)$$

$$e_2(k) = \mathbf{d}(k) + \mathbf{y}_1(k)$$

$$\mathbf{y}_1(k) = \mathbf{D}_{1,k}^{-1}\mathbf{N}_{1,k}e_1(k)$$

$$\mathbf{y}_2(k) = \mathbf{D}_{2,k}^{-1}\mathbf{N}_{2,k}e_2(k)$$

$$\mathbf{y}_2(k) = \mathbf{y}(k). \tag{9.47}$$

An elimination of $e_1(k), e_2(k), \mathbf{y}_1(k), \mathbf{y}_2(k)$ in (9.47) leads to a main relation

$$\mathbf{y}(k) = \mathbf{D}_k^{-1}\mathbf{N}_k\mathbf{r}(k) + \mathbf{D}_k^{-1}\mathbf{D}_{2,k}^{-1}\mathbf{N}_{2,k}\mathbf{d}(k) \tag{9.48}$$

where

$$\mathbf{D}_k = \mathbf{D}_{2,k}\left(\mathbf{I}_k + \mathbf{D}_{2,k}^{-1}\mathbf{N}_{2,k}\mathbf{D}_{1,k}^{-1}\mathbf{N}_{1,k}\right), \quad \mathbf{N}_k = \mathbf{N}_{2,k}\mathbf{D}_{1,k}^{-1}\mathbf{N}_{1,k}. \tag{9.49}$$

Note that by assumption, $\mathbf{D}_{2,k}$ is nonsingular. So is $\mathbf{I}_k + \mathbf{D}_{2,k}^{-1}\mathbf{N}_{2,k}\mathbf{D}_{1,k}^{-1}\mathbf{N}_{1,k}$. This matrix is an upper triangular band matrix. If even one of two matrices $\mathbf{N}_{1,k}, \mathbf{N}_{2,k}$ is singular then the element on the main diagonal of the matrix in question is equal to 0. For a case when both matrices $\mathbf{N}_{1,k}, \mathbf{N}_{2,k}$ are non-singular, the element on a main diagonal is $1 + \frac{n_{1,11}n_{2,11}}{d_{1,11}d_{2,11}} \neq 0$ because in the considered cases, one assumes that $n_{1,11}, n_{2,11}, d_{1,11}, d_{2,11} > 0$ which means that both dynamic systems have positive gains.

The block diagram in Fig. 9.13 can be rearranged to the form presented in Fig. 9.14.

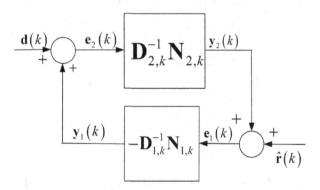

Fig. 9.14   Alternative form of the FOSSE of the closed-loop system with negative state feedback.

Note that this time signs in both adders inputs are positive. This means that the negative feedback minus sign is hidden in the controller block. A notation is introduced

$$\mathbf{P}_k = \mathbf{D}_{2,k}^{-1}\mathbf{N}_{2,k}, \ \hat{\mathbf{C}}_k = -\mathbf{C}_k = -\mathbf{D}_{1,k}^{-1}\mathbf{N}_{1,k}, \ \hat{\mathbf{r}} = -\mathbf{r}. \tag{9.50}$$

Then, from Fig. 9.14 one gets

$$\mathbf{e}_1(k) = \hat{\mathbf{r}}(k) + \mathbf{P}_k\mathbf{e}_2(k)$$
$$\mathbf{e}_2(k) = \mathbf{d}(k) + \hat{\mathbf{C}}_k\mathbf{e}_1(k). \tag{9.51}$$

The above equations can be represented in a block-matrix form

$$\begin{bmatrix} \mathbf{I}_k & -\mathbf{P}_k \\ -\hat{\mathbf{C}}_k & \mathbf{I}_k \end{bmatrix} \begin{bmatrix} \mathbf{e}_1(k) \\ \mathbf{e}_2(k) \end{bmatrix} = \begin{bmatrix} \hat{\mathbf{r}}(k) \\ \mathbf{d}(k) \end{bmatrix}. \tag{9.52}$$

From (9.52) one calculates errors $\mathbf{e}_1(k)$ and $\mathbf{e}_2(k)$

$$\begin{bmatrix} \mathbf{e}_1(k) \\ \mathbf{e}_2(k) \end{bmatrix} = \begin{bmatrix} \mathbf{I}_k & -\mathbf{P}_k \\ -\hat{\mathbf{C}}_k & \mathbf{I}_k \end{bmatrix}^{-1} \begin{bmatrix} \hat{\mathbf{r}}(k) \\ \mathbf{d}(k) \end{bmatrix}. \tag{9.53}$$

Now, one defines PML TF of the closed-loop system

$$\mathbf{G}_k = \begin{bmatrix} \mathbf{G}_{11,k} & \mathbf{G}_{12,k} \\ \mathbf{G}_{21,k} & \mathbf{G}_{22,k} \end{bmatrix} = \begin{bmatrix} \mathbf{I}_k & -\mathbf{P}_k \\ -\hat{\mathbf{C}}_k & \mathbf{I}_k \end{bmatrix}^{-1}. \tag{9.54}$$

Using direct calculations one can check that

$$\mathbf{G}_k = \begin{bmatrix} \left(\mathbf{I}_k - \mathbf{P}_k\hat{\mathbf{C}}_k\right)^{-1} & \mathbf{P}_k\left(\mathbf{I}_k - \mathbf{P}_k\hat{\mathbf{C}}_k\right)^{-1} \\ \hat{\mathbf{C}}_k\left(\mathbf{I}_k - \mathbf{P}_k\hat{\mathbf{C}}_k\right)^{-1} & \left(\mathbf{I}_k - \mathbf{P}_k\hat{\mathbf{C}}_k\right)^{-1} \end{bmatrix}$$

$$= \begin{bmatrix} \mathbf{1}_k & \mathbf{P}_k \\ \hat{\mathbf{C}}_k & \mathbf{I}_k \end{bmatrix} \begin{bmatrix} \left(\mathbf{I}_k - \mathbf{P}_k\hat{\mathbf{C}}_k\right)^{-1} & \mathbf{0}_k \\ \mathbf{0}_k & \left(\mathbf{I}_k - \mathbf{P}_k\hat{\mathbf{C}}_k\right)^{-1} \end{bmatrix}. \tag{9.55}$$

A substitution of (9.55) into (9.54) yields

$$\begin{bmatrix} \mathbf{e}_1(k) \\ \mathbf{e}_2(k) \end{bmatrix} = \begin{bmatrix} \mathbf{I}_k & \mathbf{P}_k \\ \hat{\mathbf{C}}_k & \mathbf{I}_k \end{bmatrix} \begin{bmatrix} \left(\mathbf{I}_k - \mathbf{P}_k\hat{\mathbf{C}}_k\right)^{-1} & \mathbf{0}_k \\ \mathbf{0}_k & \left(\mathbf{I}_k - \mathbf{P}_k\hat{\mathbf{C}}_k\right)^{-1} \end{bmatrix} \begin{bmatrix} \hat{\mathbf{r}}(k) \\ \mathbf{d}(k) \end{bmatrix}$$

$$= \begin{bmatrix} \mathbf{I}_k & \mathbf{D}_{2,k}^{-1}\mathbf{N}_{2,k} \\ -\mathbf{D}_{1,k}^{-1}\mathbf{N}_{1,k} & \mathbf{I}_k \end{bmatrix}$$

$$\times \begin{bmatrix} \left(\mathbf{I}_k - \mathbf{D}_{2,k}^{-1}\mathbf{N}_{2,k}\mathbf{D}_{1,k}^{-1}\mathbf{N}_{1,k}\right)^{-1} & \mathbf{0}_k \\ \mathbf{0}_k & \left(\mathbf{I}_k - \mathbf{D}_{2,k}^{-1}\mathbf{N}_{2,k}\mathbf{D}_{1,k}^{-1}\mathbf{N}_{1,k}\right)^{-1} \end{bmatrix} \begin{bmatrix} \hat{\mathbf{r}}(k) \\ \mathbf{d}(k) \end{bmatrix}. \tag{9.56}$$

Returning to the signal $\mathbf{r}(k)$ one gets the final relation

$$\begin{bmatrix} \mathbf{e}_1(k) \\ \mathbf{e}_2(k) \end{bmatrix} = \begin{bmatrix} \mathbf{I}_k & \mathbf{D}_{2,k}^{-1}\mathbf{N}_{2,k} \\ -\mathbf{D}_{1,k}^{-1}\mathbf{N}_{1,k} & \mathbf{I}_k \end{bmatrix}$$

$$\times \begin{bmatrix} -\left(\mathbf{I}_k - \mathbf{D}_{2,k}^{-1}\mathbf{N}_{2,k}\mathbf{D}_{1,k}^{-1}\mathbf{N}_{1,k}\right)^{-1} & \mathbf{0}_k \\ \mathbf{0}_k & \left(\mathbf{I}_k - \mathbf{D}_{2,k}^{-1}\mathbf{N}_{2,k}\mathbf{D}_{1,k}^{-1}\mathbf{N}_{1,k}\right)^{-1} \end{bmatrix} \begin{bmatrix} \mathbf{r}(k) \\ \mathbf{d}(k). \end{bmatrix} \quad (9.57)$$

Formula (9.55) reveals another interesting equality. Post-multiplying both sides by $\mathbf{G}_k^{-1}$ yields

$$\begin{bmatrix} \left(\mathbf{I}_k - \mathbf{P}_k\hat{\mathbf{C}}_k\right)^{-1} & \mathbf{P}_k\left(\mathbf{I}_k - \mathbf{P}_k\hat{\mathbf{C}}_k\right)^{-1} \\ \hat{\mathbf{C}}_k\left(\mathbf{I}_k - \mathbf{P}_k\hat{\mathbf{C}}_k\right)^{-1} & \left(\mathbf{I}_k - \mathbf{P}_k\hat{\mathbf{C}}_k\right)^{-1} \end{bmatrix} \mathbf{G}_k^{-1}$$

$$= \begin{bmatrix} \left(\mathbf{I}_k - \mathbf{P}_k\hat{\mathbf{C}}_k\right)^{-1} & \mathbf{P}_k\left(\mathbf{I}_k - \mathbf{P}_k\hat{\mathbf{C}}_k\right)^{-1} \\ \hat{\mathbf{C}}_k\left(\mathbf{I}_k - \mathbf{P}_k\hat{\mathbf{C}}_k\right)^{-1} & \left(\mathbf{I}_k - \mathbf{P}_k\hat{\mathbf{C}}_k\right)^{-1} \end{bmatrix} \begin{bmatrix} \mathbf{I}_k & -\mathbf{P}_k \\ -\hat{\mathbf{C}}_k & \mathbf{I}_k \end{bmatrix}$$

$$= \begin{bmatrix} \left(\mathbf{I}_k - \mathbf{P}_k\hat{\mathbf{C}}_k\right)^{-1} - \mathbf{P}_k\left(\mathbf{I}_k - \mathbf{P}_k\hat{\mathbf{C}}_k\right)^{-1}\hat{\mathbf{C}}_k \\ \hat{\mathbf{C}}_k\left(\mathbf{I}_k - \mathbf{P}_k\hat{\mathbf{C}}_k\right)^{-1} - \left(\mathbf{I}_k - \mathbf{P}_k\hat{\mathbf{C}}_k\right)^{-1}\hat{\mathbf{C}}_k \end{bmatrix}$$

$$\times \begin{bmatrix} \mathbf{P}_k\left(\mathbf{I}_k - \mathbf{P}_k\hat{\mathbf{C}}_k\right)^{-1} - \left(\mathbf{I}_k - \mathbf{P}_k\hat{\mathbf{C}}_k\right)^{-1}\mathbf{P}_k \\ \left(\mathbf{I}_k - \mathbf{P}_k\hat{\mathbf{C}}_k\right)^{-1} - \hat{\mathbf{C}}_k\left(\mathbf{I}_k - \mathbf{P}_k\hat{\mathbf{C}}_k\right)^{-1}\mathbf{P}_k \end{bmatrix} = \begin{bmatrix} \mathbf{I}_k & \mathbf{0}_k \\ \mathbf{0}_k & \mathbf{I}_k \end{bmatrix}.$$

$$(9.58)$$

A comparison of appropriate blocks in (9.58) leads to four equalities

$$\left(\mathbf{I}_k - \mathbf{P}_k\hat{\mathbf{C}}_k\right)^{-1} - \mathbf{P}_k\left(\mathbf{I}_k - \mathbf{P}_k\hat{\mathbf{C}}_k\right)^{-1}\hat{\mathbf{C}}_k = \mathbf{I}_k \quad (9.59)$$

$$\hat{\mathbf{C}}_k\left(\mathbf{I}_k - \mathbf{P}_k\hat{\mathbf{C}}_k\right)^{-1} = \left(\mathbf{I}_k - \mathbf{P}_k\hat{\mathbf{C}}_k\right)^{-1}\hat{\mathbf{C}}_k \quad (9.60)$$

$$\mathbf{P}_k\left(\mathbf{I}_k - \mathbf{P}_k\hat{\mathbf{C}}_k\right)^{-1} = \left(\mathbf{I}_k - \mathbf{P}_k\hat{\mathbf{C}}_k\right)^{-1}\mathbf{P}_k \quad (9.61)$$

$$\left(\mathbf{I}_k - \mathbf{P}_k\hat{\mathbf{C}}_k\right)^{-1} - \hat{\mathbf{C}}_k\left(\mathbf{I}_k - \mathbf{P}_k\hat{\mathbf{C}}_k\right)^{-1}\mathbf{P}_k = \mathbf{I}_k. \quad (9.62)$$

From (9.53) with (9.55) one gets

$$\mathbf{e}_2(k) = \mathbf{C}_k\left(\mathbf{I}_k - \mathbf{P}_k\hat{\mathbf{C}}_k\right)^{-1}\mathbf{r}_k(k) + \left(\mathbf{I}_k - \mathbf{P}_k\hat{\mathbf{C}}_k\right)^{-1}\mathbf{d}(k). \quad (9.63)$$

By (9.47) and (9.50)

$$\mathbf{y}(k) = \mathbf{y}_2(k) = \mathbf{P}_k \mathbf{e}_2(k)$$
$$= \mathbf{P}_k \mathbf{C}_k \left(\mathbf{I}_k + \mathbf{P}_k \mathbf{C}_k\right)^{-1} \mathbf{r}_k(k) + \mathbf{P}_k \left(\mathbf{I}_k + \mathbf{P}_k \mathbf{C}_k\right)^{-1} \mathbf{d}(k)$$
$$= \mathbf{P}_k \mathbf{C}_k \left(\mathbf{I}_k + \mathbf{P}_k \mathbf{C}_k\right)^{-1} \mathbf{P}_k \mathbf{C}_k \mathbf{r}_k(k) + \left(\mathbf{I}_k + \mathbf{P}_k \mathbf{C}_k\right)^{-1} \mathbf{P}_k \mathbf{d}(k). \qquad (9.64)$$

In the above formula transformations, equalities (9.61) and (9.62) were used. It should be realized that formula (9.64) is the same as (9.48).

Chapter 10

Fractional discrete-time PID controller

Since the early fifties of the last century PID controllers have proven their wide applicability in closed-loop systems synthesis [Åstrom and Wittenmark (1984); Astrom and Hagglund (1995); Hagglund and Astrom (1995); Tan *et al.* (2000); Astrom and Hagglund (2001); ODwyer (2009); Vilanova and Visioli (2012)]. This is because of their simplicity and relative ease of tuning where only three parameters must be tuned. Allowing the PID controller FO summation of order μ and differentiation of order ν significantly expands the possibilities of shaping the closed-loop system dynamics. However, it is associated with an increase in the number of parameters to tune. For more than a decade, intensive work has been carried out on the application of the FOPID controllers [Podlubny *at al.* (1997); Merrikh-Bayat (2011); Podlubny (1999b); Datta *et al.* (2000); Machado (2001); Xue and Chen (2002); Monje *et al.* (2004); Chen *et al.* (2004a); Chen (2006a,b); Xue *et al.* (2006); Valerio and Costa (2006a); Karanjkar *et al.* (2012); Giridharan and Karthikeyan (2014)]. The FO controllers have been successfully applied in different control systems [Chao *et al.* (2003); Xue *et al.* (2006); Hamamci (2007, 2008); Barbosa *et al.* (2008); Li *et al.* (2010); Singhal *et al.* (2012)]. As well as, important areas in robotics [Silva and Machado (2006a); Delavari *et al.* (2010)]. Different controller parameter tuning methods are applied [Lurie (1994); Podlubny (1994a); Ho *et al.* (2007); Petráš (1999); Podlubny (1999b); Raynaud and Zergaïnoh (2000); Petráš *et al.* (2002); Leu (2002); Caponetto *et al.* (2002); Xue and Chen (2002); Podlubny *at al.* (2003); Barbosa (2004a,b); Valerio and Costa (2005); Zhao *et al.* Xue and Chen (2005); Cerera *et al.* (2006); Chen (2006b); Valerio and Costa (2006); Monje*et al.* (2008); Zhao and Zhang (2008); Luo and Chen (2009a); Luo and Chen (2009b); Luo *et al.* (2007, 2011); Aldair and Wang (2010); Faieghi and Nemati (2011); Merrikh-Bayat (2011); Padula and Visioli (2011); Luo and Chen (2012)]. One should note that the FOPID controllers can also be applied to the closed-systems with IO models and vice versa [Yeroglu and Tan (2011)].

10.1   Fractional PID controller

In this chapter a PID controller characterized by fractional-order summation and discrete differentiation is discussed. The block diagram of the FOPID controller is given in Fig. 10.1.

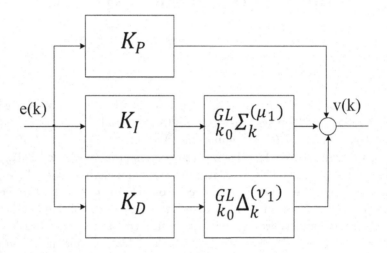

Fig. 10.1   Block diagram of the FO PID controller.

The PID action is described by an equation

$$
\begin{aligned}
v(k) &= K_P e(k) + K_I {}_{k_0}^{GL}\Sigma_k^{(\mu_1)} e(k) + K_D {}_{k_0}^{GL}\Delta_k^{(\nu_1)} e(k) \\
&= K_P e(k) + K_I {}_{k_0}^{GL}\Delta_k^{(-\mu_1)} e(k) + K_D {}_{k_0}^{GL}\Delta_k^{(\nu_1)} e(k).
\end{aligned}
\tag{10.1}
$$

where $\nu_1, \mu_1 \in \mathbf{R}_+$ are discrete differentiation and summation orders, respectively. Coefficients K_P, K_I, K_D are proportional, integration and differentiation gains.

This is formally the FO difference and summation equation. A discrete differentiation of both sides of (10.2) by ${}_{k_0}^{GL}\Delta_k^{(\mu_1)}$ gives the FODE of the form

$$
\begin{aligned}
& {}_{k_0}^{GL}\Delta_k^{(\mu_1)} v(k) \\
= K_P {}_{k_0}^{GL}\Delta_k^{(\mu_1)} e(k) &+ K_I {}_{k_0}^{GL}\Delta_k^{(\mu_1)} {}_{k_0}^{GL}\Delta_k^{(-\mu_1)} e(k) + K_D {}_{k_0}^{GL}\Delta_k^{(\mu_1)} {}_{k_0}^{GL}\Delta_k^{(\nu_1)} e(k) \\
&= K_D {}_{k_0}^{GL}\Delta_k^{(\mu_1+\nu_1)} e(k) + K_P {}_{k_0}^{GL}\Delta_k^{(\mu_1)} e(k) + K_I e(k).
\end{aligned}
\tag{10.2}
$$

Comment. One can assume that $\nu_1, \mu_1 \in \mathbf{Q}_+$ may take any positive value (including that used in the classical case when $\nu_1 = \mu_1 = 1$). Performing the procedure described in Section 6.1 one transforms equation (10.2) into the commensurate form.

Hence,

$$\mu_1 = p\nu, \ \mu_1 + \nu_1 = q\nu, \ \text{where} \nu = \frac{1}{d}, \tag{10.3}$$

and d is the least common denominator of rational orders ν_1 and μ_1. Then,

$$p < q \tag{10.4}$$

is in contradiction to the common assumption that $p \geq q$ (usually imposed on the FOBD causal systems). This important inequality (10.4) means that the FOPID controller cannot be described by the FOSSE [Kaczorek (1992)]. The FOPID controller will be further considered as an ideal FOPID controller.

The discrete TF of the ideal FOPID controller is of the form (all initial conditions are zero)

$$C_{PID}(z) = \frac{V(z)}{E(z)} = K_P + \frac{K_I}{(1 - z^{-1})^{\mu_1}} + K_D(1 - z^{-1})^{\nu_1}$$
$$= \frac{K_I + K_P(1 - z^{-1})^{\mu_1} + K_D(1 - z^{-1})^{\mu_1 + \nu_1}}{(1 - z^{-1})^{\mu_1}}. \tag{10.5}$$

This is non-proper TF. Following what is commonly applied in the classical PID control synthesis, one adds a filter in the FD part of the controller. In the simplest possible case, one adds a FO filter described by a transfer function

$$G_F(z) = \frac{a_0}{(1 - z^{-1})^{\mu_2} + a_0} \ \text{with} \ \nu_1 \leq \mu_2. \tag{10.6}$$

Then the real FOPID controller TF takes the form

$$C_{PID}(z) = \frac{V(z)}{E(z)} = K_P + \frac{K_I}{(1 - z^{-1})^{\mu_1}} + K_D a_0 \frac{(1 - z^{-1})^{\nu_1}}{(1 - z^{-1})^{\mu_2} + a_{F,0}}$$
$$= \frac{K_P v^{\mu_1 + \mu_2} + a_0 K_D v^{\mu_1 + \nu_1} + a_0 K_P v^{\mu_1} + K_I v^{\mu_2} + a_0 K_I}{v^{\mu_1} (v^{\mu_2} + a_{F,0})}. \tag{10.7}$$

where $v = 1 - z^{-1}$. For

$$\mu_1 = p_1 \nu, \ \mu_2 = p_2 \nu, \ \nu_1 = q_1 \nu, \nu_1 < \mu_2, \nu = \frac{1}{d} \tag{10.8}$$

it is a proper FOTF and it can be described by the FOSSE. To find the FOSSE of the real FOPID controller one treats it as a parallel connection of three blocks: one static and two dynamics ones. Assuming that the output matrix of the real FOPID controller equals to $\mathbf{d}_{FOPID} = [K_P]$. The second term in the sum (10.7) is described by the FOSSE

$$\,_{k_0}^{GL}\Delta_k^{(\nu)}\mathbf{x}_I(k) = \mathbf{A}_I\mathbf{x}_I(k) + \mathbf{b}_I e(k) \tag{10.9}$$

$$v_I(k) = \mathbf{c}_I\mathbf{x}_I(k) \tag{10.10}$$

where

$$\mathbf{x}_I(k) = \begin{bmatrix} \,_{k_0}^{GL}\Delta_k^{(\nu)}x_1(k) \\ \,_{k_0}^{GL}\Delta_k^{(\nu)}x_2(k) \\ \vdots \\ \,_{k_0}^{GL}\Delta_k^{(\nu)}x_{p_1-1}(k) \\ \,_{k_0}^{GL}\Delta_k^{(\nu)}x_{p_1}(k) \end{bmatrix}, \ \mathbf{A}_I = \begin{bmatrix} 0 & 1 & 0 & \cdots & 0 \\ 0 & 0 & 1 & \cdots & 0 \\ \vdots & \vdots & \vdots & & \vdots \\ 0 & 0 & 0 & \cdots & 1 \\ 0 & 0 & 0 & \cdots & 0 \end{bmatrix}, \ \mathbf{b}_I = \begin{bmatrix} 0 \\ 0 \\ \vdots \\ 0 \\ 1 \end{bmatrix} \tag{10.11}$$

$$\mathbf{c}_I = \begin{bmatrix} K_I & 0 & \cdots & 0 & 0 \end{bmatrix}. \tag{10.12}$$

The third term in the sum (10.7) is described by the FOSSE

$$\,_{k_0}^{GL}\Delta_k^{(\nu)}\mathbf{x}_D(k) = \mathbf{A}_D\mathbf{x}_D(k) + \mathbf{b}_D e(k) \tag{10.13}$$

$$v_D(k) = \mathbf{c}_D\mathbf{x}_D(k) \tag{10.14}$$

where

$$\mathbf{x}_D(k) = \begin{bmatrix} \,_{k_0}^{GL}\Delta_k^{(\nu)}x_{p_1+1}(k) \\ \,_{k_0}^{GL}\Delta_k^{(\nu)}x_{p_1+2}(k) \\ \vdots \\ \,_{k_0}^{GL}\Delta_k^{(\nu)}x_{p_1+p_2-1}(k) \\ \,_{k_0}^{GL}\Delta_k^{(\nu)}x_{p_1+p_2}(k) \end{bmatrix}, \ \mathbf{A}_D = \begin{bmatrix} 0 & 1 & 0 & \cdots & 0 \\ 0 & 0 & 1 & \cdots & 0 \\ \vdots & \vdots & \vdots & & \vdots \\ 0 & 0 & 0 & \cdots & 1 \\ -a_0 & 0 & 0 & \cdots & 0 \end{bmatrix}, \ \mathbf{b}_D = \begin{bmatrix} 0 \\ 0 \\ \vdots \\ 0 \\ 1 \end{bmatrix} \tag{10.15}$$

$$\mathbf{c}_D = \underset{1 \ \cdots \ q_1-1 \ \ q_1 \ \ q_1+1 \ \cdots \ p_1-1}{\begin{bmatrix} 0 & \cdots & 0 & K_D & 0 & \cdots & 0 \end{bmatrix}}. \tag{10.16}$$

Note: the small numbers below the elements of the matrix $\mathbf{c}_I$ indicate the only one non-zero element position. Now, according to the FOSSE building of a parallel connection rule one immediately gets the FOSSE of the real FOPID controller

$$\,_{k_0}^{GL}\Delta_k^{(\nu)}\mathbf{x}_{FOPID}(k) = \mathbf{A}_{FOPID}\mathbf{x}_{FOPID}(k) + \mathbf{b}_{FOPID}e(k) \tag{10.17}$$

$$v_{FOPID}(k) = \mathbf{c}_{FOPID}\mathbf{x}_{FOPID}(k) + \mathbf{d}_{FOPID}e(k) \tag{10.18}$$

where

$$\mathbf{x}_{FOPID}(k) = \begin{bmatrix} \,_{k_0}^{GL}\Delta_k^{(\nu)}\mathbf{x}_I(k) \\ \,_{k_0}^{GL}\Delta_k^{(\nu)}\mathbf{x}_D(k) \end{bmatrix}, \ \mathbf{A}_{FOPID} = \begin{bmatrix} \mathbf{A}_I & \mathbf{0} \\ \mathbf{0} & \mathbf{A}_D \end{bmatrix} \tag{10.19}$$

$$\mathbf{b}_{FOPID} = \begin{bmatrix} \mathbf{b}_I \\ \mathbf{b}_D \end{bmatrix}, \ \mathbf{c}_{FOPID} = \begin{bmatrix} \mathbf{c}_I & \mathbf{c}_D \end{bmatrix}, \ \mathbf{d}_{FOPID} = [K_P]. \tag{10.20}$$

From (10.7)–(10.9) one gets the FODE describing the real FOPID controller

$$
\begin{aligned}
&{}^{GL}_{0}\Delta_k^{((p_1+p_2)\nu)}v(k) + a_0 {}^{GL}_{0}\Delta_k^{(p_1\nu)}v(k)\\
&= K_P {}^{GL}_{0}\Delta_k^{((p_1+p_2)\nu)}e(k) + a_0 K_D {}^{GL}_{0}\Delta_k^{((p_1+q_1)\nu)}e(k)\\
&+ a_0 K_P {}^{GL}_{0}\Delta_k^{(p_1\nu)}e(k) + K_I {}^{GL}_{0}\Delta_k^{(p_2\nu)}e(k) + a_0 K_I e(k).
\end{aligned}
\tag{10.21}
$$

Following the procedure described in Section 6.5 one gets the FOPLE (6.116) with

$$
A_i = a^{((p_1+p_2)\nu)}(i) + a_0 a^{(p_1\nu)}(i) \tag{10.22}
$$
$$
B_i = K_P a^{((p_1+p_2)\nu)}(i) + a_0 K_D a^{((p_1+q_1)\nu)}(i)
$$
$$
+ a_0 K_P a^{(p_1\nu)}(i) + K_I a^{(p_2\nu)}(i) + a_0 K_I a^{(0\nu)}(i) \text{ for } i = 0,1,2,\dots. \tag{10.23}
$$

and $y(k) = v(k)$ and $u(k) = e(k)$. Then the main relation (6.118) describing the FOPID controller together with (6.118) uses $k_0 = 0$ to take the form

$$
\mathbf{D}_{FOPID,k}\mathbf{v}(k) = \mathbf{N}_{FOPID,k}\mathbf{e}(k) \tag{10.24}
$$

and under the permanent assumption that $A_0 = 1 + a_0 \neq 0$

$$
\mathbf{v}(k) = [\mathbf{D}_{FOPID,k}]^{-1}\mathbf{N}_{FOPID,k}\mathbf{e}(k). \tag{10.25}
$$

From four FOPID real controller description methods the most clear seems to be the FOTF one. It can be easily generalized to the form

$$
C_{PID}(z) = \frac{V(z)}{E(z)} = K_P + \frac{K_I}{(1-z^{-1})^{\mu_1}} + K_D(1-z^{-1})^{\nu}G_F(z) \tag{10.26}
$$

where $G_F(z)$ denotes the TF of any IO or FO digital filter

$$
G_F(z) = \frac{b_m z^{m-n} + b_{m-1}z^{m-n-1} + \cdots + b_1 z^{1-n} + b_) z^{-n}}{1 + a_{n-1}z^{-1} + \cdots + a_1 z^{1-n} + a_0 z^{-n}}, \quad n \geq m. \tag{10.27}
$$

10.2 FO PID controller transient characteristics

In the FOPID controller there are five parameters to tune. Three of them are already mentioned, as proportional, discrete-differentiation, and integration gains K_P, K_D, K_I. The two additional are the discrete integration and differentiation orders μ, ν, respectively. This causes a difficulty in showing the transient characteristics of the FOPID controller. The fundamental characteristics of the FOPID controller are its unit step responses. The PID controller structure consists of three independent parts: the proportional, integrating, and derivating which enable an immediate prediction of the influence of K_P, K_D, K_I on the unit step response.

Therefore the FOPID controller transient characteristics analysis will be focused on the FOs impact.

For constant parameters $K_P = 1, K_I = 0.5, K_D = 5$ and FO integration $\mu = 0.5$ the FOPID controller unit step responses are plotted in Fig. 10.2 for discrete differentiation order from a set $\nu \in \{0.1, 0.2, \ldots, 1.0\}$. In Fig. 10.3 a reverse situation is plotted. The discrete differentiation order is constant $\nu = 0.5$ and the integration order is taken form analogous set $\mu \in \{0.1, 0.2, \ldots, 1.0\}$ for the same gains.

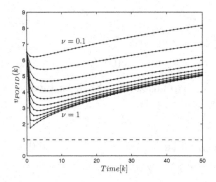
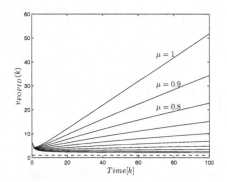

Fig. 10.2 FOPID controller unit step responses for $K_P = 1, K_I = 0.5, K_D = 5$, $\mu = 0.5$ and $\nu \in \{0.1, 0.2, \ldots, 1.0\}$.

Fig. 10.3 FOPID controller unit step responses for $K_P = 1, K_I = 0.5, K_D = 5$, $\nu = 0.5$ and $\mu \in \{0.1, 0.2, \ldots, 1.0\}$.

In both figures a vertical dotted line indicates the unit step function level. In Fig. 10.4 the FOPID controller unit step responses evaluated for $K_P = 1, K_I = 0.5, K_D = 5$ and $\nu = \mu \in \{0.1, 0.2, \ldots, 1.0\}$ are given.

The unit step responses characterize the transient behavior of the FOPID controller. One can easily find the FOPID controller differentiation and integration actions. The potency of FO differentiation and integration depends on appropriate orders. One can see that there are two typical shapes of the closed-loop system errors: an exponential decay with and without oscillations. In Figs. 10.5 and 10.6 the exponential error function and the FOPID controller output are plotted, respectively.

In the next pair of figures, the exponential error function with oscillations is considered together with the FOPID controller response. The controller parameters are the same as in the previous figures.

The integration effect in the FOPID controller is clearly visible in Fig. 10.6. This is caused by an integration of only a positive valued function. The effect of an exponential function with oscillation is much smaller because of an integration of positive and negative function values. Yet $\lim_{k \to +\infty} v_{FOPID}(k) > 0$. Now, the influence of the filter parameter a_0 in the real FOPID 10.7 will be analyzed. One starts with the ideal classical PID controller with $K_P = 1, K_I = 0.5, K_D = 5$ unit

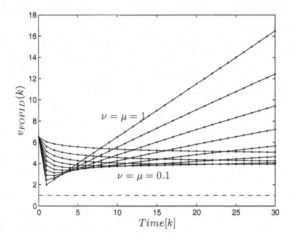

Fig. 10.4 FOPID controller unit step responses for $K_P = 1, K_I = 0.5, K_D = 5$ and $\nu = \mu \in \{0.1, 0.2, \ldots, 1.0\}$.

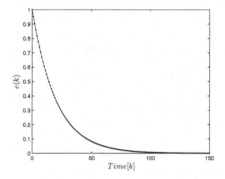

Fig. 10.5 Exponential error function.

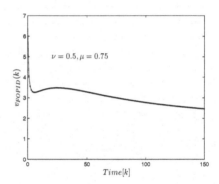

Fig. 10.6 FOPID controller exponential error function response for $K_P = 1, K_I = 0.5, K_D = 5, \nu = 0.5, \mu = 0.75$.

step response. It is plotted in Fig. 10.9. In Fig. 10.10 the IOPID controller unit step response curves for $a_0 \in \{0.1, 0.25, 1.010.0\}$ are plotted.

The plots in Fig. 10.10 reveal an influence of the filter on the discrete differentiation of the FOPID controller. Rising values of coefficients a_0 eliminate the differentiation effect noticeable in the initial period of the controller step response. A analogous effect can also be seen in the FOPID controller case. In Fig. 10.11 the FOPID controller unit step response changes, due to different filter parameters, are presented for $K_P = 1, K_I = 0.5, K_D = 5$ and $\nu = \mu = 0.5$ and $a_0 \in \{0.1, 0.25, 1.010.0\}$. Fig. 10.12 depicts similar plots simulated for the same parameters with the exception of the FOPID controller orders $\nu = \mu = 0.75$.

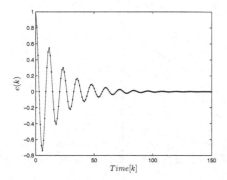

Fig. 10.7 Exponential error function with oscillations.

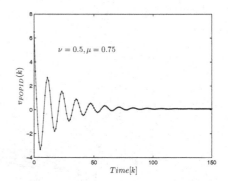

Fig. 10.8 FOPID controller exponential error function response for $K_P = 1, K_I = 0.5, K_D = 5, \nu = 0.5, \mu = 0.75$.

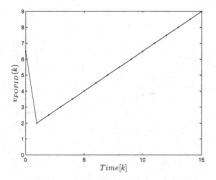

Fig. 10.9 Classical ideal PID unit step response for $K_P = 1, K_I = 0.5, K_D = 5$.

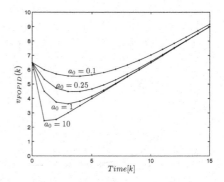

Fig. 10.10 Classical real PID unit step response for $K_P = 1, K_I = 0.5, K_D = 5$ and $a_0 \in \{0.1, 0.25, 1.0 10.0\}$.

Until now there were considered FOs within the range $\nu, \mu \in (0,1)$. There are no contraindications for use of higher discrete differentiation and integration orders. This will strengthen discrete differentiation and integration actions. In Figs. 10.13 and 10.14 the FOPID controller unit step responses for $K_P = 1, K_I = 0.5, K_D = 5$, $a_0 \in \{0.1, 0.25, 1.0 10.0\}$, $\nu = 1.2, \mu = 1$ and $\nu = 1, \mu = 1.2$ are presented, respectively.

Finally in Fig. 10.15 the FOPID controller unit step responses for $K_P = 1, K_I = 0.5, K_D = 5$, $a_0 \in \{0.1, 0.25, 1.0 10.0\}$, $\nu = \mu = 1.4$ are plotted.

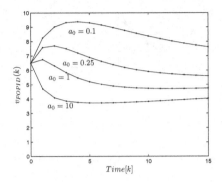

Fig. 10.11 FOPID unit step responses for $K_P = 1, K_I = 0.5, K_D = 5, \nu = \mu = 0.5$ and $a_0 \in \{0.1, 0.25, 1.010.0\}$.

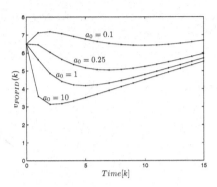

Fig. 10.12 FOPID unit step responses for $K_P = 1, K_I = 0.5, K_D = 5, \nu = \mu = 0.75$ and $a_0 \in \{0.1, 0.25, 1.010.0\}$.

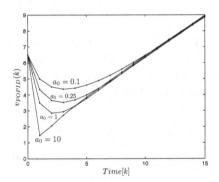

Fig. 10.13 FOPID unit step responses for $K_P = 1, K_I = 0.5, K_D = 5, a_0 \in \{0.1, 0.25, 1.010.0\}$ and $\nu = 1.2, \mu = 1$.

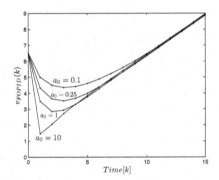

Fig. 10.14 FOPID unit step responses for $K_P = 1, K_I = 0.5, K_D = 5, a_0 \in \{0.1, 0.25, 1.010.0\}$ and $\nu = 1, \mu = 1.2$.

10.3 FO PID controller frequency characteristics

As in the case of the continuous-time system, the frequency response of the FOPID controller also plays an important role in discrete-time control system synthesis. The advantages of an application of the Bode diagrams in the closed-loop system synthesis are well-known [Ogata (1987)], [Franklin *et al.* (2002)]. A similar method can be applied to the FO discrete-time systems. When applying a classical discrete Fourier Transformation the FOPID controller TF one gets

$$C_{FOPID}(e^{j\omega_i}) = C_{FOPID}(z)|_{z=e^{j\omega_i}} = \left.\frac{V(z)}{E(z)}\right|_{z=e^{j\omega_i}}$$

$$= K_P + \frac{K_I}{(1 - e^{-j\omega_i})^{\mu_1}} + K_D(1 - e^{-j\omega_i})^{\nu_1}. \qquad (10.28)$$

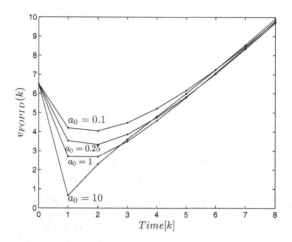

Fig. 10.15   FOPID controller unit step responses for $K_P = 1, K_I = 0.5, K_D = 5$ and $\nu = \mu = 1.4$.

Remember that separate proportional, discrete integration and differentiation actions were already considered in Section 8.1 and 8.2. It is evident that the gain introduced by the proportional part is evidently constant. One also notices, that the combined CD gain (for steady input signal) of the FOPID ideal controller

$$|C_{FOPID}(1)| = |\,C_{FOPID}(z)|_{z=e^{j0}}\,|$$

$$= |K_P + \lim_{\omega_i \to 0^+} \left\{ \frac{K_I}{(1 - e^{-j\omega_i})^{\mu_1}} \right\}| = +\infty. \tag{10.29}$$

for a normalized frequency tending to π is

$$|C_{FOPID}(-1)| = |\,C_{FOPID}(z)|_{z=e^{j\pi}}\,|$$

$$= |K_P + \frac{K_I}{2^{\mu_1}} + K_D 2^{\nu}| = K_P + \frac{K_I}{2^{\mu_1}} + K_D 2^{\nu}. \tag{10.30}$$

and depends not only on parameters K_P, K_I, K_D but also on orders ν and μ_1. Fig. 10.16 presents the discrete Nyquist plot of the IOPID ideal controller. The next three figures contain analogous plots for the FOPID ideal controllers for different combinations of differentiation and integration orders.

In Figs. 10.17 and 10.18 the discrete Nyquist plots of the FOPID ideal controller for $K_P = 1, K_I = 0.5, K_D = 5$ for two constant $\mu_1 = 0.5$ and $\mu_1 = 1$ and different $\nu \in \{0.1, 0.2, \dots, 1.0\}$ are presented, respectively.

The next cases are presented in Figs. 10.19 and 10.20 and are both $K_P = 1$, $K_I = 0.5, K_D = 5$, $\mu_1 = 1.5$ and $\nu \in \{0.1, 0.2, \dots, 1.0\}$ and $\nu_1 = 0.5$ and $\mu \in \{0.1, 0.2, \dots, 1.0\}$

The next considered plots in Figs. 10.21 and 10.22 cover the FOPID ideal controller for $\nu_1 = 1$ and $\nu_1 = 1.5$ with $\mu \in \{0.1, 0.2, \dots, 1.0\}$ and keep all other elements unchanged.

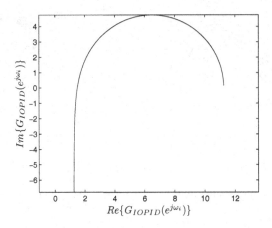

Fig. 10.16   IOPID ideal controller discrete Nyquist plot for $K_P = 1, K_I = 0.5, K_D = 5$ and $\nu = \mu = 1$.

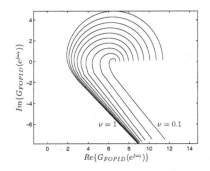

Fig. 10.17   FOPID ideal controller discrete Nyquist plot for $K_P = 1, K_I = 0.5, K_D = 5$ and $\mu_1 = 0.5$ and $\nu \in \{0.1, 0.2, \ldots, 1.0\}$.

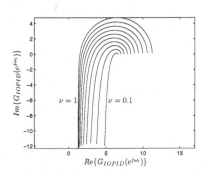

Fig. 10.18   FOPID ideal controller discrete Nyquist plot for $K_P = 1, K_I = 0.5, K_D = 5$ and $\mu_1 = 1$ and $\nu \in \{0.1, 0.2, \ldots, 1.0\}$.

Presented above the discrete Nyquist plots of the FOPID controller reveal an immense variety of frequency characteristics. This gives many more opportunities to shape the dynamics of the closed-loop systems. A change of the discrete integration orders affect the angle shape of the linear part of the characteristic, whereas a change of the discrete differentiation order, changes the radius of the circle-like part of the characteristics. Finally, in Fig. 10.23 a very strange characteristic of the FOPID ideal controller is presented. Its parameters are: $K_P = 1, K_I = 0.5, K_D = 5$ and $\nu_1 = 0.5, \mu = 0.5$. The left figure suggests the plot discontinuity, while the second figure reveals that the curve is differentiable.

Up to this point, the FOPID ideal controller frequency characteristics were

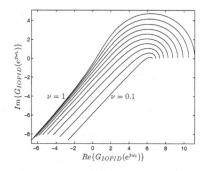

Fig. 10.19   FOPID ideal controller discrete Nyquist plot for $K_P = 1, K_I = 0.5, K_D = 5$ and $\mu_1 = 1.5$ and $\nu \in \{0.1, 0.2, \dots, 1.0\}$.

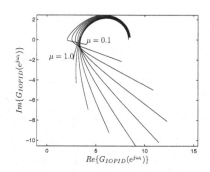

Fig. 10.20   FOPID ideal controller discrete Nyquist plot for $K_P = 1, K_I = 0.5, K_D = 5$ and $\nu_1 = 0.5$ and $\mu \in \{0.1, 0.2, \dots, 1.0\}$.

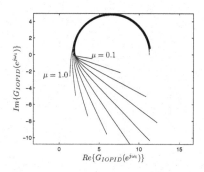

Fig. 10.21   FOPID ideal controller discrete Nyquist plot for $K_P = 1, K_I = 0.5, K_D = 5$ and $\nu_1 = 1$ and $\mu \in \{0.1, 0.2, \dots, 1.0\}$.

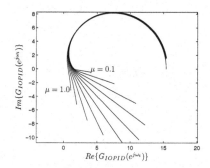

Fig. 10.22   FOPID ideal controller discrete Nyquist plot for $K_P = 1, K_I = 0.5, K_D = 5$ and $\nu_1 = 1.5$ ánd $\mu \in \{0.1, 0.2, \dots, 1.0\}$.

analyzed. The FOPID real controller (10.7) frequency characteristics are changed by the presence of the digital filter. The impact degree depends on the digital filter frequency characteristic. This influence is shown in Figs. 10.25 and 10.26. For comparisons sake both figures contain the Nyquist plot of the classical PID controller (thin line).

Figure 10.26 reveals that the time constant is too large, which means that the filter bandwidth is too small. This results in discrete Nyquist plots losing the character of the PID controller characteristic. The discrete Nyquist plot contains full information about the frequency characteristic. From it, one derives the amplitude of the frequency spectrum and the phase characteristic of the FOPID controller. Exemplary characteristics related to those presented in Fig. 10.25 are given in Figs. 10.27 and 10.28, respectively. Related Bode diagrams are given in Figs. 10.29 and 10.30.

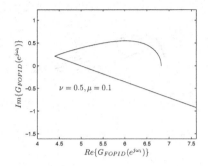

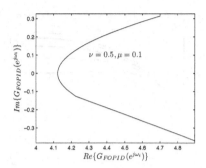

Fig. 10.23 FOPID ideal controller discrete Nyquist plot for $K_P = 1, K_I = 0.5, K_D = 5$ and $\nu_1 = 0.5, \mu = 0.1$.

Fig. 10.24 Fragment of the FOPID ideal controller discrete Nyquist plot for $K_P = 1, K_I = 0.5, K_D = 5$ and $\nu_1 = 0.5, \mu = 0.1$.

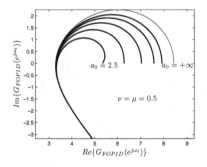

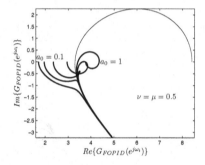

Fig. 10.25 FOPID real controller discrete Nyquist plot for $K_P = 1, K_I = 0.5, K_D = 5$ and $\nu_1 = \mu_1 = 0.5, \mu_2 = 1$ and $a_0 \in \{0.1, 0.1778, 0.3162, 0.5623, 1.0000\}$.

Fig. 10.26 FOPID real controller discrete Nyquist plot for $K_P = 1, K_I = 0.5, K_D = 5$ and $\nu_1 = \mu_1 = 0.5, \mu_2 = 1$ and $a_0 \in \{2.5, 4.4457, 7.9057, 14.0585, 25.0000\}$.

Finally the Bode diagrams of the FOPID controller, whose discrete Nyquist plot is given in Fig. 10.23, will be presented.

10.4 Closed-loop system with the FOPID controller

The synthesis of the closed-loop system with the FOPID controller proceeds in the same way as in the classical controller case. The block diagram of the system is given in Fig. 10.33 where H is a constant representing the sensor of the output signal. In the analysis, one assumes that the plant and the FOPID controller passed the order commensurating procedure. Having identified a plant treated as the FO or IO, one starts a controller tuning process. In the FOPID controller there are five parameters to tune: K_P, K_I, K_D and ν_1, μ_1. The filter parameters can be chosen separately

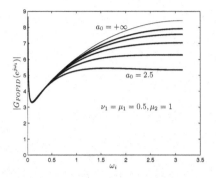

Fig. 10.27 FOPID real controller discrete amplitude frequency characteristic for $K_P = 1, K_I = 0.5, K_D = 5$ and $\nu_1 = \mu_1 = 0.5, \mu_2 = 1$ and $a_0 \in \{2.5000, 4.4457, 7.9057, 14.0585, 25.0000\}$.

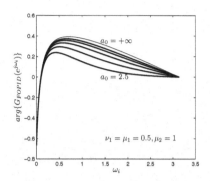

Fig. 10.28 FOPID real controller discrete phase frequency characteristic for $K_P = 1, K_I = 0.5, K_D = 5$ and $\nu_1 = \mu_1 = 0.5, \mu_2 = 1$ and $a_0 \in \{2.5000, 4.4457, 7.9057, 14.0585, 25.0000\}$.

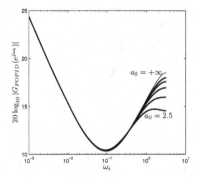

Fig. 10.29 FOPID real controller amplitude Bode plot for $K_P = 1, K_I = 0.5, K_D = 5$ and $\nu_1 = \mu_1 = 0.5, \mu_2 = 1$ and $a_0 \in \{2.5000, 4.4457, 7.9057, 14.0585, 25.0000\}$.

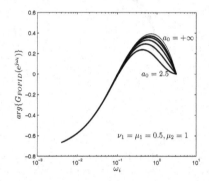

Fig. 10.30 FOPID real controller phase Bode plot for $K_P = 1, K_I = 0.5, K_D = 5$ and $\nu_1 = \mu_1 = 0.5, \mu_2 = 1$ and $a_0 \in \{2.5000, 4.4457, 7.9057, 14.0585, 25.0000\}$.

due to the noise bandwidth. The optimal FOPID parameters should be evaluated in the context of previously established closed-loop performance criteria. There are several fundamental requirements that the closed loop system must meet. Some of the requirements result in parallel optimization that gives results in conflict. In these cases, one should find a trade off solution. Generally, optimization affects the output signal $y(k)$, but through a simple scalar equation $e(k) = r(k) - Hy(k)$ derived from Fig. 10.33 all requirements imposed on $y(k)$ can be easily reformulated to the transient behavior of the closed-loop system error signal. According to various system descriptions the error signal has different forms. The error of the closed-loop system described by the FOTF is given below

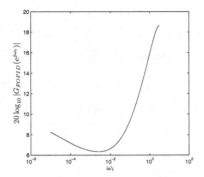

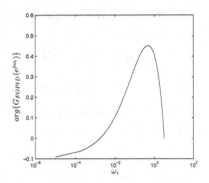

Fig. 10.31 FOPID ideal controller amplitude Bode plot for $K_P = 1, K_I = 0.5, K_D = 5$ and $\nu_1 = 0.5, \mu_1 = 0.1$.

Fig. 10.32 FOPID ideal controller phase Bode plot for $K_P = 1, K_I = 0.5, K_D = 5$ and $\nu_1 = 0.5, \mu_1 = 0.1$.

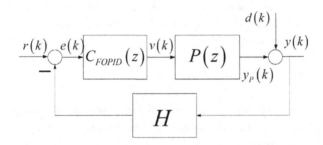

Fig. 10.33 Closed-loop system with the FOPID controller.

$$e(k) = \mathcal{Z}^{-1}\{(1 + HP(z)C_{FOPID}(z))^{-1} R(z)\}$$
$$-\mathcal{Z}^{-1}\{(1 + HP(z)C_{FOPID}(z))^{-1} HD(z)\} \qquad (10.31)$$

by the SSEs,

$$\begin{bmatrix} {}^{GL}_{k_0}\Delta_k^{(\nu)}\mathbf{x}_{FOPID}(k) \\ {}^{GL}_{k_0}\Delta_k^{(\nu)}\mathbf{x}_P(k) \end{bmatrix}$$
$$= \begin{bmatrix} \mathbf{A}_{FOPID} & -H\mathbf{b}_{FOPID}\mathbf{b}_P \\ \mathbf{b}_{FOPID}\mathbf{c}_P & \mathbf{A}_P - d_{FOPID}H\mathbf{b}_{FOPID}\mathbf{b}_P \end{bmatrix} \begin{bmatrix} \mathbf{x}_{FOPID}(k) \\ \mathbf{x}_P(k) \end{bmatrix}$$
$$+ \begin{bmatrix} \mathbf{b}_{FOPID} \\ d_{FOPID}\mathbf{b}_P \end{bmatrix} r(k)$$
$$e(k) = \begin{bmatrix} \mathbf{0} & H\mathbf{C}_P \end{bmatrix} \begin{bmatrix} \mathbf{x}_{FOPID}(k) \\ \mathbf{x}_P(k) \end{bmatrix} + r(k) - Hd(k) \qquad (10.32)$$

and finally the PLM form gives

$$\mathbf{e}(k) = \left[\mathbf{1}_k + H\mathbf{D}_{P,k}^{-1}\mathbf{N}_{P,k}\mathbf{D}_{FOPID,k}^{-1}\mathbf{N}_{FOPID,k}\right]^{-1}\mathbf{r}(k)$$

$$-H\left[\mathbf{1}_k + H\mathbf{D}_{P,k}^{-1}\mathbf{N}_{P,k}\mathbf{D}_{FOPID,k}^{-1}\mathbf{N}_{FOPID,k}\right]^{-1}\mathbf{d}(k) \qquad (10.33)$$

respectively. Among the conditions imposed on the closed-loop error function, which are mentioned below, the first one is required.

(I) Closed-loop system stability (The most important requirement. Failure to fulfill this requirement stops the synthesis procedure)

(II) Transient response specifications

(a) Delay time (The time when the closed-loop error first reaches its zero level.)

(b) Rise time (The time interval between a specified interval of the error level. It is usually taken as $[0.9e(0), 0.1e(0)]$, where $e(0)$ is a closed-loop system with an initial error value.)

(c) Peak time (The discrete time when the error reaches its first extremum.)

(d) Maximum overshoot (The system's response at an absolute maximum $\max|e(k)|$ related to the system's steady-state of value y_s. Usually, it is defined in percent as

$$M = \frac{\max_{k\in[0,+\infty)}|e(k)|}{y_s}100\% \qquad (10.34)$$

This value indicates the probability in a loss of system stability due to some plant parameter changes.)

(e) Settling time (The discrete time required to reach and stay inside a strip defined as $\pm[\epsilon, \epsilon]$) where $\epsilon = (0.03 - 0.05)y_s$

(III) Frequency characteristics specifications

(a) Maximal bandwidth of the closed-loop system

(b) Maximal peak of the magnitude characteristics (The peak or peaks are related to the closed-loop system oscillatory conditions)

(IV) Amplitude and phase margin (Measures of proximity to instability of the open-loop system in the discrete Nyquist diagram)

(V) Sums of weighted errors (These performance indices will be discussed in detail in the next subsection).

10.4.1 *Sums of weighted errors*

A measure of the controller action in the closed-loop system is represented by the performance index. This is a real scalar function indicating a discrepancy between the intended and actual transient behavior of the closed-loop system or the weighted sum of an absolute error to the power of s (WSPE)

$$I_{WS} = \sum_{i=0}^{+\infty} w(i)|e(i)|^s, \text{ for } s = 1, 2, \ldots. \tag{10.35}$$

All typical weight functions $w(i)$ must be positive and $w(i) \to +\infty$. Introducing the weight function was intended to force an end to the transient response. Moreover, the weight function must satisfy the conditions

$$\lim_{k \to +\infty} w(k)e(k) = 0, \quad \sum_{i=0}^{+\infty} w(i)|e(i)|^s = I < +\infty. \tag{10.36}$$

Please note that for even s one can only consider $e(i)$ instead of its absolute value. From a family of performance indices (10.31), one of the most popular is the sum of an absolute error (SAE) obtained for $r = 0$ and $s = 1$

$$I_{SAE} = \sum_{i=0}^{+\infty} |e(i)| \tag{10.37}$$

the sum of a squared error (SsE)

$$I_{SSE} = \sum_{i=0}^{+\infty} e^2(i) \tag{10.38}$$

or the weighted sum of a squared error (WSSE) with $w(k) = k^r$

$$I_{WSSE} = \sum_{i=0}^{+\infty} i^r e^2(i), \text{ where } r = 1, 2, \ldots. \tag{10.39}$$

As the weight function, one can use a function $a^{(\nu)}(k)$ for $\nu \in (0, 1)$. According to the graphical interpretation considered in Section 4.2 the error values which are the furthest away from the beginning of the process are given the highest weight. Hence, instead of the first-order weighted sum one can simply take FOBS of the error function to the power of s ($s = 1, 2, \ldots$.)

$$I_{FOS} = \lim_{k \to +\infty} \sum_{i=0}^{k} a^{(-\nu)}(k-i)|e(i)|^s = \lim_{k \to +\infty} \sum_{i=0}^{k} a^{(-\nu)}(i)|e(k-i)|^s$$

$$= \lim_{k \to +\infty} {}^{GL}_{0}\Delta_{k}^{(\nu)}|e(k)|^s = {}^{GL}_{0}\Delta_{+\infty}^{(\nu)}|e(k)|^s. \quad (10.40)$$

In practical numerical calculations, instead of infinite sums one takes sufficiently large sums $k_{\max}$ for which the error function equals practically zero. For the closed-loop system with the FO plant and FOPID controller described by the FOSSE the FOS may take the form

$$I_{FOS} = {}^{GL}_{0}\Delta_{k_{\max}}^{(\epsilon)} \mathbf{Q} \left[\mathbf{x}^{\mathrm{T}}(k)\mathbf{x}(k) \right]^{\frac{s}{2}} + {}^{GL}_{0}\Delta_{k_{\max}}^{(\zeta)} |v(k)|^r \quad (10.41)$$

where $r, s = 1, 2, \ldots, \epsilon, \zeta \in \mathcal{R}_+$ and $v(k)$ is the controlling signal in Fig. 10.33.

For a system represented by the block diagram in Fig. 9.14 and described by (9.53) with (9.54) one may define a quadratic performance index

$$I_{FOS} = \frac{1}{2}{}^{GL}_{0}\Delta_{k_{\max}}^{(\epsilon)} \left[\mathbf{e}_1^{\mathrm{T}}(k) \ \mathbf{e}_2^{\mathrm{T}}(k) \right] \begin{bmatrix} \mathbf{Q}_{1,k} & \mathbf{0}_k \\ \mathbf{0}_k & \mathbf{Q}_{2,k} \end{bmatrix} \begin{bmatrix} \mathbf{e}_1(k) \\ \mathbf{e}_2(k) \end{bmatrix}$$

$$= \frac{1}{2}{}^{GL}_{0}\Delta_{k_{\max}}^{(\epsilon)} \left[\mathbf{e}_1^{\mathrm{T}}(k)\mathbf{Q}_{1,k}\mathbf{e}_1(k) + \mathbf{e}_2^{\mathrm{T}}(k)\mathbf{Q}_{2,k}\mathbf{e}_2(k) \right] \quad (10.42)$$

where $\mathbf{Q}_{1,k}$ is a positive definite or a positive semidefinite Hermitian matrix of dimensions $(k+1) \times (k+1)$ and $\mathbf{Q}_{2,k}$ is a positive definite Hermitian matrix of the same dimensions. A substitution (9.53) into (10.42) yields

$$I_{FOS} = \frac{1}{2}{}^{GL}_{0}\Delta_{k_{\max}}^{(\epsilon)} \left\{ \left[\hat{\mathbf{r}}^{\mathrm{T}}(k) \ \mathbf{d}^{\mathrm{T}}(k) \right] \mathbf{G}_k^{\mathrm{T}} \begin{bmatrix} \mathbf{Q}_{1,k} & \mathbf{0}_k \\ \mathbf{0}_k & \mathbf{Q}_{2,k} \end{bmatrix} \mathbf{G}_k \begin{bmatrix} \hat{\mathbf{r}}(k) \\ \mathbf{d}(k) \end{bmatrix} \right\}. \quad (10.43)$$

In the following numerical example, the FOPID controller synthesis process is presented.

Example 10.1. Consider the FO system described by the FOTF

$$P(z) = \frac{b_1 \left(1 - z^{-1}\right)^{\nu} + b_0}{\left(1 - z^{-1}\right)^{2\nu} + a_1 \left(1 - z^{-1}\right)^{\nu} + a_0} \quad (10.44)$$

with $\nu = 0.5$, $a_1 = 1$, $a_0 = 0.25$, $b_1 = 0.25$, $b_0 = a_0$. The unit step response of the plant is presented in Fig. 10.34

The IOPID and FOPID controllers are tuned to minimize performance indices $I_{SAE}(K_P, K_I, K_D)$ and $I_{SAE}(K_P, K_I, K_D, \nu 1, \mu_1)$ with $\mu_2 = 1$ as a (constant) parameter $a_{F,0} = 0.1$. The optimal controller parameters are given in Table 10.1.

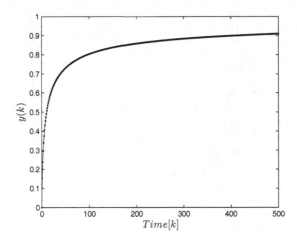

Fig. 10.34   Unit step response of the plant (10.34).

Table   10.1   FOPID   and
IOPID parameters.

	FOPID	IOPID
K_P	3.650	5.601
K_I	2.550	1.302
K_D	0.000	0.000
ν_1	0.000	0.000
μ_1	0.838	1.000
μ_2	1.000	1.000
$a_{F,0}$	0.100	0.100

The error signals plots of the closed-loop systems for both controllers are given in Fig. 10.35.

The performance index values of $I_{SAE}(K_P,\ K_I,\ K_D)$ and $I_{SAE}(K_P,\ K_I,\ K_D,\ \nu_1,\ \mu_1)$ are given in Fig. 10.36.

The advantage of the application of the FOPID controller over the IOPID can be counted as a rate of a decrease of SAE performance criteria. For FOPID, this means a 28.85% improvement of the SAE criterion.

Also, consider the same plant but with two changed parameters $\nu = 0.75, b_1 = 0$. The plant response is given in Fig. 10.37. This time, for higher order ν the response is oscillatory.

The optimal IOPID and FOPID controller parameters due to the SAE criterion are collected in Table 10.2

The error signals plots of the closed-loop systems for both controllers are given in Fig. 10.38.

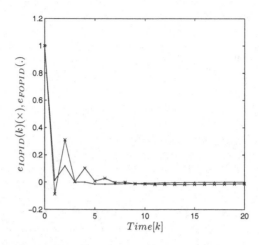

Fig. 10.35   Closed-loop system errors to the discrete unit step input for the IOPID (x) and FOPID (.) controllers.

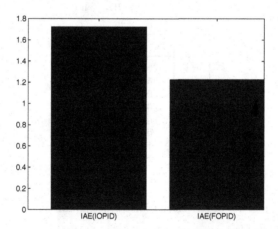

Fig. 10.36   The SAE performance criteria for the IOPID (x) and FOPID (.) controllers.

The performance index values of $I_{SAE}(K_P, K_I, K_D)$ and $I_{SAE}(K_P, K_I, K_D, \nu_1, \mu_1)$ are given in Fig. 10.39.

The advantage of the application of the FOPID controller over the IOPID can be counted as a rate of decrease of SAE performance criteria for FOPID. This means a 45.56% improvement of the SAE criterion. The oscillation plant is more difficult to control. This time an improvement was achieved showing that the FOPID controller is coping better than the classic one.

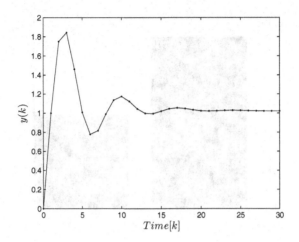

Fig. 10.37 Unit step response of the plant (10.34) with changed order.

Table 10.2 FOPID and IOPID parameters.

	FOPID	IOPID
K_P	0.000	0.500
K_I	0.529	0.281
K_D	4.360	4.310
ν_1	1.940	1.000
μ_1	1.180	1.000
μ_2	1.000	1.000
$a_{F,0}$	0.100	0.100

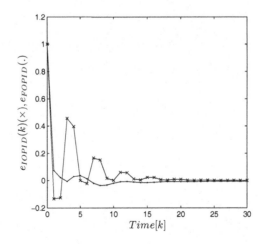

Fig. 10.38 Closed-loop system errors to the discrete unit step input for the IOPID (x) and FOPID (.) controllers.

Discrete Fractional Calculus

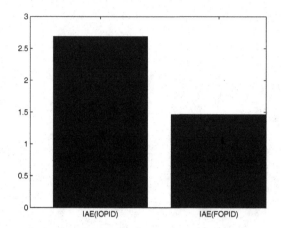

Fig. 10.39   The SAE performance criteria for the IOPID (x) and FOPID (.) controllers.

Chapter 11

FOS approximation problems

A microprocessor implementation of the FOBD/S processing is more sophisticated then in the IOBD/S based algorithms. This is due to a number of linearly growing in time multiplications and additions. It is also a necessary to store all previously measured samples of the differentiated and summated discrete-time signals. Accuracy problems also play an important role in the FOBD/S evaluation. This is caused by the rapidly decreasing function $a^{(\nu)}(k)$ for $-1 < \nu < 1$. To perform precise mathematical operations on an input signal a microprocessor system supporting satisfactory word length, floating-point or fixed-point arithmetic must be used [Angoletta (2008); Olderog and Dierks (2008); Kopetz (2011); Kuo et al. (2013)]. There are also additional limitations influencing the FOBD/S calculation effects. These limitations include, the finite operational speed of operations of the microprocessor and the related A/D converter of the microprocessor and the system limited memory. In this chapter practically realizable forms of the FOBD/S are studied in the context of its calculation accuracy. The problem is known as "The Short Memory Principle" [Tuan and Gorenflo (1995); Podlubny (1999a); Deng (2007); Deng and Li (2012); Mayoral (2012)]. The FODs are approximated by classical IODs [Chen et al. Vinagre and Podlubny (2004b)].

11.1 Simplified forms of the FOBD

It is practically impossible to calculate in a real-time the FOBD defined by formula (3.1). This requires a search for some simplified forms that can be treated as approximations of the FOTF. For simplicity, the initial time is set $k_0 = 0$. The simplest form is defined below.

Definition 11.1. Given a discrete-time, bounded real function $y(k)$ defined over a discrete-time interval $[0, k]$. Assume that this interval is divided into l contacting segments of lengths $L_1, L_{i+1} - L_i$ for $i = 1, 2, \ldots, l - 1$, A first simplified form of the Grünwald-Letnikov FOBD of the FO $0 < \nu < 1$ is defined as a sum

$$_0^{GL}\tilde{\Delta}_k^{(\nu)}y(k) = \sum_{i=0}^{k} \tilde{a}^{(\nu)}(i)y(k-i) \qquad (11.1)$$

where

$$\tilde{a}^{(\nu)}(k) = \begin{cases} a^{(\nu)}(k) & \text{for } 0 \le k < L_1 \\ a_i^{(\nu)} & \text{for } L_i \le k < L_{i+1}, i = 1, 2, \ldots, l-1 \\ 0 & \text{for } L_{i+1} \le k \end{cases} \qquad (11.2)$$

and $a_i^{(\nu)}$ is a constant satisfying inequalities $a_{(L_i)}^{(\nu)} \le a_i^{(\nu)} < a_{(L_{i+1})}^{(\nu)}$. Realize that $_0^{GL}\tilde{\Delta}_k^{(\nu)}y(k) = _0^{GL}\Delta_k^{(\nu)}y(k) = $ for $0 \le k \le L_1$. Thus, for $l = 1$ from (11.2)

$$\tilde{a}^{(\nu)}(k) = \begin{cases} a^{(\nu)}(k) & \text{for } 0 \le k < L_1 \\ 0 & \text{for } L_1 \le k \end{cases}. \qquad (11.3)$$

For $L = L_i$ and $a_i^{(\nu)} = a^{(\nu)}(L_i)$ from (11.3) one gets

$$\tilde{a}^{(\nu)}(k) = \begin{cases} a^{(\nu)}(k) & \text{for } 0 \le k < L \\ a^{(\nu)}(Li) & \text{for } Li \le k < L(i+1), i = 1, 2, \ldots, l-1 \\ 0 & \text{for } Ll \le k \end{cases}. \qquad (11.4)$$

To show the difference between the exact and approximated FOBE its vector forms are compared

$$_0^{GL}\Delta_k^{(\nu)}y(k) = \begin{bmatrix} a^{(\nu)}(0) & a^{(\nu)}(1) & \cdots & a^{(\nu)}(k) \end{bmatrix} = \begin{bmatrix} f(k) \\ f(k-1) \\ \vdots \\ f(0) \end{bmatrix} \qquad (11.5)$$

$$\tilde{G}(z, L, l) = {}^{GL}_{0}\tilde{\Delta}^{(\nu)}_{k} y(k)$$

$$= \begin{bmatrix} a^{(\nu)}(0) & \cdots & a^{(\nu)}(L-1) & \vdots & a^{(\nu)}(L) & \cdots & a^{(\nu)}(L) & \vdots & \cdots & a^{(\nu)}(lL) \end{bmatrix} \begin{bmatrix} f(k) \\ \vdots \\ f(k-L+1) \\ \cdots \\ f(k-L) \\ \vdots \\ f(k-2L+1) \\ \cdots \\ \vdots \\ \cdots \\ f(k-lL) \\ \vdots \\ f(lL) \end{bmatrix}$$

$$= \begin{bmatrix} a^{(\nu)}(0) & \cdots & a^{(\nu)}(L-1) & \vdots & a^{(\nu)}(L) & a^{(\nu)}(2L) & \cdots & a^{(\nu)}(lL) \end{bmatrix}$$

$$\times \begin{bmatrix} f(k) \\ \vdots \\ f(k-L+1) \\ f(k-L) + f(k-L-1) + \cdots + f(k-2L+1) \\ f(k-2L) + f(k-2L-1) + \cdots + f(k-3L+1) \\ \vdots \\ f(k-lL-l) + f(k-lL-l-1) + \cdots + f(k-lL+1) \end{bmatrix}$$

$$= \begin{bmatrix} a^{(\nu)}(0) & \cdots & a^{(\nu)}(L-1) & \vdots & a^{(\nu)}(L) & a^{(\nu)}(2L) & \cdots & a^{(\nu)}(lL) \end{bmatrix} \begin{bmatrix} f(k) \\ \vdots \\ f(k-L+1) \\ \sum_{i=1}^{L} f(k-L-i+1) \\ \sum_{i=1}^{L} f(k-2L-i+1) \\ \vdots \\ \sum_{i=1}^{L} f(k-lL-i+1) \end{bmatrix}$$

$$(11.6)$$

The one-sided $\mathcal{Z}$-Transform of the FOBD (5.35) is of the form $\mathcal{Z}\{{}^{GL}_{0}\Delta^{(\nu)}_{k} y(k)\} = \sum_{i=0}^{+\infty} a^{(\nu)}(i) z^{-i} = (1 - z^{-1})^{\nu}$ while

$$\mathcal{Z}\left\{{}_{0}^{GL}\tilde{\Delta}_{k}^{(\nu)}y(k)\right\}$$

$$=\left[a^{(\nu)}(0)\cdots a^{(\nu)}(L-1)\,\vdots\,a^{(\nu)}(L)\,a^{(\nu)}(2L)\cdots a^{(\nu)}(lL)\right]\begin{bmatrix}1\\\vdots\\z^{-L+1}\\\sum_{i=1}^{L}z^{-L-i+1}\\\sum_{i=1}^{L}z^{-2L-I+1}\\\vdots\\\sum_{i=1}^{L}z^{-lL-i+1}\end{bmatrix}Y(z)$$

$$(11.7)$$

is an IO classical discrete TF. One should note that the choice of coefficients $a_i^{(\nu)}$ in 11.2 serve to preserve the stability of the FODE with 11.1. The idea of the FOBD approximations is based on dividing a series of consecutive values of $a^{(\nu)}(k)$ into l parts and a substitution by constant values $a_i^{(\nu)}$. The approximation idea is clarified by Figs. 11.1–11.4 where four coefficient versions: (11.3), (11.2) with $a_i^{(\nu)}=a^{(\nu)}(Li)$, $a_i^{(\nu)}=a^{(\nu)}(Li+L)$ and $a_i^{(\nu)}=a^{(\nu)}(Li+0.5\lfloor L\rfloor)$ against the background of the exact values $a^{(\nu)}(k)$ (in a continuous line) for $u=0.5$, $L=50$ and $l=5$ are presented. For a better clarity of plots, the common logarithms of the absolute values of $a^{(\nu)}(k)$ and $a^{(\nu)}(k)$ are plotted. It can be seen that the approximation becomes better for consecutive intervals. In Figs. 11.5–11.7, plots of the differences between $\log_{10}\tilde{a}^{(\nu)}$ and $\log_{10}a^{(\nu)}$ for $a_i^{(\nu)}=a^{(\nu)}(Li)$, $a_i^{(\nu)}=a^{(\nu)}(Li+L)$ and $a_i^{(\nu)}=a^{(\nu)}(Li+0.5\lfloor L\rfloor)$ are displayed.

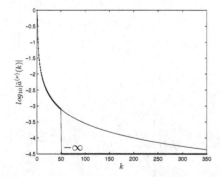

Fig. 11.1   Logarithms of absolute values of the function $a^{(\nu)}(k)$ for (11.3).

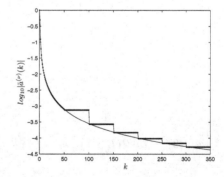

Fig. 11.2   Logarithms of absolute values of the function $a^{(\nu)}(k)$ for (11.4).

For the FOBD in the Horner form, the approximation function $\tilde{c}^{(\nu)}(k)$ values must be calculated from (1.54) while taking into account (11.2).

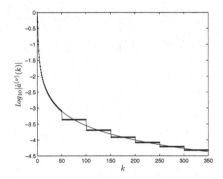

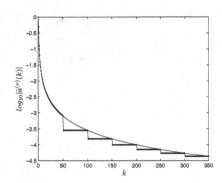

Fig. 11.3 Logarithms of absolute values of coefficient function (11.4) for $a_i^{(\nu)} = a^{(\nu)}(Li + 0.5 \lfloor L \rfloor)$.

Fig. 11.4 Logarithms of absolute values of coefficient function (11.4) for $a_i^{(\nu)} = a^{(\nu)}(Li + L)$.

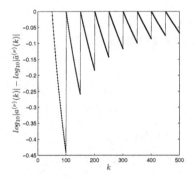

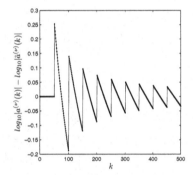

Fig. 11.5 Difference between logarithms of absolute values of coefficient function (11.4) and its approximation $a_i^{(\nu)} = a^{(\nu)}(Li)$.

Fig. 11.6 Difference between logarithms of absolute vales of coefficient function (11.4) for $a_i^{(\nu)} = a^{(\nu)}(Li + 0.5 \lfloor L \rfloor)$.

11.2 Simplified forms of the FOBS

In general, to simplify the FOBS one can use analogous approximation coefficients as described in the previous section. The Horner form of the FOBS provides another interesting possibility. Applying the property of the function $c^{(-\nu)}(k)$ for $\nu \in \mathbf{Z}_+$ (1.56) an approximation

$$\tilde{c}^{(-\nu)}(k) = \begin{cases} c^{(-\nu)}(k) & \text{for } k_0 \le k < L \\ 1 & \text{for } k \le L \end{cases} \tag{11.8}$$

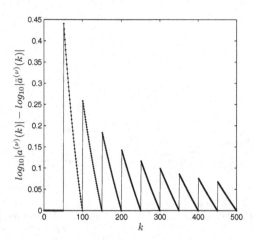

Fig. 11.7   Difference between logarithms of absolute values of coefficient function (11.4) for $a_i^{(\nu)} = a^{(\nu)}(Li + L)$.

is admissible. The approximated Horner form of the FOBS is as follows

$$
\begin{aligned}
{}^{H}_{k_0}\tilde{\Delta}^{(-\nu)}_{k} f(k) &= f(k) \\
&+ c^{(-\nu)}(1)\Big[f(k-1) + c^{(-\nu)}(2)\Big[f(k-2) + c^{(-\nu)}(3)\left[f(k-3) + \cdots\right. \\
&+ c^{(-\nu)}(f-L+1)\ [f(L-1) + f(L) + \cdots + f(k_0+1) + f(k_0)]\cdots]]] \\
&= f(k) + c^{(-\nu)}(1)\Big[f(k-1) + c^{(-\nu)}(2)\Big[f(k-2) + c^{(-\nu)}(3)\left[f(k-3) + \cdots\right. \\
&\qquad\qquad + c^{(-\nu)}(f-L+1)\ \left[\sum_{i=k_0}^{L-1} f(i)\right]\cdots]]].
\end{aligned}
$$

$$(11.9)$$

One should note that for (11.8) a related function $a^{(-\nu)}(k)$ has the form

$$
\tilde{a}^{(-\nu)}(k) = \begin{cases} a^{(-\nu)}(k) & \text{for } 0 \le k < L_1 \\ a^{(-\nu)}(L) & \text{for } L_1 \le k \end{cases} \tag{11.10}
$$

which is slightly different from the one defined by (11.3).

11.3   The "Short Memory Principle"

To achieve an assumed accuracy of approximated FOBD one should chose appropriate lengths of parameters $L_1, L_2, \ldots, L_l$ in (11.2) or L and l for (11.4) [Podlubny (1999a); Deng (2007)]. However, there is a trade-off between these lengths and a calculation error. These considerations will only be concentrated on the simplest case (11.3). The FOBD realization error absolute value is defined by

$$e_L(k) = {}_{k_0}^{GL}\Delta_k^{(\nu)}f(k) - {}_{k_0}^{GL}\tilde{\Delta}_k^{(\nu)}f(k). \tag{11.11}$$

This error function will serve as a tool for evaluating the optimal length of L. The optimality of L is considered a minimal value that is guaranted not to exceed the assumed maximal absolute error. The L may be treated as a finite "'Calculation Tail'. Its calculation rule formulates the Theorem.

Theorem 11.1. *Consider a bounded discrete-variable function $f(k)$ such that $|f(k)| < F$ for $k \in [k_0 + \infty]$. For $\nu \in (0,1)$ and assumed maximal error e_{max} the length L for which $e_L(k) \le e_{max}$ for $k \in [k_0 + \infty]$ must satisfy equation*

$$L_{opt} = \left\{ L : \min_{L \in \mathbf{Z}_+} a^{(\nu-1)}(L) < \frac{e_{max}}{F} \right\}. \tag{11.12}$$

Proof. It is obvious that $e_L(k) = 0$ for $k \in [k_0, L-1]$. For $k \ge L$ from (11.11) one has

$$|e_L(k)| = |\sum_{i=k_0}^{k} a^{(\nu)}(i-k_0)f(k-i+k_0) - \sum_{i=k_0}^{k} \tilde{a}^{(\nu)}(i-k+L)f(k-i+k_0)|$$

$$= |\sum_{i=0}^{k-k_0} a^{(\nu)}(i)f(k-i) - \sum_{i=k-k_0}^{k} a^{(\nu)}(i-k+L)f(2k-i-L)|$$

$$= |\sum_{i=0}^{k-k_0} a^{(\nu)}(i)f(k-i) - \sum_{i=0}^{L} a^{(\nu)}(i)f(k-i)| \le F|\sum_{i=0}^{k-k_0} a^{(\nu)}(i) - \sum_{i=0}^{L} a^{(\nu)}(i)|$$

$$= F|a^{(\nu-1)}(k-k_0) - a^{(\nu-1)}(L)| < F|a^{(\nu-1)}(L)|. \tag{11.13}$$

The last inequality is justified by properties of the function $a^{(\nu-1)}(k)$ for which $a^{(\nu-1)}(L) = const > 0$, $a^{(\nu-1)}(k-k_0) > 0$ is monotonically decreasing and $a^{(\nu-1)}(L) > a^{(\nu-1)}(k)$ for $k > L$. $\square$

The result of Theorem 11.1 serves for establishing L_{opt} on a base of inequality (11.12). The steps are formulated in an Algorithm.

Algorithm 11.1.

(I) For a given maximal error e_{max} evaluate $F > |f(k)|$
(II) Find a minimal L_{opt} satisfying inequality

$$\frac{(\nu-1)(\nu-2)\cdots(\nu-L)}{L!} < \frac{e_{max}}{F}. \tag{11.14}$$

The simplest way to find L_{opt} is to solve the inequality graphically: plot a function $a^{(\nu-1)}(L)$ vs. L and find a cross point with a constant function $g(L) = \frac{e_{max}}{F}$. The following numerical example confirms the result of Theorem 11.1.

Example 11.1. Consider a discrete-variable function defined on an interval $[0, k_{max}] = [0, 500]$

$$f(k) = \begin{cases} e^{-0.01k} \cos\left(\frac{\pi}{6}\right) & \text{for } 0 \geq k \geq 0.5k_{max} \\ e^{-0.01(0.5k_{max}-k)} \cos\left(\frac{0.5k_{max}-k}{6}\right) & \text{for } 0.5k_{max} < k \geq 0.5k_{max} \end{cases} . \quad (11.15)$$

Plot of (11.15) is presented in Fig. 11.8.

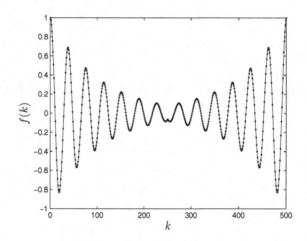

Fig. 11.8   Plot of function (11.15).

In Fig. 11.9 two plots are given: $_0^{GL}\Delta_k^{(\nu)} f(k)$ and $_0^{GL}\tilde{\Delta}_k^{(\nu)} f(k)$ defined for $\nu = 0.5$ and $L = 50$. The differences between both plots are practically invisible. Hence, in the next Fig. 11.10 an error function (11.10) related to (11.15) is plotted.

In Fig. 11.11 the maximum error function e_{max} vs. "the calculation tail" is plotted.

In modern microprocessor systems the multiplication and addition operations times may be assumed to be equal. Therefore, it is worth evaluating the total operations number needed to perform the FOBD calculation due to the exact and simplified forms. The maximal numbers are collected in Table 11.1.

In Fig. 11.12 multiplications and additions numbers of the FOBD calculation process vs. discrete time are due to methods indicated in Table 11.1. In the first (exact) method there is no steady-state number of calculations and multiplications. Whereas, such values are present in methods 2 and 3. Its values are given in the fifth column.

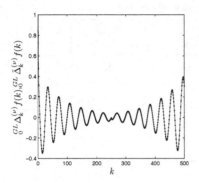

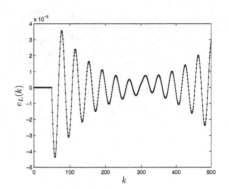

Fig. 11.9 Plots of $_0^{GL}\Delta_k^{(\nu)}f(k)$ and $_0^{GL}\tilde{\Delta}_k^{(\nu)}f(k)$ defined for $\nu = 0.5$ and $L = 50$.

Fig. 11.10 Error function (11.10) related to (11.15).

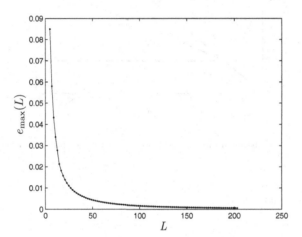

Fig. 11.11 Plot of error function $e_{\max}(L)$.

Two parameters L, l in (11.3) and (11.4) may be selected when comparing the frequency characteristics. Defining the frequency approximation errors as

$$e_{IAE}(\omega_i, L, l) = |G_D(j\omega_i) - \tilde{G}_D(j\omega_i, L, l)| \qquad (11.16)$$

$$e_{ISE}(\omega_i, L, l) = |G_D(j\omega_i) - \tilde{G}_D(j\omega_i, L, l)|^2 \qquad (11.17)$$

where $G_D(j\omega_i)$ and $\tilde{G}_D(j\omega_i, L, l)$ are defined by (8.9) and (11.6), respectively. On the basis of above frequency approximation errors one defines frequency IAE and

Table 11.1   FOBD versions operation numbers.

Method	FOBD number	Summations number	Multiplications number	Total	Formula
1	${}^{GL}_{0}\Delta_{k}^{(\nu)}f(k)$	k	k	$2k$	(11.1)
2	${}^{GL}_{0}\tilde{\Delta}_{k}^{(\nu)}f(k)$	L	L	$2L$	(11.3)
3	${}^{GL}_{0}\tilde{\Delta}_{k}^{(\nu)}f(k)$	lL	$L+1$	$(l+1)L+1$	(11.4)

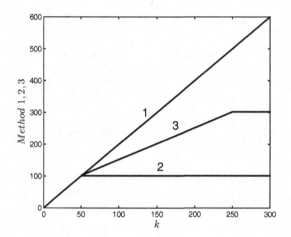

Fig. 11.12   Number of multiplications and additions vs. discrete-time for methods 1, 2 and 3.

ISE FOBD approximation criteria

$$J_{IAE}(L,l) = \sum_{k=0}^{k_{\max}} |G_D(j\omega_{i,k}) - \tilde{G}_D(j\omega_{i,k}, L, l)| \tag{11.18}$$

$$J_{ISE}(L,l) = \sum_{k=0}^{k_{\max}} |G_D(j\omega_{i,k}) - \tilde{G}_D(j\omega_{i,k}, L, l)|^2 \tag{11.19}$$

where $\omega_{i,k} \in [0, \pi]$.

The shapes of approximated FOBD due to different L and l are presented in a numerical example.

Example 11.2. Given a fractional order $\nu = 0.5$. Figs. 11.13 and 11.14 contain discrete Nyquist diagrams $G_D(j\omega)$ (1), $G_D(j\omega, 50, 1)$ (2) and $G_D(j\omega, 50, 5)$ (3) plotted for two frequency ranges $\omega \in [0, \pi]$ and $\omega \in [0, \frac{\pi}{6}]$.

In Fig. 11.15, absolute values of $e_{IAE}(\omega_i, 50, 1)$ and $e_{IAE}(\omega_i, 50, 5)$ are presented.

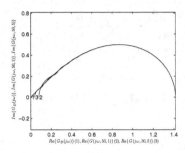

Fig. 11.13 Nyquist diagrams $G_D(j\omega)$ (1), $G_D(j\omega, 50, 1)$ (2) and $G_D(j\omega, 50, 5)$ (3) for $\omega \in [0, \pi]$.

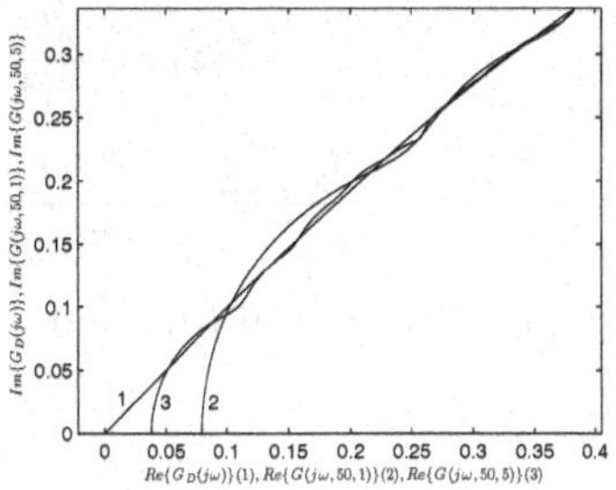

Fig. 11.14 Nyquist diagrams $G_D(j\omega)$ (1), $G_D(j\omega, 50, 1)$ (2) and $G_D(j\omega, 50, 5)$ (3) for $\omega \in [0, \frac{\pi}{6}]$.

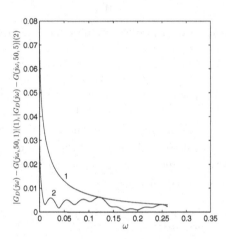

Fig. 11.15 Absolute values of $e_{IAE}(\omega_i, 50, 1)$ and $e_{IAE}(\omega_i, 50, 5)$ for $\omega \in [0, \frac{\pi}{6}]$.

11.4 Simplified form of the FOTF

The fundamental form of the FOTF, for a permanent assumption that $a_p = 1$

$$G(z) = \frac{\sum_{i=0}^{q} b_i (1 - z^{-1})^{(i\nu)}}{\sum_{i=0}^{p} a_i (1 - z^{-1})^{(i\nu)}}$$

$$= \frac{b_q(1 - z^{-1})^{q\nu} + b_{q-1}(1 - z^{-1})^{(q-1)\nu} + \cdots + b_1(1 - z^{-1})^{\nu} + b_0}{a_p(1 - z^{-1})^{p\nu} + a_{p-1}(1 - z^{-1})^{(p-1)\nu} + \cdots + a_1(1 - z^{-1})^{\nu} + a_0}. \quad (11.20)$$

may be expanded into an infinite series $(1 - z^{-1})^{\nu}$ (see formula (1.17))

$$G(z) = \frac{b_q(1 - z^{-1})^{q\nu} + \cdots + b_1(1 - z^{-1})^\nu + b_0}{a_p(1 - z^{-1})^{p\nu} + \cdots + a_1(1 - z^{-1})^\nu + a_0}$$

$$= \frac{b_q \sum_{i=0}^{+\infty} a^{(q\nu)}(i)z^{-i} + \cdots + b_1 \sum_{i=0}^{+\infty} a^{(\nu)}(i)z^{-i} + b_0}{a_p \sum_{i=0}^{+\infty} a^{(p\nu)}(i)z^{-i} + \cdots + a_1 \sum_{i=0}^{+\infty} a^{(\nu)}(i)z^{-i} + a_0}. \tag{11.21}$$

Recalling formulas (6.107) and (6.117) the last form is also equivalent to

$$G(z) = \frac{\sum_{i=0}^{+\infty} B_i z^{-i}}{\sum_{i=0}^{+\infty} A_i z^{-i}} \text{ with } A_0 \neq 0. \tag{11.22}$$

In the last form of the FOTF, the numerator and denominator are infinite polynomials in complex variable z^{-1}. Such polynomials have an infinite number of roots. Now, by properties discussed in Sections 1.3.3 and 6.4 for $\nu \in (0, 1]$

$$A_i \to 0, B_i \to 0. \tag{11.23}$$

(see Fig. 1.16). This suggests a simplification of (11.22). Define two integers L_N, L_D being sufficiently large that

$$G_{L_N,L_D}(z) = \frac{\sum_{i=0}^{L_N} B_i z^{-i}}{\sum_{i=0}^{L_D} A_i z^{-i}} \approx G(z). \tag{11.24}$$

One must remember that the roots of polynomials are very sensitive to the coefficient's variations ([Ralston (1965)]). It is obvious that the approximation $G_{L_N,L_D}(z)$ gets better for increasing L_N and L_D. Value L_N and L_D selection can be made taking into account the error TF

$$E_{L_N,L_D}(z) = [G(z) - G_{L_N,L_D}(z)] U(z). \tag{11.25}$$

A steady-state error is calculated using the final value theorem of the $\mathcal{Z}$-Transform [Ostalczyk (2002)]

$$e_{L_N,L_D}(\infty) = \lim_{z \to +\infty} E_{L_N,L_D}(z) = \lim_{z \to 1} [G(z) - G_{L_N,L_D}(z)] U(z)(z - 1). \tag{11.26}$$

For the discrete unit step function with $\mathcal{Z}\{1(k)\} = \frac{z}{z-1}$ one gets

$$e_{L_N,L_D}(\infty) = G(1) - G_{L_N,L_D}(1) = \frac{b_0}{a_0} - \frac{\sum_{i=0}^{L_N} B_i}{\sum_{i=0}^{L_D} A_i}. \tag{11.27}$$

Usually $e_{L_N,L_D}(\infty) \neq 0$. To preserve a zero steady-state a gain correction coefficient (GCC) should be added to (11.24)

$$G_{L_N,L_D}(z) = K_{L_N,L_D} \frac{\sum_{i=0}^{L_N} B_i z^{-i}}{\sum_{i=0}^{L_D} A_i z^{-i}} \text{ where } K_{L_N,L_D} = \frac{b_0 \sum_{i=0}^{L_D} A_i}{a_0 \sum_{i=0}^{L_N} B_i}. \tag{11.28}$$

If a stable FOTF is subjected to an input signal satisfying the condition $\lim_{k \to +\infty} u(k) = 0$ then a steady-state response is zero. There is no need to add this coefficient. In the next numerical example a discussion concerning the choice of L_N, L_D will be performed.

Example 11.3.

$$G(z) = \frac{b_1(1 - z^{-1})^\nu + b_0}{(1 - z^{-1})^{2\nu} + a_1(1 - z^{-1})^\nu + a_0} \tag{11.29}$$

where $\nu = 0.5, a1 = -1, a0 = 0.4181, b1 = 0.2, b0 = 0.35$. In Figs. 11.20 and 11.21 the exact and approximated with $L_N = L_D = 1000$ responses are given, respectively.

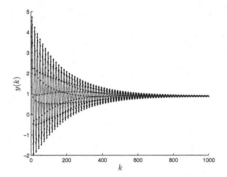

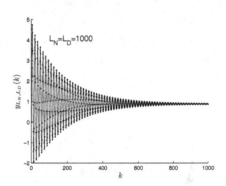

Fig. 11.16   Response of 11.29.

Fig. 11.17 Approximated response for $L_N = L_D = 1000$.

The difference between the exact solution and its approximation is almost invisible. However, the difference exists. The shape of $e_{L_N,L_D}(k)$ for two cases: with and without the GCC (11.28) will be considered. In Figs. 11.18 and 11.19 plots of considered errors are presented, respectively. Note that the black areas depict strong oscillations of errors.

The plots reveal that the GCC introduction leads to much higher maximal error which further tends (however slowly) towards zero. Whereas, in the second analyzed case the error values are much smaller, but there is a steady-state non-zero error. Such a situation may be important in the closed-loop dynamic system synthesis, when the value of a constant gain influences the closed-loop system stability. Also note that for $K_{L_N,L_D} = 1$ for $k \in [0, L_D]$ the error equals to zero.

In the Example 11.3 it was assumed $L_N = L_D$. Now, one abandons this assumption and checks the influence of the relation L_N to L_D to the $e_{L_N,L_D}(k)$. This is shown in the numerical example.

Example 11.4. 11.4 Compare $e_{L_N,2L_N}(k)$ with $e_{0.5L_N,L_N}(k)$ and check the influence of $L_N = L_D$ on the error function.

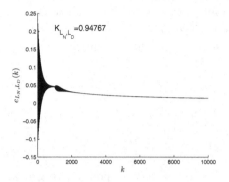

Fig. 11.18   Error $e_{L_N,L_D}(k)$ vs. discrete time k with the correction coefficient.

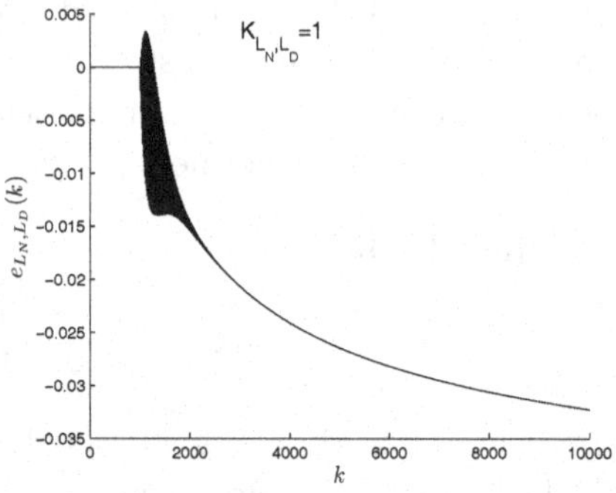

Fig. 11.19   Error $e_{L_N,L_D}(k)$ vs. discrete time k without the correction coefficient.

Solution. The plots of errors $e_{L_N,2L_N}(k)$ and $e_{2L_N,L_N}(k)$ are given in Figs. 11.20 and 11.21, respectively.

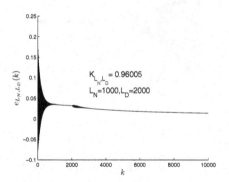

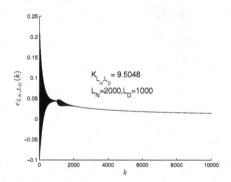

Fig. 11.20   Error $e_{L_N,2L_N}(k)$ vs. discrete time k with the correction coefficient.

Fig. 11.21   Error $e_{2L_N,L_N}(k)$ vs. discrete time k with the correction coefficient.

A careful viewing of the above plots reveals that a better approximation is achieved for a case when $L_D > L_N$. This is caused by the FOTF form considered in the numerical example where $2 = p > q = 1$.

In Fig. 11.22 the error functions for different lengths of $L_N = L_D\{1000, 2000, \dots, 9000\}$ for $K_{L_N,L_D} = 1$ are plotted. The graphs show relatively rapid disappearance of maximal errors. One should note that for $k \in [0 L_D]$ there is a zero error, hence there are horizontal line sectors.

There is also another measure for $G(z)$ and $G_{L_N,L_D}(z)$ matching. One compares its frequency characteristics. Consider (11.21) with a discrete Dirac pulse as the input $U(z) = \mathcal{Z}\{\delta(k)\} = 1$. Then as a performance criterion one takes

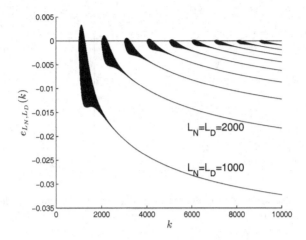

Fig. 11.22   Error function for $L_N = L_D\{1000, 2000, \ldots, 9000\}$.

$$I(L_N, L_D) = \min_{L_N, L_D \in \mathbf{Z}_+} \left[\max_{\omega \in \mathbf{R}_+} |e_{L_N, L_D}(j\omega)|^i \right]$$

$$= \min_{L_N, L_D \in \mathbf{Z}_+} \left[\max_{\omega \in \mathbf{R}_+} |G(j\omega) - G_{L_N, L_D}(j\omega)|^i \right] \text{ for } i = 1, 2. \qquad (11.30)$$

In Example 11.5 the applicability of this measure will be checked.

Example 11.5. 11.5 Plot $e_{L_N, L_D}(j\omega)$ and $|e_{L_N, L_D}(j\omega)|$ for the FOTF and its approximation considered in the previous example for $L_N = L_D = 1000$.

Solution. The Nyquist plots of $G(j\omega_i)$ and $G_{L_N, L_D}(j\omega_i)$ are given in Fig. 11.23. The plots of its magnitude vs. chosen interval of frequency reveal a strong peak, which is responsible for oscillations. This is presented in Fig. 11.24.
The differences between the plot are almost invisible. To derive the approximation effects in Fig. 11.25 the Nyquist plot of the difference $G(j\omega_i) - G_{L_N, L_D}(j\omega_i)$ is presented. The magnitudes of $|G(j\omega_i) - G_{L_N, L_D}(j\omega_i)|$ vs. frequency for $L_N = L_D = 1000, 2000, 3000$ are plotted in Fig. 11.26.

As was mentioned earlier, with increasing $G(j\omega_i)$ and $G_{L_N, L_D}(j\omega_i)$ the peaks decrease. The FOTF considered in numerical examples investigated above is rather difficult to approximate. A specific shape of the magnitude plot with "thin" and high peak causes this. The highest approximation errors occur only just in the peak range.

The most difficult FOS to approximate is a system being stable in a Lapunov sense but not asymptotically stable (the marginally stable system). Then an approximation should also posses this property. Recall that the approximated system

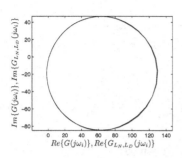

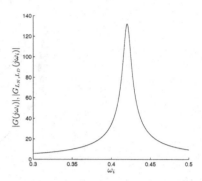

Fig. 11.23 The Nyquist plots of $G(j\omega_i)$ and $G_{L_N,L_D}(j\omega_i)$ for $L_N = L_D = 1000$.

Fig. 11.24 The magnitude plots $|G(j\omega_i)|$ and $|G_{L_N,L_D}(j\omega_i)|$ for $L_N = L_D = 1000$.

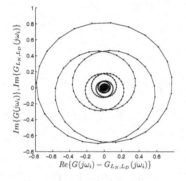

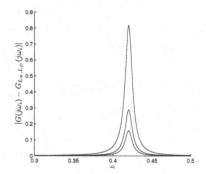

Fig. 11.25 The Nyquist plots of $G(j\omega_i) - G_{L_N,L_D}(j\omega_i)$ for $L_N = L_D = 1000$.

Fig. 11.26 The magnitude plots $|G(j\omega_i)| - |G_{L_N,L_D}(j\omega_i)|$ for $L_N = L_D = 1000, 2000, 3000$.

described by $G_{L_N,L_D}(z)$ is in fact one that is a classical discrete-time one. There are several simple stability criteria. In the $G_{L_N,L_D}(z)$ case there is a problem of its rather high order. In the next numerical example one analyses the marginally stable FOTF and its approximation.

Example 11.6. Consider the FOTF (11.29) with coefficients $a_1 = -1, a_0 = 0.4142135476330133, b_1 = 0.2, b_0 = 0.35$. Find its IO approximation.

Solution. The discrete unit step response for two time ranges (a) [0 1000] and (b) [1 100000] is presented in Figs. 11.27(a) and 11.27(b), respectively.

Now, one simulates two approximations. For $L_N = L_D = 1000$ one obtains an unstable solution whereas for $L_N = L_D = 5000$ one obtains asymptotically stable ones. Considered responses are plotted in Figs. 11.28(a) and 11.28(b) and 11.29(a) and 11.29(b) for two ranges, respectively.

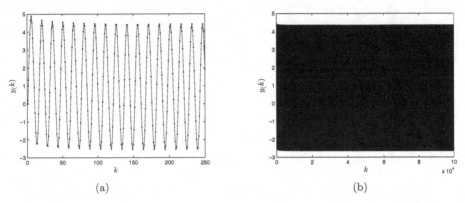

Fig. 11.27   The discrete Dirac pulse response of (11.29) a time range (a) [0 250] and (b) [0 100000].

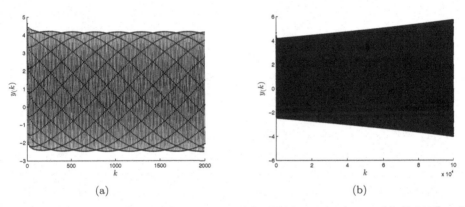

Fig. 11.28   The discrete Dirac pulse response of (11.29) for a time range (a) [0 2000] and (b) [0 100000].

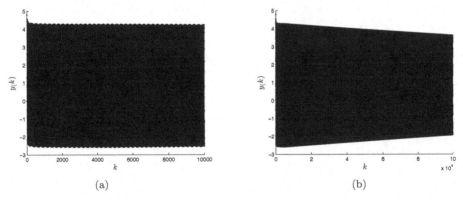

Fig. 11.29   The discrete Dirac pulse response of (11.29) for a time range (a) [0 10000] and (b) [0 100000].

Chapter 12

Fractional potential

In this Chapter an application of the fractional potential [Oustaloup (1995); Poty *et al.* (2004a,b); Melchior *et al.* (2009)] to the path planning in a mobile vehicle will be shown. First, a notion of the discrete distance will be defined. Based on that, one defines discrete curves and figures [Hearn and Baker (1994); Salomon (1999)]. Next, to every discrete distance there is a defined surjection relating every discrete fractional potential value to points on a discrete plane. Application of the fractional additivity property one creates a final potential plane which will be useful, among others, in the mobile devices path planning.

12.1 Discrete circle

In this section, a short introduction to a discrete curve in a discrete plane is presented. The main focus will be on the discrete circle. It will be a sort of Bresenham's circle [Bresenham (1965)]. First, one defines a discrete square.

Definition 12.1. A discrete square $S(i,j)$ is defined by four corners of coordinates $s_A(i - \frac{1}{2}, j + \frac{1}{2})$, $s_A(i - \frac{1}{2}, j - \frac{1}{2})$, $s_A(i + \frac{1}{2}, j - \frac{1}{2})$, $s_A(i + \frac{1}{2}, j + \frac{1}{2})$.

An area of $S(i,j)$ is equal to 1.

A discrete square containing a fragment of a circle $x^2 + y^2 = r^2$, where $x, y \in \left[-r - \frac{1}{2}, r + \frac{1}{2}\right] \in \mathbb{R}$ will be denoted as $S(i,j,r)$. A single such square is presented in Fig. 12.1.

Definition 12.2. A discrete circle of radius $r(0,0)$ centered at the origin $C(r)$ is defined as a set of discrete squares $S(r)$

$$C\left[r(k_1, k_2)\right] = \bigcup_{i,j=1,2,\dots r} S(i,j,r). \tag{12.1}$$

The discrete circle of radius $r = 5$ is given in Fig. 12.2.

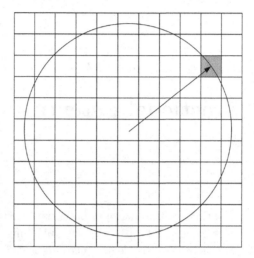

Fig. 12.1   Discrete circle element.

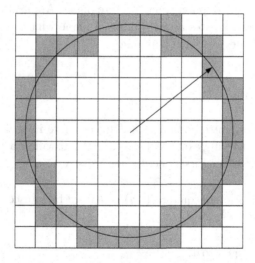

Fig. 12.2   Discrete circle of radius $r = 4$.

Note that the radius r satisfies inequalities

$$\sqrt{\left(i - \frac{1}{2}\right)^2 + \left(j - \frac{1}{2}\right)^2} < r < \sqrt{\left(i + \frac{1}{2}\right)^2 + \left(j + \frac{1}{2}\right)^2} \tag{12.2}$$

for $i, j = 1, 2, \ldots, r$. Discrete circles for different chosen radii are presented in Figs. 12.3–12.10. Here, one should note that consecutive circle elements are collected in matrices of dimensions $(2r + 1) \times (2r + 1)$ so the circle center is placed at $(r + 1) \times (r + 1)$.

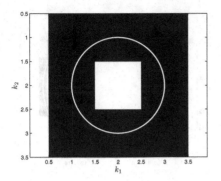

Fig. 12.3 Discrete circle of radius $r = 1$.

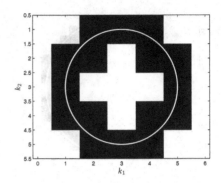

Fig. 12.4 Discrete circle of radius $r = 2$.

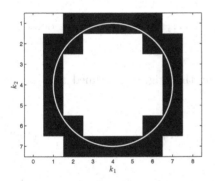

Fig. 12.5 Discrete circle of radius $r = 3$.

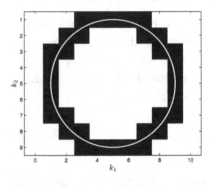

Fig. 12.6 Discrete circle of radius $r = 4$.

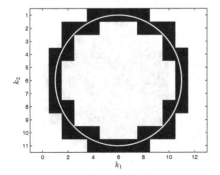

Fig. 12.7 Discrete circle of radius $r = 5$.

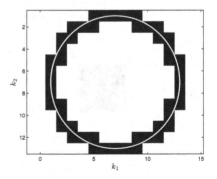

Fig. 12.8 Discrete circle of radius $r = 6$.

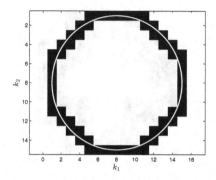

Fig. 12.9   Discrete circle of radius $r = 7$.

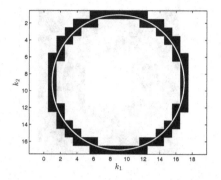

Fig. 12.10   Discrete circle of radius $r = 8$.

12.1.1   *Discrete mask*

Now one collects all circles with radii $r \in \{1, 2, \ldots, r_{\max}\}$. Such a set defines a discrete disk characterized by radius $r_{\max}$.

Definition 12.3. A discrete disk (centered at the origin) is defined as a set of discrete circles of radii $r \in \{1, 2, \ldots, r_{\max}\}$

$$D(r_{\max}) = \bigcup_{r=1,2,\ldots r_{\max}} C(r). \tag{12.3}$$

One should note that in the discrete circles there is an undefined circle with radius $r = 0$. In the discrete disk definition, one should supplement a set (12.3) by a square $S(0, 0, 0)$. Selected discrete disks are presented in Figs. 12.11–12.13.

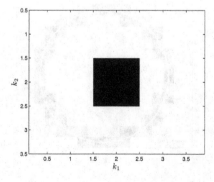

Fig.   12.11   Discrete   disk   of   radius $r_{\max} = 1$.

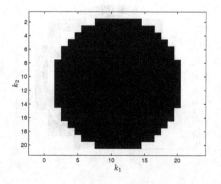

Fig.   12.12   Discrete   disk   of   radius $r_{\max} = 10$.

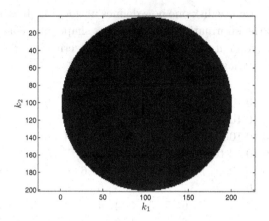

Fig. 12.13 Discrete disk of radius $r_{\max} = 100$.

The discrete disk may also be represented by a matrix containing information about presence 0 or absence 10. An exemplary disk matrix of dimensions $(2r+1) \times (2r+1)$ for $r = 10$ is given below.

$$
\mathbf{D}\left[D\left(r_{\max}\right)\right] =
\begin{bmatrix}
10 & 10 & 10 & 10 & 10 & 10 & 10 & 10 & 10 & 10 & 10 & 10 & 10 & 10 & 10 & 10 & 10 & 10 & 10 & 10 & 10 \\
10 & 10 & 10 & 10 & 10 & 10 & 10 & 0 & 0 & 0 & 0 & 0 & 0 & 0 & 10 & 10 & 10 & 10 & 10 & 10 & 10 \\
10 & 10 & 10 & 10 & 10 & 0 & 0 & 0 & 0 & 0 & 0 & 0 & 0 & 0 & 0 & 0 & 10 & 10 & 10 & 10 & 10 \\
10 & 10 & 10 & 10 & 0 & 0 & 0 & 0 & 0 & 0 & 0 & 0 & 0 & 0 & 0 & 0 & 0 & 10 & 10 & 10 & 10 \\
10 & 10 & 10 & 0 & 0 & 0 & 0 & 0 & 0 & 0 & 0 & 0 & 0 & 0 & 0 & 0 & 0 & 0 & 10 & 10 & 10 \\
10 & 10 & 0 & 0 & 0 & 0 & 0 & 0 & 0 & 0 & 0 & 0 & 0 & 0 & 0 & 0 & 0 & 0 & 0 & 10 & 10 \\
10 & 10 & 0 & 0 & 0 & 0 & 0 & 0 & 0 & 0 & 0 & 0 & 0 & 0 & 0 & 0 & 0 & 0 & 0 & 10 & 10 \\
10 & 0 & 0 & 0 & 0 & 0 & 0 & 0 & 0 & 0 & 0 & 0 & 0 & 0 & 0 & 0 & 0 & 0 & 0 & 0 & 10 \\
10 & 0 & 0 & 0 & 0 & 0 & 0 & 0 & 0 & 0 & 0 & 0 & 0 & 0 & 0 & 0 & 0 & 0 & 0 & 0 & 10 \\
10 & 0 & 0 & 0 & 0 & 0 & 0 & 0 & 0 & 0 & 0 & 0 & 0 & 0 & 0 & 0 & 0 & 0 & 0 & 0 & 10 \\
10 & 0 & 0 & 0 & 0 & 0 & 0 & 0 & 0 & 0 & 0 & 0 & 0 & 0 & 0 & 0 & 0 & 0 & 0 & 0 & 10 \\
10 & 0 & 0 & 0 & 0 & 0 & 0 & 0 & 0 & 0 & 0 & 0 & 0 & 0 & 0 & 0 & 0 & 0 & 0 & 0 & 10 \\
10 & 0 & 0 & 0 & 0 & 0 & 0 & 0 & 0 & 0 & 0 & 0 & 0 & 0 & 0 & 0 & 0 & 0 & 0 & 0 & 10 \\
10 & 0 & 0 & 0 & 0 & 0 & 0 & 0 & 0 & 0 & 0 & 0 & 0 & 0 & 0 & 0 & 0 & 0 & 0 & 0 & 10 \\
10 & 10 & 0 & 0 & 0 & 0 & 0 & 0 & 0 & 0 & 0 & 0 & 0 & 0 & 0 & 0 & 0 & 0 & 0 & 10 & 10 \\
10 & 10 & 0 & 0 & 0 & 0 & 0 & 0 & 0 & 0 & 0 & 0 & 0 & 0 & 0 & 0 & 0 & 0 & 0 & 10 & 10 \\
10 & 10 & 10 & 0 & 0 & 0 & 0 & 0 & 0 & 0 & 0 & 0 & 0 & 0 & 0 & 0 & 0 & 0 & 10 & 10 & 10 \\
10 & 10 & 10 & 10 & 0 & 0 & 0 & 0 & 0 & 0 & 0 & 0 & 0 & 0 & 0 & 0 & 0 & 10 & 10 & 10 & 10 \\
10 & 10 & 10 & 10 & 10 & 0 & 0 & 0 & 0 & 0 & 0 & 0 & 0 & 0 & 0 & 0 & 10 & 10 & 10 & 10 & 10 \\
10 & 10 & 10 & 10 & 10 & 10 & 10 & 0 & 0 & 0 & 0 & 0 & 0 & 0 & 10 & 10 & 10 & 10 & 10 & 10 & 10 \\
10 & 10 & 10 & 10 & 10 & 10 & 10 & 10 & 10 & 10 & 10 & 10 & 10 & 10 & 10 & 10 & 10 & 10 & 10 & 10 & 10
\end{bmatrix}.
$$

$$(12.4)$$

On the basis of a discrete disk one now defines now a discrete mask. The discrete mask in addition to the information concerning its shape, also contains information about the distances from its center. A matrix describing the mask defined by $M(r_{\max})$ for $r_{\max} = 10$ has the form

$$
\mathbf{M}\left[D\left(r_{\max}\right)\right] =
\begin{bmatrix}
10 & 10 & 10 & 10 & 10 & 10 & 10 & 10 & 10 & 10 & 10 & 10 & 10 & 10 & 10 & 10 & 10 & 10 & 10 & 10 & 10 \\
10 & 10 & 10 & 10 & 10 & 10 & 10 & 9 & 9 & 9 & 9 & 9 & 9 & 9 & 10 & 10 & 10 & 10 & 10 & 10 & 10 \\
10 & 10 & 10 & 10 & 10 & 9 & 9 & 8 & 8 & 8 & 8 & 8 & 8 & 8 & 9 & 9 & 10 & 10 & 10 & 10 & 10 \\
10 & 10 & 10 & 10 & 9 & 8 & 8 & 7 & 7 & 7 & 7 & 7 & 7 & 7 & 8 & 8 & 9 & 10 & 10 & 10 & 10 \\
10 & 10 & 10 & 9 & 8 & 8 & 7 & 7 & 6 & 6 & 6 & 6 & 6 & 7 & 7 & 8 & 8 & 9 & 10 & 10 & 10 \\
10 & 10 & 9 & 8 & 8 & 7 & 6 & 6 & 5 & 5 & 5 & 5 & 5 & 6 & 6 & 7 & 8 & 8 & 9 & 10 & 10 \\
10 & 10 & 9 & 8 & 7 & 6 & 5 & 5 & 4 & 4 & 4 & 4 & 4 & 5 & 5 & 6 & 7 & 8 & 9 & 10 & 10 \\
10 & 9 & 8 & 7 & 7 & 6 & 5 & 4 & 3 & 3 & 3 & 3 & 3 & 4 & 5 & 6 & 7 & 7 & 8 & 9 & 10 \\
10 & 9 & 8 & 7 & 6 & 5 & 4 & 3 & 3 & 2 & 2 & 2 & 3 & 3 & 4 & 5 & 6 & 7 & 8 & 9 & 10 \\
10 & 9 & 8 & 7 & 6 & 5 & 4 & 3 & 2 & 1 & 1 & 1 & 2 & 3 & 4 & 5 & 6 & 7 & 8 & 9 & 10 \\
10 & 9 & 8 & 7 & 6 & 5 & 4 & 3 & 2 & 1 & 0 & 1 & 2 & 3 & 4 & 5 & 6 & 7 & 8 & 9 & 10 \\
10 & 9 & 8 & 7 & 6 & 5 & 4 & 3 & 2 & 1 & 1 & 1 & 2 & 3 & 4 & 5 & 6 & 7 & 8 & 9 & 10 \\
10 & 9 & 8 & 7 & 6 & 5 & 4 & 3 & 3 & 2 & 2 & 2 & 3 & 3 & 4 & 5 & 6 & 7 & 8 & 9 & 10 \\
10 & 9 & 8 & 7 & 7 & 6 & 5 & 4 & 3 & 3 & 3 & 3 & 3 & 4 & 5 & 6 & 7 & 7 & 8 & 9 & 10 \\
10 & 10 & 9 & 8 & 7 & 6 & 5 & 5 & 4 & 4 & 4 & 4 & 4 & 5 & 5 & 6 & 7 & 8 & 9 & 10 & 10 \\
10 & 10 & 9 & 8 & 8 & 7 & 6 & 6 & 5 & 5 & 5 & 5 & 5 & 6 & 6 & 7 & 8 & 8 & 9 & 10 & 10 \\
10 & 10 & 10 & 9 & 8 & 8 & 7 & 7 & 6 & 6 & 6 & 6 & 6 & 7 & 7 & 8 & 8 & 9 & 10 & 10 & 10 \\
10 & 10 & 10 & 10 & 9 & 8 & 8 & 7 & 7 & 7 & 7 & 7 & 7 & 7 & 8 & 8 & 9 & 10 & 10 & 10 & 10 \\
10 & 10 & 10 & 10 & 10 & 9 & 9 & 8 & 8 & 8 & 8 & 8 & 8 & 8 & 9 & 9 & 10 & 10 & 10 & 10 & 10 \\
10 & 10 & 10 & 10 & 10 & 10 & 10 & 9 & 9 & 9 & 9 & 9 & 9 & 9 & 10 & 10 & 10 & 10 & 10 & 10 & 10 \\
10 & 10 & 10 & 10 & 10 & 10 & 10 & 10 & 10 & 10 & 10 & 10 & 10 & 10 & 10 & 10 & 10 & 10 & 10 & 10 & 10
\end{bmatrix}.
$$

$$(12.5)$$

Such mask matrices can be treated as an image. It is presented in Fig. 12.14 after an appropriate re-scaling.

It is also possible to represent the mask matrix by its constant values contours and 3D image. These representations are plotted in Figs. 12.15 and 12.16, respectively.

Now, a transformed mask should be defined. For any bounded discrete one-variable discrete function $f(r)$ one defines a generalized mask $M[f(r)]$ characterized by a matrix $\mathbf{M}[f(r)]$

$$
\mathbf{M}[f(r)] =
\begin{bmatrix}
& \vdots & \vdots & \vdots & \\
\cdots & f(1) & f(1) & f(1) & \cdots \\
\cdots & f(1) & f(0) & f(1) & \cdots \\
\cdots & f(1) & f(1) & f(1) & \cdots \\
& \vdots & \vdots & \vdots &
\end{bmatrix}.
$$

$$(12.6)$$

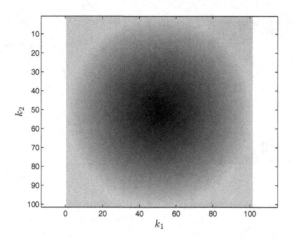

Fig. 12.14   Mask matrix image for $r_{\max} = 10$.

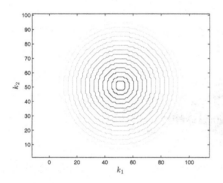

Fig. 12.15   Discrete mask constant values contours.

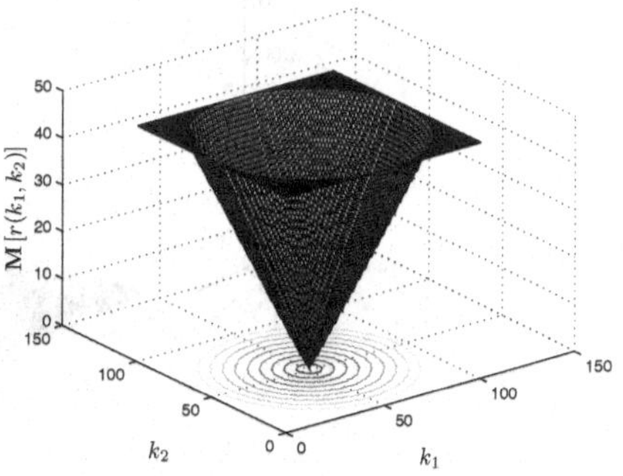

Fig. 12.16   Discrete mask 3D plot.

Below, some examples of generalized masks are proposed.

Example 12.1. At the outset one establishes the range of the generalized mask $r_{\max} = 50$. Next one considers a discrete variable function

$$f(r) = e^{ar} \text{ where } a = -0.05, \ r = \sqrt{k_1^2 + k_2^2}, \ k_1, k_2 \in [-r_{\max}, r_{\max}]. \quad (12.7)$$

In relation to the function given above one gets the mask depicted in Fig. 12.17. Changing the parameter a one can easily change the shape of the generalized mask. Next, one considers very important discrete variable function as defined on the discrete plane. This function describes the distribution of the potential generated by the unit charge. Similar plots to those presented in the above three figures are given in Figs. 12.20–12.23, respectively.

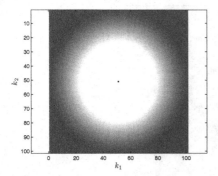

Fig. 12.17   Generalized mask matrix image for $r_{\max} = 50$.

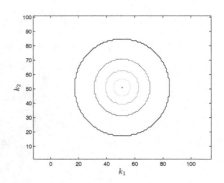

Fig. 12.18   Discrete mask constant values contours.

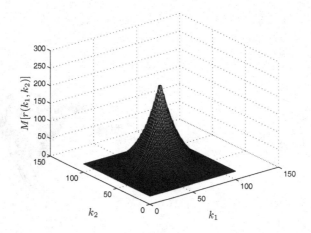

Fig. 12.19   Generalized discrete mask 3D plot.

Fig. 12.20   Generalized mask matrix image for the potential.

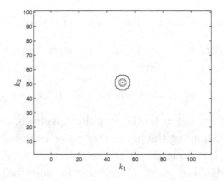

Fig. 12.21   Discrete potential mask constant value contours.

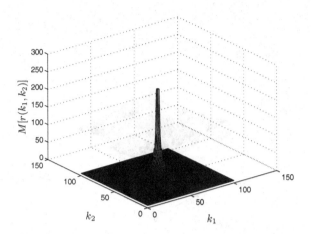

Fig. 12.22 Generalized discrete mask of the potential 3D plot.

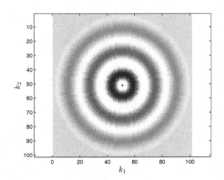

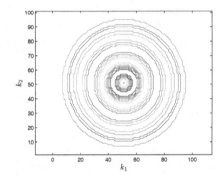

Fig. 12.23 Generalized mask matrix image for the potential.

Fig. 12.24 Discrete potential mask constant value contours.

$$f(r) = \frac{255}{r} \text{ where } r = \sqrt{k_1^2 + k_2^2}, \ k_1, k_2 \in [-r_{\max}, r_{\max}]. \tag{12.8}$$

There is an immense number of function choice that can be used to generate the discrete mask. To confirm the last statement one considers the function

$$f(r) = 50 \cos\left(\frac{\pi r}{8}\right) e^{-0.04r} + 50 \text{ where } r = \sqrt{k_1^2 + k_2^2}, \ k_1, k_2 \in [-r_{\max}, r_{\max}]. \tag{12.9}$$

Appropriate plots are given in Figs. 12.23–12.25, respectively.

The discrete function can be defined by its values collected in the Table 12.1.

A plot of the function is given in Fig. 12.26.

Appropriate plots are given in Figs. 12.23–12.25, respectively.

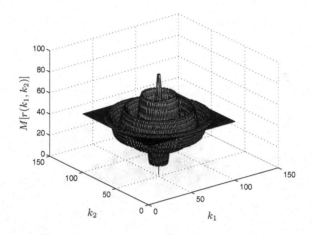

Fig. 12.25   Generalized discrete mask of the potential 3D plot.

Table   12.1   Discrete
function values.

r	$f(r)$
0	254
1	253
2	252
$\vdots$	$\vdots$
13	239
14	240
15	208
16	206
$\vdots$	$\vdots$
33	168
34	170
35	62
36	59
$\vdots$	$\vdots$
49	23
50	20

12.1.2   *Discrete fractional potential*

One considers an isolated charged sphere of radius R, which is located in a homogenous isotropic environment with no other charges. One assumes a positive charge Q. Then the electric field-lines associated with a positive point charge are a set of unbroken, evenly spaced straight-lines which radiate from the charge. Since there is full spherical symmetry the electric potential is described by the Laplace's equation

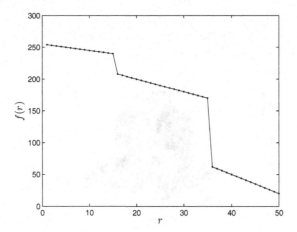

Fig. 12.26   Plot of the discrete function tabularized in Table 12.1.

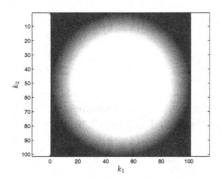

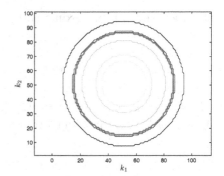

Fig. 12.27   Generalized mask matrix image for the potential.

Fig. 12.28   Disrete potential mask constant value contours.

$$\frac{d^2V(r)}{dr^2} + \frac{2}{r}\frac{dV(r)}{dr} = 0. \tag{12.10}$$

Since the zero of potential is arbitrary, it is reasonable to choose the zero of potential at infinity: the standard practice with localized charges. Hence, the solution is of the form

$$V(r) = -\frac{1}{4\pi\varepsilon_0}\int_\infty^r \frac{1}{x^2}dx = \frac{1}{4\pi\varepsilon_0}\frac{Q}{r}, \text{ for } r > R \tag{12.11}$$

where ϵ_0 is the electric constant (permittivity of free space). This is known as the Coulomb potential. In practice, one evaluates a simplified version of the potential

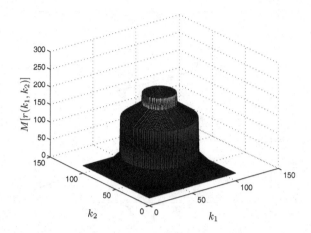

Fig. 12.29   Generalized discrete mask of the potential 3D plot.

where the integral is approximated by a finite sum

$$V(r) = -\frac{Q}{4\pi\varepsilon_0} \sum_{i=L}^{r} \frac{1}{(ih)^2} h. \tag{12.12}$$

Now, one considers a discrete-variable function $f(k)$ with a coordinate transformation $k \to k'$. The mentioned transformation is clarified in Fig. 12.30.

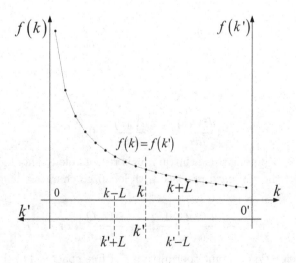

Fig. 12.30   Discrete-variable function coordinate transformation.

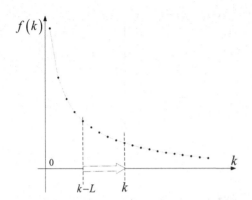

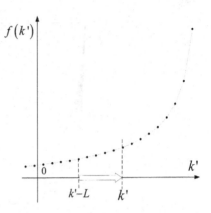

Fig. 12.31   Discrete-function $f(k)$ plot.

Fig. 12.32   Discrete function $f(k')$ plot.

For any $k = k'$ one defines the FOBS of orders $\mu \in [-2, -1]$ over the ranges $[k - L, k]$ and $k' - L$. Appropriate functions with the FOBDs calculation ranges are given in Figs. 12.31 and 12.32, respectively.

$$\substack{GL \\ k-L}\Delta_k^{(-\mu)} f(k) = \sum_{i=0}^{L} a^{(-\nu)}(i) f(kh - ih) h^\nu \tag{12.13}$$

$$\substack{GL \\ k'-L}\Delta_{k'}^{(-\mu)} f(k') = \sum_{i=0}^{L} a^{(-\nu)}(i) f(k'h - ih) h^\nu. \tag{12.14}$$

Taking into account that from Fig. 12.30

$$f(k) = f(k') \tag{12.15}$$
$$f(k + i) = f(k' - i)$$

one obtains

$$\substack{GL \\ k+L}\Delta_k^{(-\mu)} f(k) = \sum_{i=0}^{L} a^{(-\nu)}(i) f(kh + ih) h^\nu. \tag{12.16}$$

The generalized discrete potential (GFP) generated by a single charge is defined as

$$V^{(-\mu)}(kh) = Q'\substack{GL \\ k+L}\Delta_k^{(-\mu)} f(k) = \sum_{i=0}^{L} a^{(-\nu)}(i) f(kh + ih) h^\nu \text{ for } kh > R. \tag{12.17}$$

where $Q' = Q/(4\epsilon)$ is a charge value and h is a value defining a distance kh. The shapes of the generalized (fractional) potential are plotted in Fig. 12.33 for $\mu \in [1, 2]$.

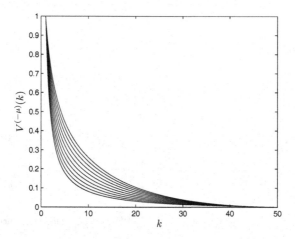

Fig. 12.33 Discrete generalized potentials for $\nu \in [1,2]$ for $k \geq R = 1$.

In the above curves, there are two ones calculated for $\mu = 1$ using formula (12.16) and

$$V^{(-1)}(kh) = \frac{Q'}{kh}.\qquad(12.18)$$

In Fig. 12.34 both plots are presented. There is no noticeable difference between them so one can consider that the GDP numerical method is satisfactory. The difference between plots is given in Fig. 12.34.

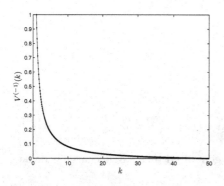

Fig. 12.34 Discrete mask for $r_{\max} = 50$ image.

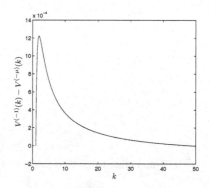

Fig. 12.35 Discrete mask contours.

The individual shape of the GDP is preserved for every straight discrete path radiating from the charge. Hence, there is an established relationship between the distance $r(k_1, k_2)$ from the charge and the GDP

$$\mathbf{M}\left[V^{(-\mu)}(r)\right] = \begin{bmatrix} \vdots & \vdots & \vdots \\ \cdots V^{(-\mu)}(1) & V^{(-\mu)}(1) & V^{(-\mu)}(1) \cdots \\ \cdots V^{(-\mu)}(1) & m(0) & V^{(-\mu)}(1) \cdots \\ \cdots V^{(-\mu)}(1) & V^{(-\mu)}(1) & V^{(-\mu)}(1) \cdots \\ \vdots & \vdots & \vdots \end{bmatrix}. \tag{12.19}$$

The GDP calculation procedure is explained in the numerical example.

Example 12.2. For a mask length $r_{\max} = 300$ and the FP $\mu = 1.5$ evaluate the GDP. In Figs. 12.36–12.38 the mask, its contours and 3D plot are plotted, respectively.

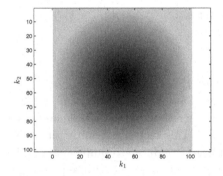

Fig. 12.36   Discrete mask for $r_{\max} = 50$ image.

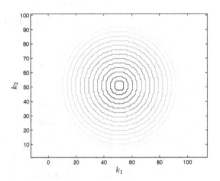

Fig. 12.37   Discrete mask contours.

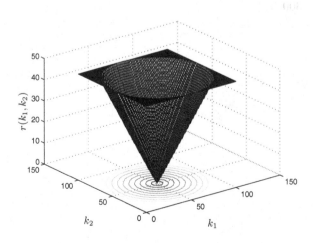

Fig. 12.38   Discrete mask 3D plot.

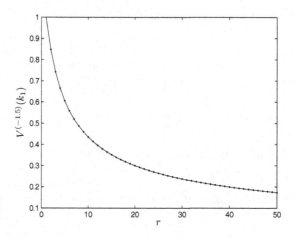

Fig. 12.39   GDP values.

In Fig. 12.39 the GDP for $\mu = 1.5$ is presented.

The related GDP image, contour and 3D plots are given in Figs. 12.40–12.42.

Due to a configuration of point charges the generalized potential is equal to the sum of the point charges' individual potentials. This fact significantly simplifies calculations, since addition of scalar potential fields is reduced to simple mask additions. In the numerical example selected obstacles and related GDPs are presented.

Example 12.3. Evaluate the GDP of selected 2D obstacles.

First one presents two separated charges of the same $Q = 1$ placed at coordinates $(300, 250)$ and $(300, 350)$

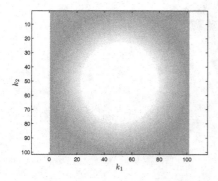

Fig. 12.40   GDP image.

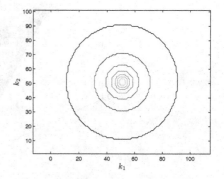

Fig. 12.41   GDP contours.

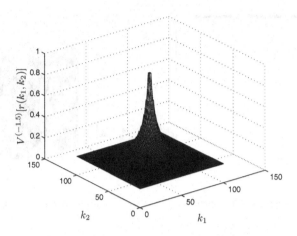

Fig. 12.42   GDP 3D plot.

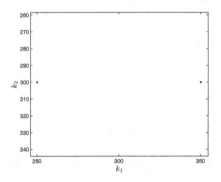

Fig. 12.43   Two separated charges.

Fig. 12.44   GDP generated by two charges.

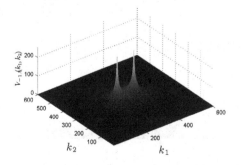

Fig. 12.45   3D plot of the GDP.

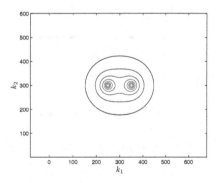

Fig. 12.46   Equipotential curves of the GDP.

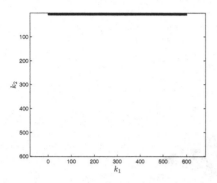

Fig. 12.47   Charged sided wall.

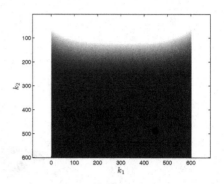

Fig. 12.48   GDP generated by sided wall.

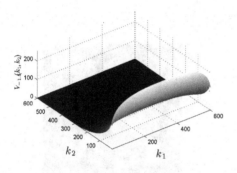

Fig. 12.49   3D plot of the GDP.

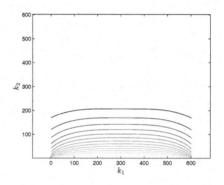

Fig. 12.50   Equipotential curves of the GDP.

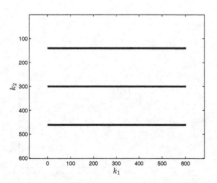

Fig. 12.51   Charged three parallel walls.

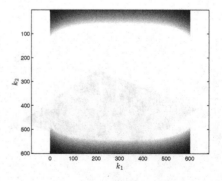

Fig. 12.52   GDP generated by three parallel wall.

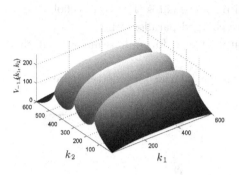

Fig. 12.53   3D plot of the GDP.

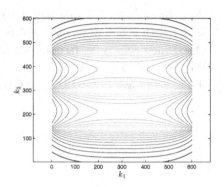

Fig. 12.54   Equipotential curves of the GDP.

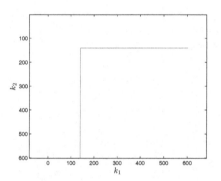

Fig. 12.55   L-shaped walls.

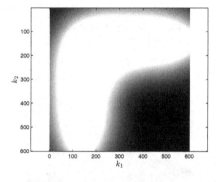

Fig. 12.56   GDP generated by the L-shaped walls.

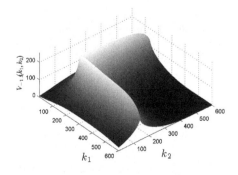

Fig. 12.57   3D plot of the GDP.

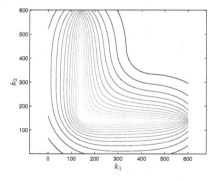

Fig. 12.58   Equipotential curves of the GDP.

Next one considers a side wall described by coordinates $[1 - 10, 1 - 600]$.

The fourth group of images depicts three charged parallel walls.

A mobile device may face an obstacle in a letter L form. It is modeled by the GDP presented in Figs. 12.55–12.58.

Finally, in Figs. 12.59–12.62 the GDP of the path formed in Z-letter is presented.

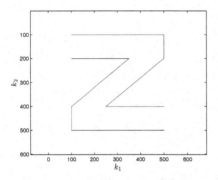

Fig. 12.59   Z-shaped walls.

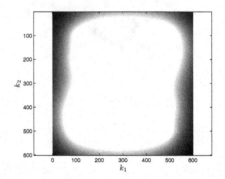

Fig. 12.60   GDP generated by the Z-shaped walls.

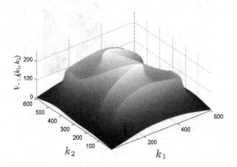

Fig. 12.61   3D plot of the GDP.

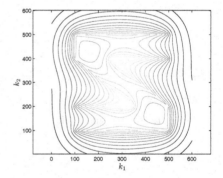

Fig. 12.62   Equipotential curves of the GDP.

Chapter 13

FO image filtering and edge detection

The most important characterization of the objects that create a discrete image is the object's edges. In the computer vision process, the individual object's edges detection is a fundamental procedure. The discrete image edges extraction procedures have been investigated for over forty years and several successful methods have been invented [Roberts (1963); Prewitt (1970); Sobel (1978); Marr and Hildreth (1980); Canny (1986); Aldair and Wang (2010); Alonso et al. (2011)]. Most of them, as a fundamental mathematical tool, apply the derivative, or more precisely the classical first order or 2D difference notion in the edge detection. However, this process strongly reinforces usually present noise in the image. The noise and the sharp edges in the image are related to high frequency components in the image frequency characteristics. This forces of a preliminary image filtering, resulting in a trade-off between a noise reduction and edge detection accuracy. As was mentioned earlier in the Chapters on frequency characteristics of the FOD. Contrary to the FOD, which amplifies the high frequencies, the FOI serves as a low-frequency filter.

Several methods have been recognized to be useful in detecting the edges in digital images. It is worth mentioning the methods using the wavelet transforms [Yitzhaky and Peli (2003); Zhang et al. (2009); Alonso et al. (2011); Morita and Sato (2014)] or the fuzzy logic [Talai and Talai (1997); Liang et al. (2001); Liang and Looney (2003); Becerikli and Karan (2005); Madasu (2009)].

The earliest works originate from France, invented by scientific researchers conducted by Allain Oustaloup form Universitae Bordeaux where a fractional-order edge detector named CRONE was developed [Oustaloup et al. (1991a,b); Oustaloup (1995); Mathieu et al. (2003)]. Relating to the classic IODs and IOSs the FOBD/S can be applied to the edge detection and image filtering. Generalized masks based on the FOBDs and FOBSs are now successfully used in the image edge detection and filtering methods [Pu et al. (2008, 2010); Sparavigna (2009); Yang et al. (2010); Gao et al. (2011); Chen and Fei (2012); Hou et al. (2012); Tian et al. (2014); Han and Liu (2014)].

13.1   Edge detection based on the fractional differentiation

To evaluate the FOBD at the discrete-time instant $k > 0$ one needs the order $\nu \in \mathbb{R}_+$ and $k+1$ function samples. In the simplified form one takes into account $L > 0$ last samples of analyzed function $f(k_1, k_2)$. This leads to a generalized mask defined by a matrix

$$\mathbf{M}\left[a^{(\nu)}(r)\right] = s_c$$

$$\times \begin{bmatrix}
& \cdots & v_{-2,-L}a^{(\nu)}(L) & v_{-1,-L}a^{(\nu)}(L) & v_{0,-L}a^{(\nu)} & v_{1,-L}a^{(\nu)}(L) & v_{2,-L}a^{(\nu)}(L) & \cdots & \\
& \vdots & \vdots & \vdots & \vdots & \vdots & \vdots & & \vdots \\
v_{-L,-2}a^{(\nu)}(L) & \cdots & v_{-2,-2}a^{(\nu)}(3) & v_{-1,-2}a^{(\nu)}(2) & v_{0,-2}a^{(\nu)}(2) & v_{1,-2}a^{(\nu)}(2) & v_{2,-2}a^{(\nu)}(3) & \cdots & v_{L,-2}a^{(\nu)}(L) \\
v_{-L,-1}a^{(\nu)}(L) & \cdots & v_{-2,-1}a^{(\nu)}(2) & v_{-1,-1}a^{(\nu)}(1) & v_{0,-1}a^{(\nu)}(1) & v_{1,-1}a^{(\nu)}(1) & v_{2,-1}a^{(\nu)}(2) & \cdots & v_{L,-1}a^{(\nu)}(L) \\
v_{-L,0}a^{(\nu)}(L) & \cdots & w_{-2,0}a^{(\nu)}(2) & v_{-1,0}a^{(\nu)}(1) & v_{0,0}a^{(\nu)}(0) & v_{1,0}a^{(\nu)}(1) & w_{2,0}a^{(\nu)}(2) & \cdots & v_{L,0}a^{(\nu)}(L) \\
v_{-L,1}a^{(\nu)}(L) & \cdots & v_{-2,1}a^{(\nu)}(2) & v_{-1,1}a^{(\nu)}(1) & v_{0,1}a^{(\nu)}(1) & v_{1,1}a^{(\nu)}(1) & v_{2,1}a^{(\nu)}(2) & \cdots & v_{L,1}a^{(\nu)}(L) \\
v_{-L,2}a^{(\nu)}(L) & \cdots & v_{-2,2}a^{(\nu)}(3) & v_{-1,2}a^{(\nu)}(2) & v_{0,2}a^{(\nu)}(2) & v_{1,2}a^{(\nu)}(2) & v_{2,2}a^{(\nu)}(3) & \cdots & v_{L,2}a^{(\nu)}(L) \\
& \vdots & \vdots & \vdots & \vdots & \vdots & \vdots & & \vdots \\
& \cdots & v_{-2,L}a^{(\nu)}(L) & v_{-1,L}a^{(\nu)}(L) & v_{0,L}a^{(\nu)}(L) & v_{1,L}a^{(\nu)}(L) & v_{2,L}a^{(\nu)}(L) & \cdots &
\end{bmatrix}$$

$$(13.1)$$

where s_c is a pixel scaling coefficient and $v_{i,j}$ are direction coefficients. If $v_{i,j} = 1$ for $i, j \in \{-L, -L+1, \ldots, L-1, L\}$ then no special direction is preferred and the mask (13.1) takes the form

$$\mathbf{M}\left[a^{(\nu)}(r)\right] = s_c \begin{bmatrix}
& \cdots & a^{(\nu)}(L) & a^{(\nu)}(L) & a^{(\nu)}(L) & a^{(\nu)}(L) & a^{(\nu)}(L) & \cdots & \\
& \vdots & \vdots & \vdots & \vdots & \vdots & \vdots & & \vdots \\
a^{(\nu)}(L) & \cdots & a^{(\nu)}(3) & a^{(\nu)}(2) & a^{(\nu)}(2) & a^{(\nu)}(2) & a^{(\nu)}(3) & \cdots & a^{(\nu)}(L) \\
a^{(\nu)}(L) & \cdots & a^{(\nu)}(2) & a^{(\nu)}(1) & a^{(\nu)}(1) & a^{(\nu)}(1) & a^{(\nu)}(2) & \cdots & a^{(\nu)}(L) \\
a^{(\nu)}(L) & \cdots & a^{(\nu)}(2) & a^{(\nu)}(1) & 1 & a^{(\nu)}(1) & a^{(\nu)}(2) & \cdots & a^{(\nu)}(L) \\
a^{(\nu)}(L) & \cdots & a^{(\nu)}(2) & a^{(\nu)}(1) & a^{(\nu)}(1) & a^{(\nu)}(1) & a^{(\nu)}(2) & \cdots & a^{(\nu)}(L) \\
a^{(\nu)}(L) & \cdots & a^{(\nu)}(3) & a^{(\nu)}(2) & a^{(\nu)}(2) & a^{(\nu)}(2) & a^{(\nu)}(3) & \cdots & a^{(\nu)}(L) \\
& \vdots & \vdots & \vdots & \vdots & \vdots & \vdots & & \vdots \\
& \cdots & a^{(\nu)}(L) & a^{(\nu)}(L) & a^{(\nu)}(L) & a^{(\nu)}(L) & a^{(\nu)}(L) & \cdots &
\end{bmatrix}.$$

$$(13.2)$$

For $v_{0,-i} = 1$ for $i = 0, 1, \ldots, L-1, L$ and all remaining coefficients equal to zero one gets the upper mask

$$
\mathbf{M}_u\left[a^{(\nu)}(r)\right] = s_c
\begin{bmatrix}
0 \cdots & 0\ 0\ a^{(\nu)}(L)\ 0\ 0 & \cdots\ 0 \\
\vdots & \vdots\ \vdots \quad \vdots\quad \vdots\ \vdots & \vdots \\
0 \cdots & 0\ 0\ a^{(\nu)}(2)\ 0\ 0 & \cdots\ 0 \\
0 \cdots & 0\ 0\ a^{(\nu)}(1)\ 0\ 0 & \cdots\ 0 \\
0 \cdots & 0\ 0 \quad 1 \quad 0\ 0 & \cdots\ 0 \\
0 \cdots & 0\ 0 \quad 0 \quad 0\ 0 & \cdots\ 0 \\
0 \cdots & 0\ 0 \quad 0 \quad 0\ 0 & \cdots\ 0 \\
\vdots & \vdots\ \vdots \quad \vdots\quad \vdots\ \vdots & \vdots \\
0 \cdots & 0\ 0 \quad 0 \quad 0\ 0 & \cdots\ 0
\end{bmatrix}.
\tag{13.3}
$$

For $v_{-i,0} = 1$ for $i = 0, 1, \ldots, L-1, L$ and all remaining coefficients equal to zero one gets the left mask

$$
\mathbf{M}_l\left[a^{(\nu)}(r)\right] = s_c
\begin{bmatrix}
0 & \cdots & 0 & 0 & 0\ 0\ 0\ \cdots\ 0 \\
\vdots & & \vdots & \vdots & \vdots\ \vdots\ \vdots \quad \vdots \\
0 & \cdots & 0 & 0 & 0\ 0\ 0\ \cdots\ 0 \\
0 & \cdots & 0 & 0 & 0\ 0\ 0\ \cdots\ 0 \\
a^{(\nu)}(L) & \cdots & a^{(\nu)}(2) & a^{(\nu)}(1) & 1\ 0\ 0\ \cdots\ 0 \\
0 & \cdots & 0 & 0 & 0\ 0\ 0\ \cdots\ 0 \\
0 & \cdots & 0 & 0 & 0\ 0\ 0\ \cdots\ 0 \\
\vdots & & \vdots & \vdots & \vdots\ \vdots\ \vdots \quad \vdots \\
0 & \cdots & 0 & 0 & 0\ 0\ 0\ \cdots\ 0
\end{bmatrix}.
\tag{13.4}
$$

For $v_{-i,-i} = 1$ for $i = 0, 1, \ldots, L-1, L$ and all remaining coefficients equal to zero one gets the left upper mask

$$
\mathbf{M}_{lu}\left[a^{(\nu)}(r)\right] = s_c
\begin{bmatrix}
a^{(\nu)}(L) & \cdots & 0 & 0 & 0\ 0\ 0\ \cdots\ 0 \\
\vdots & & \vdots & \vdots & \vdots\ \vdots\ \vdots \quad \vdots \\
0 & \cdots\ a^{(\nu)}(2) & 0 & 0\ 0\ 0 \cdots\ 0 \\
0 & \cdots & 0 & a^{(\nu)}(1)\ 0\ 0\ 0 \cdots\ 0 \\
0 & \cdots & 0 & 0 & 1\ 0\ 0\ \cdots\ 0 \\
0 & \cdots & 0 & 0 & 0\ 0\ 0\ \cdots\ 0 \\
0 & \cdots & 0 & 0 & 0\ 0\ 0\ \cdots\ 0 \\
\vdots & & \vdots & \vdots & \vdots\ \vdots\ \vdots \quad \vdots \\
0 & \cdots & 0 & 0 & 0\ 0\ 0\ \cdots\ 0
\end{bmatrix}.
\tag{13.5}
$$

In a similar way, one defines a left down $\mathbf{M}_{ld}$, a down $\mathbf{M}_d$, a right down $\mathbf{M}_{rd}$, a right $\mathbf{M}_r$ and a right upper $\mathbf{M}_{ru}$ mask, respectively. Admitting fractionality of the mask one reaches immense possibilities of mask creation. Yet, one should keep in mind that there is a need to establish mask dimensions larger then typical 3×3.

Here one should mention that the commonly used typical masks are special cases of mask defined by formula (13.1). For instance, the typical low-pass average filters are described by the masks

$$
\mathbf{M}\left[a^{(-1)}(r)\right] = \begin{bmatrix} 1 & 1 & 1 \\ 1 & 1 & 1 \\ 1 & 1 & 1 \end{bmatrix}. \tag{13.6}
$$

It is a special case of the mask (13.2) with $L = 1, s_c = 1, \nu = -1$. Another typical mask

$$
\mathbf{M}\left[a^{(-1)}(r)\right] = \begin{bmatrix} 1 & 1 & 1 \\ 1 & 2 & 1 \\ 1 & 1 & 1 \end{bmatrix} = 2 \begin{bmatrix} 0.5 & 0.5 & 0.5 \\ 0.5 & 1 & 0.5 \\ 0.5 & 0.5 & 0.5 \end{bmatrix} \tag{13.7}
$$

means (13.2) with $L = 1, s_c = 2, \nu = -0.5$. So called Roberts operators

$$
\mathbf{M}_u\left[a^{(1)}(r)\right] = \begin{bmatrix} 0 & -1 & 0 \\ 0 & 1 & 0 \\ 0 & 0 & 0 \end{bmatrix}, \ \mathbf{M}_{ur}\left[a^{(1)}(r)\right] = \begin{bmatrix} 0 & 0 & -1 \\ 0 & 1 & 0 \\ 0 & 0 & 0 \end{bmatrix}, \ \mathbf{M}_r\left[a^{(1)}(r)\right] = \begin{bmatrix} 0 & 0 & 0 \\ 0 & 1 & -1 \\ 0 & 0 & 0 \end{bmatrix}
$$

$$\tag{13.8}$$

are special cases of these defined by formulas (13.3), $\mathbf{M}_{u,r}\left[a^{(1)}(r)\right]$ and $\mathbf{M}_r\left[a^{(1)}(r)\right]$. The well-known Laplace operators

$$
\mathbf{M}_{u,d,l,r}\left[a^{(0.25)}(r)\right] = -4 \begin{bmatrix} 0 & -0.25 & 0 \\ -0.25 & 1 & -0.25 \\ 0 & -0.25 & 0 \end{bmatrix}
$$

$$
\mathbf{M}_{u,lu,l,ld,d,rd,r,ru}\left[a^{(0.125)}(r)\right] = -8 \begin{bmatrix} -0.125 & -0.125 & -0.125 \\ -0.125 & 1 & -0.125 \\ -0.125 & -0.125 & -0.125 \end{bmatrix} \tag{13.9}
$$

are also the special cases of $\mathbf{M}_{u,d,l,r}\left[a^{(0.25)}(r)\right]$ and $\mathbf{M}_{u,d,l,r}\left[a^{(0.125)}(r)\right]$.

Combinations of masks defined for different FOs - positive and/or negative further enlarges the variety of masks. An exemplary mask is presented below

$$\mathbf{M}_{lu,l,ld,d,rd,r,ru,u}\left[a^{(\nu_{lu},\nu_l,\nu_{ld},\nu_d,\nu_{rd},\nu_r,\nu_{ru},\nu_r)}(r)\right]=s_c$$

$$\times\begin{bmatrix}a^{(\nu_{ul})}(L) & \cdots & 0 & 0 & a^{(\nu_u)}(L) & 0 & 0 & \cdots & a^{(\nu_{ur})}(2)\\ \vdots & & \vdots & \vdots & \vdots & \vdots & \vdots & & \vdots\\ 0 & \cdots & a^{(\nu_{ul})}(2) & 0 & a^{(\nu_u)}(2) & 0 & a^{(\nu_{ur})}(2) & \cdots & 0\\ 0 & \cdots & 0 & a^{(\nu_{ul})}(1) & a^{(\nu_u)}(1) & a^{(\nu_{ur})}(1) & 0 & \cdots & 0\\ a^{(\nu_l)}(L) & \cdots & a^{(\nu_l)}(2) & a^{(\nu_l)}(1) & 1 & a^{(\nu_r)}(1) & a^{(\nu_r)}(2) & \cdots & a^{(\nu_r)}(L)\\ 0 & \cdots & 0 & a^{(\nu_{dl})}(1) & a^{(\nu_d)}(1) & a^{(\nu_{dr})}(1) & 0 & \cdots & 0\\ 0 & \cdots & a^{(\nu_{dl})}(2) & 0 & a^{(\nu_d)}(2) & 0 & a^{(\nu_{dr})}(2) & \cdots & 0\\ \vdots & & \vdots & \vdots & \vdots & \vdots & \vdots & & \vdots\\ a^{(\nu_{dl})}(L) & \cdots & 0 & 0 & a^{(\nu_d)}(L) & 0 & 0 & \cdots & a^{(\nu_{dr})}(L)\end{bmatrix}.$$

$$(13.10)$$

One should note that usually one preserves a horizontal and/or vertical symmetry of orders, ie. $\nu_{ul}=\nu_{ur}$, $\nu_{dl}=\nu_{dr}$, $\nu_u=\nu_d$, $\nu_l=\nu_r$ for $L=1$. As special cases of presented masks may serve the Gauss filter evaluated for $\delta=1$

$$\mathbf{M}_{u,lu,l,ld,d,rd,r,ru}\left[a^{(-\frac{3}{8},-\frac{5}{8},\dots,-\frac{3}{8},-\frac{5}{8})}(r)\right]=\frac{4}{25}\begin{bmatrix}\frac{3}{8} & \frac{5}{8} & \frac{3}{8}\\ \frac{5}{8} & 1 & \frac{5}{8}\\ \frac{3}{8} & \frac{5}{8} & \frac{3}{8}\end{bmatrix}.\qquad(13.11)$$

The application of the FO masks will be clarified in an numerical example.

Example 13.1. Consider a discrete image formed by two crosses rotated relative to each other by $\frac{\pi}{4}$. The image is heavily noised. The image and its 3D plot are presented in Figs. 13.1 and 13.2, respectively.

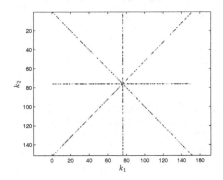

Fig. 13.1 Test image.

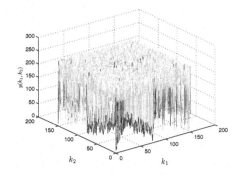

Fig. 13.2 3D plot of the test image.

Because it is difficult to see the noise image in the following two images the two different points of view of the Fig. 13.2 are presented in Figs. 13.3 and 13.4.

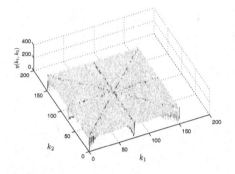

Fig. 13.3   3D test image 3D first second viewing angle.

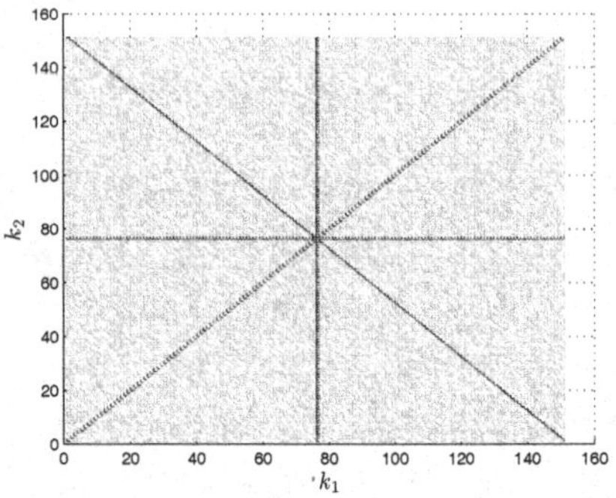

Fig. 13.4   3D test image third viewing angle.

Extract the image edge using different FO masks.

Solution. For $l = 7$ and $\nu = 0.8$ results of masks (13.2) applications are presented in consecutive Fig. 13.5 and with an unoptimized threshold in Fig. 13.6.

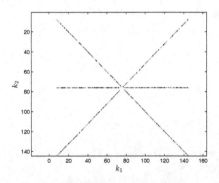

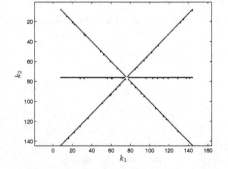

Fig. 13.5   Noised image of Fig. 13.1 transformed due to the upper mask (13.3) with $L = 7$ and $\nu = 0.8$.

Fig. 13.6   Noised image of Fig. 13.1 transformed due to the upper mask (13.3) with $L = 7$ and $\nu = 0.8$ with a threshold.

An application of another mask leads to the following results.

For comparative purposes in Fig. 13.13, image of Fig. 13.12 is presented in the 3D plot. One realizes the shape of the transformed image from Fig. 13.2 is much

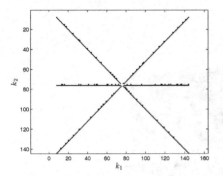

Fig. 13.7 Noised image of Fig. 13.1 transformed due the down mask with $L = 7$, $\nu = 0.8$ and a threshold.

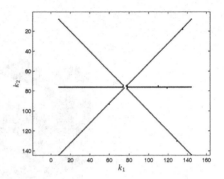

Fig. 13.8 Noised image of Fig. 13.1 transformed due the down and upper mask (13.3) with $L = 7$, $\nu = 0.8$ and a threshold.

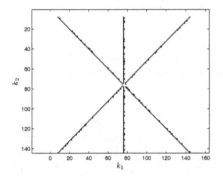

Fig. 13.9 Noised image of Fig. 13.1 transformed due to the backward mask with $L = 7$, $\nu = 0.8$ and a threshold.

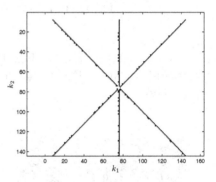

Fig. 13.10 Noised image of Fig. 13.1 transformed due to the forward mask with $L = 7$, $\nu = 0.8$ and a threshold.

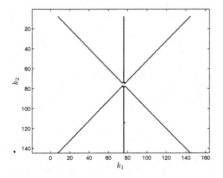

Fig. 13.11 Noised image of Fig. 13.1 transformed due to the backward and forward mask with $L = 7$, $\nu = 0.8$ and a threshold.

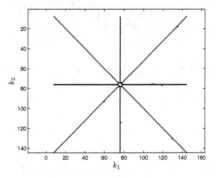

Fig. 13.12 Noised image of Fig. 13.1 transformed due the backward-forward and upper-downward mask with $L = 7$, $\nu = 0.8$ and a threshold.

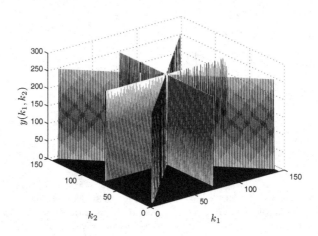

Fig. 13.13   Transformed by ul-lr mask image 3D plot.

clearer than that presented in Fig. 13.2. Finally, there are given plots of edges detected using Canny (Fig. 13.14, Prewitt Fig. 13.15 and Sobel Fig. 13.16 filters, respectively).

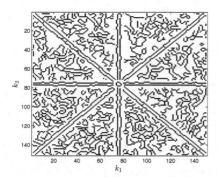

Fig. 13.14   Noised image of Fig. 13.1 transformed due to the Canny mask with an automatic Matlab threshold.

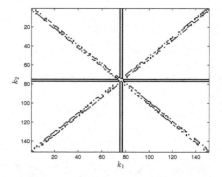

Fig. 13.15   Noised image of Fig. 13.1 transformed due to the Prewitt mask with an automatic Matlab threshold.

The next example shows combined operations one can perform on the image. The application of the FO integration playing a role of the filter will serve to extract a noise.

Example 13.2. A digital image given in Fig. 13.17 of the "Small dog" is considered. Its 3D plot is presented in Fig. 13.18.

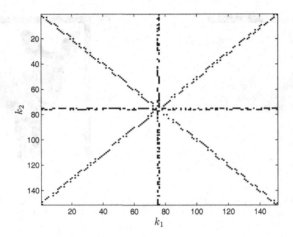

Fig. 13.16 Noised image of Fig. 13.1 transformed due to the Roberts mask with an automatic Matlab threshold.

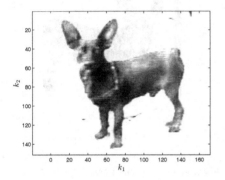

Fig. 13.17 Digital image of "Small Dog".   Fig. 13.18 3D plot of "Small Dog".

First one applies the upper mask and the upper mask with a threshold for $L = 7$ and $\nu = -1.5$. Note: in this example the FO is negative which means that the mask works as the FOI having filtering property. The results are given in Figs. 13.19 and 13.20, respectively.

In Figs. 13.21 and 13.26 the effects of downward, upward-downward, backward, forward, backward-forward and upward-downward-backward-forward mask applications are presented, respectively.

In Fig. 13.27 the 3D plot related to Fig. 13.26 is given.

As was mentioned earlier, a combined operations may lead to new results. The classical Canny, Prewitt and Roberts edge detection methods, applied to the upward mask, give images presented in Figs. 13.28–13.30, respectively.

Finally, the effects of applying the considered classical edge detection methods directly to "Small Dog" image are presented in Figs. 13.31–13.33, respectively.

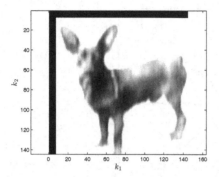

Fig. 13.19 "Small Dog" filtered due to the upward mask with $L = 7$ and $\nu = -1.5$.

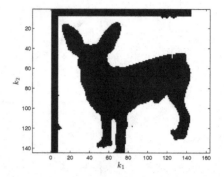

Fig. 13.20 "Small Dog" filtered due to the upward mask with $L = 7$, $\nu = -1.5$ and a threshold.

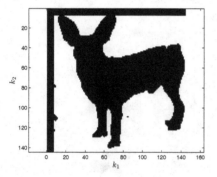

Fig. 13.21 "Small Dog" filtered due to the downward mask with $L = 7$, $\nu = -1.5$ with a threshold.

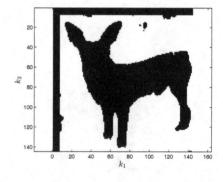

Fig. 13.22 "Small Dog" filtered due to the downward-upward mask with $L = 7$, $\nu = -1.5$ and a threshold.

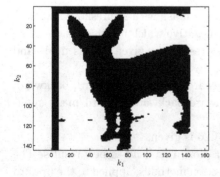

Fig. 13.23 "Small Dog" filtered due to the backward mask with $L = 7$, $\nu = -1.5$ with a threshold.

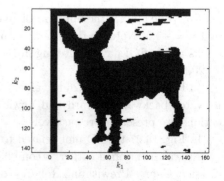

Fig. 13.24 "Small Dog" filtered due to the forward mask with $L = 7$, $\nu = -1.5$ and a threshold.

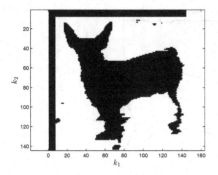

Fig. 13.25 "Small Dog" filtered due to the backward-forward mask with $L = 7$, $\nu = -1.5$ with a threshold.

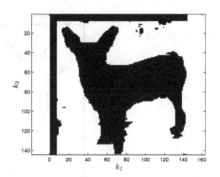

Fig. 13.26 "Small Dog" filtered due to the upward-downward-backward-forward mask with $L = 7$, $\nu = -1.5$ and a threshold.

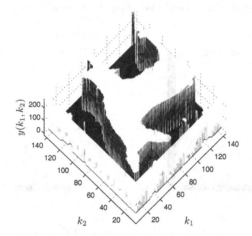

Fig. 13.27 3D plot of "Small Dog" filtered due to the upward-downward-backward-forward mask with $L = 7$, $\nu = -1.5$ and a threshold.

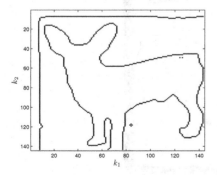

Fig. 13.28 Canny mask of "Small Dog" filtered due to the upward mask with $L = 7$, $\nu = -1.5$ with a threshold.

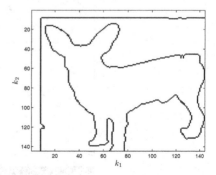

Fig. 13.29 Prewitt mask of "Small Dog" filtered due to the upward mask with $L = 7$, $\nu = -1.5$ and a threshold.

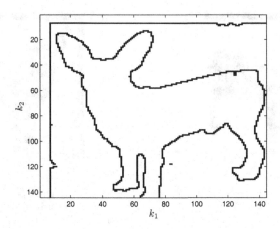

Fig. 13.30   Roberts mask of "Small Dog" filtered due to the upward mask with $L = 7$, $\nu = -1.5$ and a threshold.

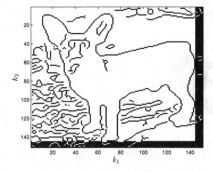

Fig. 13.31   Canny mask of "Small Dog" with a threshold.

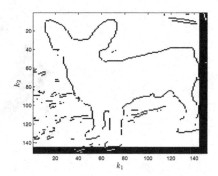

Fig. 13.32   Prewitt mask of "Small Dog" with a threshold.

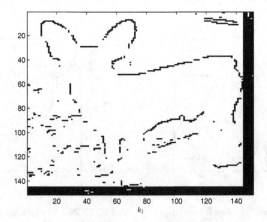

Fig. 13.33   Roberts mask of "Small Dog" with a threshold.

Appendix A

Selected linear algebra formulae and discrete-variable special functions

A.1 Discrete-variable Euler Gamma function

The continuous-, complex-variable $z \in \mathbb{C}$ Euler Gamma function is defined by so-called Euler integral of the second kind for $\mathcal{R}(z) > 0$

$$\Gamma(z) = \int_0^\infty x^{z-1} e^{-x} dx. \tag{A.1}$$

The integral is convergent for all complex $z \in \mathbb{C}$ satisfying inequality $\mathcal{R}(z) > 0$. Very useful is the recursive formula

$$\Gamma(z+1) = z\Gamma(z) \quad \text{for } \mathcal{R}(z) > 0. \tag{A.2}$$

It can be generalized to the form

$$\Gamma(z+n) = z(z+1)\cdots(z+n-1)\Gamma(z) \quad \text{for } n \in \mathbb{Z}_0 > 0. \tag{A.3}$$

As a special case for $z = k, n = 1$

$$\Gamma(k+1) = k!. \tag{A.4}$$

A.2 Involutory matrix

A square matrix $\mathbf{P}$ is called involutory [Ashrafi and Gibson (2004)] matrix if for $k \in \mathbb{Z}_+$ it satisfies

$$\mathbf{P}^n = \begin{cases} \mathbf{P} & \text{for } n - \text{even} \\ \mathbf{I} & \text{for } n - \text{odd} \end{cases} \tag{A.5}$$

This means that the involutory matrix is equal to its inverse i.e. $\mathbf{P} = \mathbf{P}^{-1}$. In the next Subsections some related involutory matrices are presented.

A.2.1 *The involutory Pascal matrix*

In this Section some properties of the involutory matrices are given.

A square matrix $\mathbf{P}_k$ of dimension $(k+1) \times (k+1)$ is called the Pascal matrix [Bernstein (2009)] if its elements satisfy

$$p_{ij} = \begin{cases} 0 & \text{for } 1 \leqslant i < j \leqslant k+1 \\ a^{(i-1)}(j-1) & \text{for } 1 \leqslant j \leqslant i \leqslant k+1 \end{cases} \tag{A.6}$$

Remark A.1. The Pascal matrix $\mathbf{P}_k$ elements may be also defined as

$$p_{ij} = \begin{cases} 0 & \text{for } 1 \leqslant i < j \leqslant k \\ (-1)^{j+1} a^{(-j)}(i-j) & \text{for } 1 \leqslant j \leqslant i \leqslant k \end{cases}. \tag{A.7}$$

Hence,

$$
\mathbf{P}_k
$$

$$
= \begin{bmatrix}
a^{(0)}(0) & 0 & 0 & \cdots & 0 & 0 & 0 \\
a^{(1)}(0) & a^{(0)}(1) & 0 & \cdots & 0 & 0 & 0 \\
a^{(2)}(0) & a^{(2)}(1) & a^{(2)}(2) & \cdots & 0 & 0 & 0 \\
\vdots & \vdots & \vdots & & \vdots & \vdots & \vdots \\
a^{(k-2)}(0) & a^{(k-2)}(1) & a^{(k-2)}(2) & \cdots & a^{(k-2)}(k-2) & 0 & 0 \\
a^{(k-1)}(0) & a^{(k-1)}(1) & a^{(k-1)}(2) & \cdots & a^{(k-1)}(k-2) & a^{(k-1)}(k-1) & 0 \\
a^{(k)}(0) & a^{(k)}(1) & a^{(k)}(2) & \cdots & a^{(k)}(k-2) & a^{(k)}(k-1) & a^{(k)}(k)
\end{bmatrix}
$$

$$
= \begin{bmatrix}
a^{(-1)}(0) & 0 & 0 & \cdots & 0 & 0 & 0 \\
a^{(-1)}(1) & a^{(-2)}(0) & 0 & \cdots & 0 & 0 & 0 \\
a^{(-1)}(2) & a^{(-2)}(1) & a^{(-3)}(0) & \cdots & 0 & 0 & 0 \\
\vdots & \vdots & \vdots & & \vdots & \vdots & \vdots \\
a^{(-1)}(k-2) & a^{(-2)}(k-3) & a^{(-3)}(k-4) & \cdots & a^{(2-k)}(0) & 0 & 0 \\
a^{(-1)}(k-1) & a^{(-2)}(k-2) & a^{(-3)}(k-3) & \cdots & a^{(2-k)}(1) & a^{(1-k)}(0) & 0 \\
a^{(-1)}(k) & a^{(-2)}(k-1) & a^{(-3)}(k-2) & \cdots & a^{(2-k)}(2) & a^{(1-k)}(1) & a^{(-k)}(0)
\end{bmatrix} \tag{A.8}
$$

and

$$a^{(i-1)}(j-1) = a^{(-j)}(i-j) \quad \text{for } 1 \leqslant j \leqslant i \leqslant k \tag{A.9}$$

For instance, for $k = 5$ the Pascal matrix is of the form

$$\mathbf{P}_5 = \begin{bmatrix}
1 & 0 & 0 & 0 & 0 & 0 & 0 \\
1 & -1 & 0 & 0 & 0 & 0 & 0 \\
1 & -2 & 1 & 0 & 0 & 0 & 0 \\
1 & -3 & 3 & -1 & 0 & 0 & 0 \\
1 & -4 & 6 & -4 & 1 & 0 & 0 \\
1 & -5 & 10 & -10 & 5 & -1 & 0 \\
1 & -6 & 15 & -20 & 15 & -6 & 1
\end{bmatrix} \tag{A.10}$$

Theorem A.1. *The Pascal matrix $\mathbf{P}_k$ is an involutory matrix.*

Proof. To prove that the Pascal matrix is involutory it suffices to prove that

$$\mathbf{P}_k^2 = \mathbf{1}_k. \tag{A.11}$$

Consider matrices $\mathbf{A}_k^{(1)}$ and $\mathbf{A}_k^{(-1)}$ defined by formulas (2.17) and (2.60) for $k_0 = 0$, respectively. Then,

$$\mathbf{A}_k^{(-1)}\mathbf{A}_k^{(1)} = \mathbf{1}_k \tag{A.12}$$

and

$$\left[\mathbf{A}_k^{(-1)}\mathbf{A}_k^{(1)}\right]^{\mathrm{T}} = [\mathbf{1}_k]^{\mathrm{T}} = \mathbf{1}_k. \tag{A.13}$$

Performing a transposition operation one gets

$$\left[\mathbf{A}_k^{(-1)}\mathbf{A}_k^{(1)}\right]^{\mathrm{T}} = \left[\mathbf{A}_k^{(1)}\right]^{\mathrm{T}}\left[\mathbf{A}_k^{(-1)}\right]^{\mathrm{T}} = [\mathbf{1}_k]^{\mathrm{T}} = \mathbf{1}_k \tag{A.14}$$

A pre-multiplication of formula (A.11) by the matrix $\left[\mathbf{A}_k^{(1)}\right]^{\mathrm{T}}$ and a post-multiplication by the matrix $\left[\mathbf{A}_k^{(-1)}\right]^{\mathrm{T}}$ yields

$$\left[\mathbf{A}_k^{(1)}\right]^{\mathrm{T}}\mathbf{P}_k^2\left[\mathbf{A}_k^{(-1)}\right]^{\mathrm{T}} = \left\{\left[\mathbf{A}_k^{(1)}\right]^{\mathrm{T}}\mathbf{P}_k\right\}\left\{\mathbf{P}_k\left[\mathbf{A}_k^{(-1)}\right]^{\mathrm{T}}\right\} = \mathbf{1}_k. \tag{A.15}$$

Now, one checks products

$$\left[\mathbf{A}_k^{(1)}\right]^{\mathrm{T}}\mathbf{P}_k = \begin{bmatrix} \mathbf{1}_1 & \mathbf{0}_{k-1} \\ \mathbf{0}_{k-1}^{\mathrm{T}} & \mathbf{P}_{k-1} \end{bmatrix} = \begin{bmatrix} 1 & \mathbf{0}_{k-1} \\ \mathbf{0}_{k-1}^{\mathrm{T}} & \mathbf{P}_{k-1} \end{bmatrix} \tag{A.16}$$

$$\mathbf{P}_k\left[\mathbf{A}_k^{(-1)}\right]^{\mathrm{T}} = \begin{bmatrix} 1 & \mathbf{0}_{k-1} \\ \mathbf{0}_{k-1}^{\mathrm{T}} & \mathbf{P}_{k-1} \end{bmatrix} \tag{A.17}$$

Therefore, one derives

$$\mathbf{P}_{k-1}^2 = \mathbf{1}_{k-1}. \tag{A.18}$$

A repetition of this procedure $k - 1$ times more ends the proof.

□

A multiplication of equality (A.11) by the matrix $\mathbf{P}_k$ confirms the belonging of the Pascal matrix to the involutory matrices set defined by formula (A.5).

A.2.2   *The involutory unit anti-diagonal matrix*

Definition A.1. A $(k+1) \times (k+1)$ anti-diagonal unit matrix

$$\mathbf{N}_k = \begin{bmatrix} 0 & 0 & \cdots & 0 & 1 \\ 0 & 0 & \cdots & 1 & 0 \\ \vdots & \vdots & & \vdots & \vdots \\ 0 & 1 & \cdots & 0 & 0 \\ 1 & 0 & \cdots & 0 & 0 \end{bmatrix} \tag{A.19}$$

is the involuntary matrix.

Elementary multiplications show that $\mathbf{N}_k \mathbf{N}_k = \mathbf{1}_k$.

A.3   The Toeplitz matrix

Definition A.2.

A $(k+1) \times (k+1)$ matrix is Toeplitz one if its (i, j) entry depends only on the value of $i - j$. Therefore, matrix diagonals: main, upper and lower are constant.

$$\mathbf{T}_k = \begin{bmatrix} t_0 & t_1 & \cdots & t_{k-1} & t_k \\ t_{-1} & t_0 & \cdots & 1 & 0 \\ \vdots & \vdots & & \vdots & \vdots \\ t_{-k+1} & 1 & \cdots & t_0 & t_1 \\ t_{-k} & t_{-k+1} & \cdots & 0 & t_0 \end{bmatrix} \tag{A.20}$$

Bibliography

Ahmed, E., El-Sayed, A. A. and Hala, A. E, (2006), On some Routh-Hurwitz conditions for fractional order differential equations and their applications in Lorenz, Rössler, Chua and Chen systems, *Physics Letters A* **358** pp. 1–4.

Al-Alaoui M. A, B. W. (1993). Novel digital integrator and differentiator, *Electronic Letters* **29**, 4, pp. 376–378.

Aoun, M., Malti. R., Levron, F. and Oustaloup, A. (2004), Numerical simulations of fractional systems: An overview of existing methods and improvements. *Nonlinear Dynamics* **38** pp. 117–131.

Aoun, M., Malti, R., Levronc, F. and Oustaloup, A. (2007), Synthesis of fractional Laguerre basis for system approximation, *Automatica* **43** pp. 1640–1648.

Aldair, A. A. and Wang, W. J. (2010). Design of Fractional Order Controller Based on Evolutionary Algorithm for a Full Vehicle Nonlinear Active Suspension Systems, *International Journal of Control and Automation* **3**, 4, pp. 33–46.

Alonso, M., T., Lopez-Martinez, C., Mallorqui, J. J. and Salembier, P. (2011). Edge Enhancement Algorithm Based on the Wavelet Transform for Automatic Edge Detection in SAR Images, *IEEE Transactions on Geoscience and Remote Sensing* **49**, 1, pp. 222–235.

Angoletta, M. E. (2008). Digital signal processor fundamentals and system design, in D. Brandt (ed.), *CAS - CERN Accelerator School: Course on Digital Signal Processing* (Sigtuna, Sweden), pp. 167–229.

P. Arena, R. Caponetto, Fortuna, L. and Porto, D. (2000). *Nonlinear Noninteger Order Circuits and Systems - An Introduction*, (World Scientific, Singapore)

Ashrafi, A. and Gibson, P. M. (2004). An involutory Pascal matrix, *Linear Algebra and its Applications* **387**, pp. 277–286.

Åström, K. J. and Wittenmark B. (1984). *Computer Controlled Systems: Theory and Design* (Prentice-Hall, Inc., Englewood Cliffs).

Åström, K. J. and Hagglund T. (1995). *PID controllers: theory, design, and tuning* (Instrument Society of America, USA).

Åström, K. J. and Hagglund T. (2001). The future of {PID} control, *Control Engineering Practice* **9**, 11, pp. 1163–1175.

Axtell, M. and Bise, M. (1990). Fractional calculus applications in control systems, in *Proceedings of the IEEE 1990 National Aerospace and Electronics Conference* (New York, USA), pp. 563–566.

Bagley R. L. and Torvik. P. (1984) On the appearance of the fractional derivative in the behavior of real materials, *Journal of Applied Mechanics* **51**, pp. 294–298.

Bagley, R. L. and Calico, R. A. (1991) Fractional order state equations for the control of viscoelastic damped structures, *Journal of Guidance, Control and Dynamics* **14**, 2, pp. 304–311.

Baleanu, D., Machado, J. T. and Luo, A. J. (2011). *Fractional Dynamics and Control* (Springer).

Baleanu, D., Diethelm, K. and Trujillo, J. J. (2012). *Fractional Calculus: Models and Numerical Methods. Series on Complexity, Nonlinearity and Chaos* (World Scientific Publishing Company, Singapore).

Barbosa, R. S., Machado, T. A. and Ferreira, I. M. (2004a). Tuning of PID Controllers Based on Bode's Ideal Transfer Function, *Nonlinear Dynamics* **38**, pp. 305–321.

Barbosa, R. S., Machado, T. A. and Ferreira, I. M. (2004b). PID controller tuning using fractional calculus concepts, *Fractional Calculus & Applied Analysis,* **7**, 2, pp. 119–134.

Barbosa, R S., Machado, T. A. and Jesus, I. S. (2008). On the Fractional PID Control of a Laboratory Servo System, in *Proceedings of the 17th World Congress. The International Federation of Automatic Control* (Seoul, Korea), pp. 15273-15278.

Barbosa, R S., Machado, T. A. and Jesus, I. S. (2010). Effect of fractional orders in the velocity control of a servo system, *Computers and Mathematics with Applications* **59**, pp. 1679–1686.

Barkai, E. (2002). CTRW pathways to the fractional diffusion equation, *Chemical Physics* **284**, pp. 13–27.

Bayin, Ş. S. (2006). *Mathematical methods in science and engineering.* A John Wiley & Sons, Inc.

Becerikli, Y., and Karan, T., K. (2005). A New Fuzzy Approach for Edge Detection, in *Proceedings of 8th International Work-Conference on Artificial Neural Networks* (Barcelona, Spain), pp. 943–951.

Ben Adda, F. (1997). Geometric interpretation of the fractional derivative, *Journal of Fractional Calculus* **11**, pp. 21–52.

Bernstein. D. S. (2009). *Matrix Mathematics: Theory, Facts, and Formulas,* 2nd edn. (Princeton University Press).

Bhattacharya, S. K. (2008). *Control Systems Engineering* (Pearson Education, India).

Bonnet, C. and Partington, J. R. (2000), Coprime factorizations and stability of fractional differential systems, *Systems and Control Letters* **41** pp. 167–174.

Bose, T. (2004). *Digital Signal and Image Processing* (John Wiley & Sons).

Bresenham, J. (1965). Algorithm for Computer Control of a Digital Plotter, *IBM Systems Journal* **4**, 1, pp. 25–30.

Brzeziński, D. and Ostalczyk P. (2012). The Grünwald-Letnikov Formula and Its Equivalent Horners Form Accuracy Comparison and Evaluation for Application to Fractional Order PID Controllers, *Proc. 17th Int. Conference on Methods and Models in Automation and Robotics (MMAR'12)* (Midzyzdroje, Poland). pp. 579–584.

Busłowicz, M. (2008). Stability of linear continuous-time fractional-order systems with delays of the retarded type, *Bulletin of the Polish Academy of Sciences* **56**, pp. 319–324.

Busłowicz, M. and Kaczorek, T. (2009). Simple conditions for practical stability of positive fractional discrete time linear systems, *International Journal of Applied Mathematics and Computer Science* **19**, 2, pp. 263–269.

Butt, R. (2009). *Introduction to Numerical Analysis Using MATLAB®* (Jones & Bartlett Learning).

Canny, J. (1986). A computational Approach to Edge Detection, *IEEE Transactions on Pattern Analysis and Machine Intelligence* **6**, 8, pp. 679–698.

Caputo, M. and Mainardi, F. (1971), A new dissipation model based on memory mechanism, *Pure and Applied Geophysics* **91**, 8, pp. 134–147.

Chao, H., Luo, Y., Di, L. and Chen, Y. (2003). Roll-channel fractional order controller design for a small fixed-wing unmanned aerial vehicle, *Control Engineering Practice* **18**, pp. 761–772.

Caponetto, R., Fortuna, L. and Porto, D. (2002). Parameter Tuning od a Non Integer Order PID Controller, in *Proceedings of the 15th international symposium on mathematical theory of networks and systems* (Notre Dame, USA),

Carlson, G. E. and Halijak, C. A. (1964). Approximation of fractional capacitors $(\frac{1}{s})^{-n}$ by a regular Newton process, *IRE Transactions on Circuit Theory* **CT-11**, 2, pp. 210–213.

Caponetto, R., Dongola, G., Fortuna, L. and Petras, I. (2010). *Fractional Order Systems: Modeling and Control Applications* (World Scientific Series on Nonlinear Science: Series , VOL. 72, Singapore).

Cerera, J., Banos, A., Monje, C. A. and Vinagre B. M. (2006). Tuning of Fractional PID Controllers by using QFT, in *IEEE Industrial Electronics (IECON2006)* Paris, France, pp. 5402–5407.

Chao, H., Luo, Y., Di, L. and Chen, Y. (2010) Roll-channel fractional order controller design for a small fixed-wing unmanned aerial vehicle, *Control Engineering Practice* **18**, 7, pp. 761–772.

Chen, Y. and Moore, K. L. (2002a) Analytical stability bound for a class of delayed fractional order dynamic systems, *Nonlinear Dynamics* **29**, 7, pp. 191–200.

Chen, Y., and Moore, K. L. (2002b). Discretization schemes for fractional-order differentiators and integrators, *IEEE Transactions on Circuits and Systems-I: Fundamental Theory and Applications* **49**, 3, pp. 363–367.

Chen, Y. and Vinagre, B. M. (2003). A new IIR-type digital frqactional order differentiator, *Signal Processing* **83**, pp. 2359–2365.

Chen, Y., Moore, K. L., Vinagre, B. M. and Podlubny, I. (2004a). Robust PID controller autotuning with a phase shaper, in *Proceedings of the First IFAC Symposium on Fractional Differentiation and its Applications* (Bordeaux, France), pp. 481–492.

Chen, Y., Vinagre, B. M. and Podlubny, I. (2004b). Continued Fraction Expansion Approaches to Discretizing Fractional Order Derivatives an Expository Review, *Nonlinear Dynamics* **38**, 1-4, pp. 155–170.

Chen Y. (2006a). Ubiquitous fractional order controllers?, in *Proceedings of the Second IFAC Symposium on Fractional Differentiation and its Applications* (Porto, Portugal), pp. 481–492.

Chen, Y., Dou, H., Vinagre, B. M. and Monje C. A. (2006b). A robust tuning methods for fractional order PI controllers, in *Proceedings of the Second IFAC Symposium on Fractional Differentiation and its Applications* (Porto, Portugal), pp. 83–88.

Chen, Y., Petráš, I. and Xue, D. (2009). Fractional Order Control - A Tutorial, in *Proceedings of the American Control Conference* (St. Louis, USA), pp. 1397–1411.

Chen, X. and Fei, X. (2012). Improving Edge-detection Algorithm based on Fractional Differential Approach, in *1012 International Conference on Image, Vision and Computing (ICIVC 2012)* (IACSIT Press, vol. 50, Singapore), pp. 1–6.

Dahlquist, G. and Björk, Å. (1974). *Numerical Mehtods.* Prentice-Hall.

Das, S. (2009). *Functional Fractional Calculus for System Identification and Controls* (Springer-Verlag, Berlin-Heidelberg).

Das, S. and Pan, I. (2012). *Fractional Order Signal Processing: introductory Concepts and Applications. Springer Briefs in Applied Sciences and Technology* (Springer, Heidelberg).

Datta, M., Ho, M., and Bhattacharyya, S. P. (2000) *Structure and Synthesis of PID Controllers* (Springer Verlag, London).

De Espíndola, J. J., Bavastri, C. A., and De Oliveira Lopes, E. M. (2008). Design of optimum systems of viscoelastic vibration absorbers for a given material based on the fractional calculus model, *Journal of Vibration and Control* **14**, 9-10, pp. 1607–1630.

Deng, W., Li, Ch. and Lu, J. (2007), Stability analysis of linear fractional differential system with multiple time delays, *Nonlinear Dynamics* **48** pp. 409–416.

Delavari, H., Ghaderi, R., Ranjbar, A. N., and HosseinNia, H. S. and Momani, S. (2010). Adaptive Fractional PID Controller for Robot Manipulator, in *Proceedings of the 4th IFAC Workshop on Fractional Differentiation and its Applications (FDA'10)* (Badajoz, Spain), pp. 1–7.

Deng, W. (2007). Short memory principle and a predictor-corrector approach for fractional differential equations, *Journal of Computational and Applied Mathematics* **206**, 1, pp. 174–188.

Deng, W. and Li, C. (2012). Numerical Schemes for Fractional Ordinary Differential Equations, Miidla, P. (ed.), *Numerical Modelling* (InTech Europe), pp. 355–374. http://www.intechopen.com/books/numerical-modelling/numerical-schemes-for-fractional-ordinary-differential-equations

Diaz, J. B. and Osler, T. J. (1974). Differences of fractional order, *Mathematics of Computation* **28**, pp. 185–202.

Diethelm, K. (1995). An algorithm for the numerical solution of differential equations of fractional order, *Electronic Transactions on Numerical Analysis* **5**, pp. 1–6.

Diethelm, K. (2010). *The Analysis of Fractional Differential Equations: An Application-Oriented Exposition Using Differential Operators of Caputo Type*. Lecture Notes in Mathematics (Springer, Heidelberg).

Diethelm, K. (2011). An efficient parallel algorithm for the numerical solution of fractional differential equations, *Fractional Calculus and Applied Analysis* **14**, pp. 475–490.

Domek, S. (2013). *Fractional Calculus in the predictive control (in Polish)* (Wydawnictwo Uczelniane Zachodniopomorskiego Uniwersytetu Technologicznego w Szczecinie, Szczecin).

Dorčák, L.,J. Lesko, V. and Kostial, I. (1997). Identification of Fractional - Order Dynamical Systems, in *Proceeding of the 12th International Conference on Process Control and Simulation - (ASRTP'96)* (Kosice, Slovak Republic), pp. 62–68.

Dorčák, L., Prokop, J. and Koštial I. (1994). Investigations of the properties of fractional-order dynamical systems, in *Proc.of the 11th International Conference on Process Control and Simulation (ASRTP94)* (Kosice Zlata Idka, Slovakia), pp. 58–66.

Dorčák, L., Petráš, I., Koštial, I. and Terpák (2002). Fractional-order state space models, in *Proc.of the International Carpatian Control Conference (ICCC2002)* (Malenovice, Slovakia), pp. 193–198.

Dorčák, Ľ., Gonzalez, E., Terpak, J., Valsa, J. and Pivka, L. (2013). Identification of fractional-order dynamical systems based on nonlinear function optimization, *International Journal of Pure and Applied Mathematics* **89**, 2, pp. 225–250.

Dorf, R. C. and Bishop, R. H. (2005). *Modern Control Systems* (Pearson Prentice Hall, Upper Saddle River).

Duarte, F. and Machado, J. T. (2009). Describing function of two masses with backlash. *Nonlinear Dynamics,* **56**, 4, pp. 409–413.

Duarte, F. and Machado, J. T. (2009). Fractional describing function of systems with nonlinear friction. *Intelligent Engineering Systems and Computational Cybernetics*, (Springer, The Netherlands), pp. 257–266.

Dzieliński, A. and Sierociuk, D. (2005). Adaptive feedback control of fractional order discrete state - space systems, in *Proceedings of International Conference on Computational Intelligence for Modelling Control and Automation, CIMCA'2005* (Vienna, Austria), pp. 804–809.

Dzieliński, A. and Sierociuk, D. (2006a). Observer for discrete fractional order state - space systems, in *Proceedings of the 2th IFAC Workshop on Fractional Differentiation and its Applications (FDA'06)* (Porto, Portugal), pp. 524–529.

Dzieliński, A. and Sierociuk, D. (2006). Stability of discrete fractional state - space systems, in *Proceedings of the 2th IFAC Workshop on Fractional Differentiation and its Applications (FDA'06)* (Porto, Portugal), pp. 518–523.

Dzieliński, A. and Sierociuk, D. (2007a). Controllability and observability of fractional order discrete state - space systems., in *Proceedings of the 13th IEEE/IFAC International Conference on Methods and Models in Automation and Robotics (MMAR'07)* (Szczecin, Poland), pp. 129–134.

Dzieliński, A. and Sierociuk, D. (2007b). Ultracapacitor modelling and control using discrete fractional order state - space model, in *Proceedings of the 8th International Carpathian Control Conference (ICCC'07)* (Štrbské Pleso, Slovakia), pp. 127–130.

Dzieliński, A. and Sierociuk, D. (2007c). Ultracapacitor modelling and control using discrete fractional order state - space models and fractional kalman filter. in *Proceedings of European Control Conference (ECC'07)* (Kos, Greece), pp. 2916–2922.

Dzieliński, A. and Sierociuk, D. (2008). Simulation and experimental tools for fractional order control education, Proceedings of the 17th World Congress. The International Federation of Automatic Control (Seoul, Korea), pp. 11654–11659.

Dzieliński, A. and Sierociuk, D. (2010). Some applications of fractional order calculus, *Bulletin of the Polish Academy of Sciences. Technical Sciences*, **58**, 4, pp. 2916–2922.

Dzieliński, A., Sarwas, G. and Sierociuk, D. (2011). Comparison and validation of integer and fractional order ultracapacitor models, *Advances in Difference Equations a SpringerOpen Journal* **11**, 1, pp. 1–15, http://www.advancesindifferenceequations.com/content/2011/1/11.

El-Salam, S. A. and El-Sayed, A. A. (2007). On the stability of some fractional-order nonautonomous systems. *Electr. J. of Qualitative Theory of Differential Equations*, **6** pp. 1–14.

Faieghi, M. R. and Nemati, A. (2011). On Fractional-Order PID Design , in T. Michałowski (ed.), *Applications of MATLAB in Science and Engineering* (InTech, London), pp. 273–292.

Feliu, V., Vinagre, B. M. and Monje, C. A. (2007). Fractional-order Control of a Flexible Manipulator, in Sabatier, J., Agrawal, O. and Machado, J. T. (eds.), *Advances in Fractional Calculus* (Springer Netherlands),

Fiedler, M. (1986). *Special matrices and their applications in numerical mathematics*, (Martinus Nijhoff Publishers, Dodrecht).

Franklin, G. F., Powell, D. J. and Workman, Michael, L. (1998). *Digital Control of Dynamic Systems* 3rd edn. (Adison-Wesley).

Franklin, G. F., Powell J. D. and Emami-Naeini A. (2002). *Feedback Control of Dynamic Systems*, (Prentice-Hall, Inc).

Freed, A., Diethelm K. and Luchko Y. (2002). *Fractional-Order Viscoelasticity (FOV): Constitutive Development Using the Fractional Calculus: First Annual Report*, (National Aeronautics and Space Administration, Glenn Research Center, Cleveland).

Fulford, G., Forrester, P. and Jones A. (1997). *Modelling with Differential and Difference Equations* (Cambridge University Press, Cambridge).

Gao, C. B., Zhou, J. L., Hu, J. R. and Lang, F. N. (2011). Edge detection of color image based on quaternion fractional differential, *IET Image Processing* **5**, 3, pp. 261–272.

Garg, V. and Singh, K. (2012). An Improved Grünwald-Letnikov Fractional Differential Mask for Image Enhancement, *International Journal of Advanced Computer Science and Applications* **3**, 3, pp. 130–135.

Giridharan, K. and Karthikeyan A (2014). Fractional Order PID Controller Based Bidirectional Dc/Dc Converter, *International Journal of Innovation Research in Science, Engineering and Technology* **3**, 3, pp. 557–564.

Gorenflo, R. and Mainardi F. (1997a). Fractional Calculus: Integral and Differential Equations of Fractional Order, in A. Carpinteri and F. Mainardi (eds.), *Fractals and Fractional Calculus in Continuum Mechanics* (Springer Verlag, Wien and New York), pp. 223–276.

Gorenflo, R. and Mainardi F. (1997b). Fractional Calculus: Some Numerical Methods, in A. Carpinteri and F. Mainardi (eds.), *Fractals and Fractional Calculus in Continuum Mechanics* (Springer Verlag, Wien and New York), pp. 277–290.

Günter, M. (1988). *Zeitdiscrete Steuerungssysteme* (VEB Verlag Technik, Berlin).

Guermah, S., Djennoune, S. and Bettayeb M. (2010). Discrete-Time Fractional-Order Systems: Modeling and Stability Issues, in Mahmoud M. S. (ed.), *Advances in Discrete Time Systems* (InTech), pp. 183–212.

Hagglund, T. and Åström, K. J. (1995). *PID controllers: Theory, design, and tuning.* 2nd edn. (ISA The Instrumentation, Systems, and Automation Society).

Halanay, A. and Samuel, J. (1997). *Differential Equations, Discrete Systems and Control.* (Kluwer Academic Publishers, Dodrecht).

Hamamci, S. E. (2007). Analgorithm for stabilization of fractional order time delay systems using fractional order PID controllers. *IEEE Transactions on Automatic Control*, **52**, 10, pp. 1964–1969.

Hamamci, S. E. (2008). Stabilization using fractional-oreder PI and PID controllers, *Nonlinear Dynamics* **51**, pp. 329–343.

Han, Q., Liu, K. (2014). *A Novel Edge Detection Operator Based On Fractional Gaussian Differential* (National Science Foundation), http://www.paper.edu.cn.

T. T. Hartley, C. F. Lorenzo, and H. K. Qammar. (2003). Chaos in a fractional order Chuas system, *IEEE Transactions Circuits and Systems I* **42**, 8, pp. 485–490.

Donald Hearn, D. and Baker, M. P. (1994. *Computer Graphics* (Prentice-Hall, Upper Saddle River).

Heleschewitz, D. and Matignon, D. (1998). Diffusive realisations of fractional integrodifferential operators: Structural analysis under approximation, in *Proceedings of IFAC Conference on System Structure and Control, vol. 2* (Nantes, France), pp. 243–248.

Henry, B. I. and Wearne, S. L. (2000). Fractional reactiondiffusion, *Physica A: Statistical Mechanics and its Applications* **276**, 3, pp. 448–455.

Henry, B. I., Langlandsand, T. M. and Wearne, S. L. (2006). Anomalous diffusion with linear reaction dynamics: From continuous time random walks to fractional reaction-diffusion equations, *Physical Review E* **73**, 3, pp. 1–34.

Herrmann, R. (2011). *Fractional Calculus: An Introduction for Physicists* (Scientific Publishing Company, Singapore).

Heymans, N. and Podlubny, I. (2006). Physical interpretation of initial conditions for fractional differential equations with Riemann-Liouville fractional derivatives, *Rheol Acta* **45**, pp. 765–771.

Heymans, N. (2008). Dynamic measurements in long - memory materials: fractional calculus evaluation of approach to steady state, *Journal of Vibration and Control* **14**, 9-10 pp. 1587-1596.

Hilfer, R. (2000). *Applications of Fractional Calculus in Physics* (World Scientific Publishing Company, Singapore).

Ho, W. K., Gan, O. P., Tay, E. B. and Ang, E. L. (1996). Performance and gain and phase margins of well-known PID tuning formulas. *IEEE Transactions on Control Systems Technology*, 4 pp. 473–477.

Hofreiter, M. and Zítek P. (2003). Argument increment stability criterion for linear delta models, *International Journal on Applied Mathematics and Computer Science* 13, 13, pp. 485–491.

Hou, M., Liu, Y. and Fu, Y. (2012). The Improved Fractional Order Differential Image Information Extraction Algorithm, in *Proc. Fourth International Conference on Computational and Information Sciences (CPS12)* (Chongqing, China), pp. 651–654.

Ifeachor, E. C. and Jervis, B. W. (1993). *Digital Signal Processing: A Practical Approach*, 6th edn. (Addison-Wesley, Harlow, England).

Isermann, R. (1988). *Digitale Regelsysteme. Band I, Grundlagen Deterministische Regelungen* (Springer-Verlag, Berlin).

Jacquot, R. G. (1995). *Modern Digital Systems* (Marcel Dekker, Inc., New York).

Jesus, I. S. and Machado, T. A. (2008). An Evolutionary Approach for the Synthesis of Fractional Potentials, *Journal of Fractional Calculus & Applied Analysis* 11, 3, pp. 1–12.

Jesus, I. S. and Machado, T. J. (2012). Application of Integer and Fractional Models in Electrochemical Systems, *Hindawi Publishing Company, Mathematical Problems in Engineering* 2012, ID 248175, pp. 1–17. doi:10.1155/2012/248175

Jifeng, W. and Yuankai L. (2005). Frequency Domain analysis and Applications for Fractional Order Control Systems, *Journal of Physics: Conference Series* 13, pp. 268–273.

Jury, E. I. (1964). *Theory and Application of the Z-Transform Method* (John Wiley & Sons, Inc., New York).

Kaczorek, T. (1992). *Linear Control Systems: Analysis of Multivariable Systems*, (John Wiley & Sons, Inc.).

Kaczorek, T. (2004). Reachability and controllability to zero of positive fractional discrete-time systems, *Machine Intelligence and Robotic Control* 6, 4, pp. 139–143.

Kaczorek, T. (2008). Practical stability of positive fractional discrete-time linear systems, *Bulletin of the Polish Academy of Sciences. Technical Sciences* 56, 4, pp. 313–317.

Kaczorek, T. (2009). Realization problem for positive fractional hybrid 2D linear systems, *Fractional Calculus and Applied Analysis* 11, 3, pp. 1–16.

Kaczorek, T. (2009). Positivity and stabilization of fractional discrete-time systems with delays, in *IEEE International Conference on Control and Automation* (Christchurch, New Zealand), pp. 1163–1167.

Kaczorek, T. (2010). Tests for practical stability of positive fractional discrete-time linear systems, *Pomiary Automatyka Robotyka* 2, pp. 463–469.

Kaczorek, T. (2011a). Positive fractional linear systems, *Pomiary Automatyka Robotyka* 2, pp. 504–511.

Kaczorek, T. (2011b). Positive fractional 2D continuous-discrete linear systems, *Bulletin of the Polish Academy of Sciences: Techniucal Sciences* 59, 4, pp. 575–580.

Kaczorek, T. (2011c). Singular fractional discrete-time linear systems, *Control and Cybernetics* 40, 3, pp. 1–9.

Kaczorek, T. (2011). *Selected Problems of Fractional Systems Theory*, (Springer, Berlin Heidelberg).

Kaczorek, T. (2012). Computation of initial conditions and inputs for given outputs of fractional and positive discrete-time linear systems, *Archives of Control Sciences* **22**, 2, pp. 145–159.

Kaczorek, T. and Sajewski Ł (2013). *Realization Problem for Positive and Fractional Systems*, (Printing House of Bialystok University of Technology, Bialystok).

Kaczorek, T. (2014). Minimum energy control of fractional positive continuous-time linear systems with bounded inputs, *International Journal of Applied Mathematics and Computer Science* **24**, 2, pp. 335-340.

Kaczorek, T. (2014). Minimum energy control of fractional positive discrete-time linear systems with bounded inputs, in *Proceedings of the IFAC World Congress in Cape Town* (Cape Town, South Africa), pp. 2909–2914.

Kailath, S. (1980). *Linear Systems*, (Prentice-Hall, Inc.).

Kalia, R. N. (1993). *Recent Advances in Fractional Calculus* (Global Publishing Company).

Kalia R. N., Srivastava H. M. (1999). Fractional Calculus and Its Applications Involving Functions of Several Variables, *Applied Mathematics Letters* **12**, pp. 19–23.

Karanjkar, D. S., Chatterji, S. and Venkateswaran P. R. (2012). Trends in Fractional Order Controllers, *International Journal of Emerging Technology and Advanced Engineering* **2**, 3, pp. 383–389.

Keisler, H. J. (2013). *Elementary Calculus. An Infinitesimal Approach*, (University of Wisconsin),

Khalil, H. K. (2002). *Nonlinear Systems* 3rd edn. (Prentice Hall, Upper Saddle River).

Kilbas, A. A., Srivastava, H. M., Trujillo, J. J. (2006a). *Theory and Applications of Fractional Differential Equations: North-Holland Mathematics Studies 204* (Elsevier, Amsterdam).

Kilbas, A. and Rogosin S. V. (2009). *Analytic Methods of Analysis and Differential Equations: AMADE 2009* (Cambridge Scientific Publishers, Cambridge).

Kiryakova, V. S. (1994). *Generalized Fractional Calculus and Applications* (Longman Sci. Tech. & J. Wiley, New York).

Klafter, J., Lim, S. C. and Metzler, R. (2011). *Fractional Dynamics: Recent Advances* (World Scientific Publ. Co., Singapore).

Klamka, J. (2006). *Controllability of Dynamical Systems*, (Kluver Academic Publ., Dodrecht).

Klamka, J. (2008). Controllability of fractional discrete-time systems with delay, *Zeszyty Naukowe Politechniki Śląskiej. Seria Automatyka* **151**, pp. 67-72.

Klamka, J. (2010). Controllability and Minimum Energy Control Problem of Fractional Discrete-Time Systems, in Baleanu, D., Guvenc, Z. B., Machado, J. T. (eds.), *New Trends in Nanotechnology and Fractional Calculus Applications* (Springer, Netherlands), pp. 503–509.

Klamka, J. (2011). Local controllability of fractional discrete-time semilinear systems, *Acta Mechanica et Automatica* **15**, 2, pp. 55–58.

Klimek, M. (2009). *On Solutions of Linear Fractional Equations of a Variational Type* (The Publishing Office of Częstochowa University of Technology, Serie: Monographs no. 172, Częstochowa).

Kopetz, H. (2011). *Real-time systems: design principles for distributed embedded applications* (Springer, New York).

Kuo, S. M., Lee, B. H. and Tian, W. (2013). *Real-time digital signal processing: fundamentals, implementations and applications* (Wiley, Chichester).

Ladaci, S.and Charefa A. (2006). On fractional adaptive control., *Nonlinear Dynamics* **43**, 4, pp. 365–378.

Lakshmikantham, V., Leela, S. and Devi, J. V. (2009). *Theory of Fractional Dynamic Systems* (Cambridge Scientific Publishers, Cambridge).

Langlands, T. M. and Henry, B. I. (2001). The accuracy and stability of an implicit solution method for the fractional diffusion equation, . *Comput. Phys* **205**, 2, pp. 267–270.

Lanusse, P., Oustaloup, A. and Mathieu, B. (1993). Third generation CRONE control, in *Conference Proceedings of the International Conference on Systems, Man and Cybernetics:* (Le Touquet, France), pp. 149–155.

Lanusse, P. Pommier, V. and Oustaloup, A. (2000) Robust control of LTI square MIMO plants using two CRONE control design approaches. in *Proceedings of the IFAC Symposium on Robust Control Design (ROCOND'2000)*, (Prague, Czech Republic).

Lanusse, P. Pommier, V. and Oustaloup, A. (2000) Fractional control system design for a hydraulic actuator. in *Proceedings of the First IFAC Conference on Mechatronics Systems*, (Darmstadt, Germany).

Leszczyński, J. S. (2011). *An Introduction to Fractional Mechanics* (Częstochowa University of Technology, Częstochowa).

Linz, P. and Wang, R. (2003). *Exploring Numerical Methods. An Introduction to Scientific Computing Using MATLAB* (Jones and Bartlett Publishers, Inc., Sudbury).

Leu, J. F., Tsay, S. Y. and Hwang, C. (2002). Design of optimal fractional order PID controllers, *Journal of the Chinese Institute of Chemical Engineers*, **33**, 2, pp. 193–202.

Li, H. S., Luo, Y. and Chen, Y. (2010). A fractional order proportional and derivative (FOPD) motion controller: tuning rule and experiments. *IEEE Transactions on Control Systems Technology*, **18** 2 pp. 516–520,

Li, H., Huang, J., Liu, D., Zhang, J. and Teng, F. (2011). Design of fractional order iterative learning control on frequency domain. in *Proceedings of the International Conference on Mechatronics and Automation*, (Beijing, China) pp. 2056–2060.

Liang, L., R., Basallo, E., and Looney, C., G. (2001). Image edge detection with fuzzy classifier, in *Proceedings of the ISCA 14th International Conference* (Las Vegas, USA), pp. 279–283.

Liang, L., R. and Looney, C., G. (2003). Competitive fuzzy edge detection, *Applied Soft Computing* **3**, pp. 123–137.

Liung, L. (1999). *System Identyfication. Theory for the User* (Prentice Hall PTR, Upper Saddle River).

Lubich, C. H. (1986). Discretized fractional calculus, *SIAM Journal of Mathematical Analysis* **17**, 3, pp. 704–719.

Ludovic, L. L., Oustaloup, A., Trigeassou, J. C. and Levron, F. (1998) Frequency identification by non integer model. in *Proc. of IFAC Symposium on System Structure and Control*, (Nantes, France)

Luo, Y. and Chen, Y. (2009a). Fractional order [proportional derivative] controller for a class of fractional order systems, *Automatica* **45**, pp. 2446-2450.

Luo, Y. and Chen, Y. (2009b). Fractional order [proportional derivative] controller for Robust Motion Control: Tuning Procedure and Validation, in *Proceedings of the American Control Conference* (St. Louis, USA), pp. 1412-1417.

Luo, Y. and Chen, Y. (2012) Stabilizing and Robust FOPI Controller Synthesis for First Order Plus Time Delay Systems. *Automatica*, **48**, 9, pp. 2159–2167.

Luo, Y., Wang, C. Y., Chen, Y. and Pi, Y. G. (2010). Tuning fractional order proportional integral controllers for fractional order systems. *Journal of Process Control*, **20** 7 pp. 823–831,

Luo, Y., Li, H. and Chen, Y. (2011). Experimental study of fractional order proportional derivative controller synthesis for fractional order systems, *Mechatronics* **21**, pp. 204–214.

Lurie, B. J. (1994). Three-parameter tunable tilt-integral-derivative (TID) controller, *US Patent US5371670*

Ma, C. and Hori, Y. (2004). Backlash vibration suspension control of torsional system by novel fractional order PID controller, *IEEE Transactions on Industry Applications* **124**, 3, pp. 312–317.

Ma, C. and Hori, Y. (2007). Fractional order control: Theory and applications in motion control [past and present], *IEEE Industrial Electronics Magazine* **1**, 4, pp. 6–16.

Machado, J. T. (1997). Analysis and design of fractional-order digital control systems, *Journal of Systems Analysis-Modeling-Simulation* **27**, 2-3, pp. 107–122.

Machado, J. T. (1998). Fractional-Order Derivative Approximation in Discrete-Time Control Systems, *Journal of Systems Analysis-Modeling-Simulation* **34**, 4, pp. 419–434.

Machado, J. T. (2001). Discrete-time fractional-order controllers, *Journal of Fractional Calculus & Applied Analysis* **4**, 1, pp. 47–66.

Machado, J. T., Silva, M. F., Barbosa, R. S., Jesus, I. S., Reis, C. M., Marcos, M. G. and Galhano A. F. (2010). Some Applications of Fractional Calculus in Engineering, *Mathematical Problems in Engineering* **2010**, ID 639801, pp. 1–34.

Machado, J. T., Kiryakova, V. and Mainardi, F. (2011). Recent history of fractional calculus, *Communications in Nonlinear Science and Numerical Simulation* **16**, 3, pp. 1140–1153.

Machado, J. T., Galhano, A.M. and Trujillo, J. J. (2013). Science metrics on fractional calculus development since 1966, *Fractional Calculus & Applied Analysis* **16**, 2, pp. 479–500.

Madasu, H., Verma, O., P., Gangwar, P. and Vasikarla, S. (1997). Fuzzy Edge and Corner Detector for Color Images, in *Proc. Proceedings of the Sixth International Conference on Information Technology: New Generations* (Las Vegas, USA), pp. 1301–1306.

Magin, R. L. (2006). *Fractional calculus in bioengineering* (Begell House Inc., Redding, CT).

Magin, R. L. and Ovadia, M. (2008). Modeling the cardiac tissue electrode interface using fractional calculus, *Journal of Vibration and Control* **14**, 9-10, pp. 1431–1442.

Magin, R. L. (2012). Fractional calculus in bioengineering: A tool to model complex dynamics, in *Proc. of the 13th International Carpathian Control Conference (ICCC'12)* (High Tatras, Podbansk, Slovak Republic) pp. 464–469.

Mainardi, F. (2010). *Fractional Calculus and Waves in Linear Viscoelasticity* (Imperial College Press, London).

Malinowska, A. B. and D.F.M. Torres, D. F. (2012). *Introduction to the Fractional Calculus of Variations.* (Imperial College Press, London).

Manabe, S. (1961). The non-integer integral and its application to control systems, *Japanese Institute of Electrical Engineers Journal* **6**, 3-4, pp. 83–87.

Manabe, S. (1963). The System Design by the Use of a Model Consisting of a Saturation and Non-integer Integrals, *ETJ of Japan* **8**, 3-4, pp. 147–150.

Mathieu, B., Melchior, P., Oustalopu, A. and Ceyral, C. (2003). Fractional differentiation for edge detection, *Signal Processing* **8**, pp. 2421–2432.

Matignon, D. (1996) Stability results for fractional differential equations with applications to control processing. in *Proceedings of IMACS-SMC* , (Lille, France), pp. 963–968.

Matignon, D. (1998) Stability properties for generalized fractional differential systems in *ESAIM: Proceedings Fractional Differential Systems: Models, Methods and Applications*, pp. 145158. URL: http://www.emath.fr/proc/Vol.5/

Matušů, R. (2011). Application of fractional order calculus to control theory, *International Journal of Mathematical Models and Methods in Applied Sciences* **5**, 7, pp. 1162-1169.

Marr, D., Hildreth E., (1980). Theory of edge detection, *Proceeding of the Royal Society of London. Series B* **207**, 1167, pp. 187–217.

Mayoral, L. (2012). Testing for Fractional Integration Versus Short Memory with Structural Breaks, *Oxford Bulletin of Economics and Statistics* **74**, 2, pp. 278–305.

Mbodje, B. and Montseny, G. (1995), Boundary fractional derivative control of the wave equation. *IEEE Transactions on Automatic Control* **40**, 2, pp. 378–382.

Méhauté, Machado, T. J., Trigeassou, J. C. and Sabatier, J. (2005). *Fractional Differentiation and its Applications.* (Ubooks Verlag, Neusäss).

Melchior, P., Metoui, B., Najar, S. A, Naceur, M. and Oustaloup, A. (2009). Robust path planning for mobile robot based on fractional attractive force, in *Proceedings of the American Control Conference* (Hyatt Regency Riverfront, St. Louis, USA), pp. 1424-1429.

Meral, F. C., Royston, T. J. and Magin, R. L. (2010). Fractional calculus in viscoelasticity: An experimental study, *Communications in Nonlinear Science and Numerical Simulation* **15**, 4, pp. 939-945.

Merrikh-Bayat, F., Afshar, M. and K.-Ghartemani, M. (2008a), Extension of the root-locus method to a certain class of fractional-order systems. *ISA Transactions* **48**, 1, pp. 4853.

Merrikh-Bayat F. and Afshar, M. (2008b), Extending the Root-Locus Method to Fractional-Order Systems, *Journal of Applied Mathematics, Hindawi Publ. Corp.,* **ID 528934**, pp. 1–13. (doi:10.1155/2008/528934).

Merrikh-Bayat, F. (2011). Optimal Tuning Rules of the Fractional-Order PID Controllers with Application to First-Order Plus Time Delay Processes, in *Proceedings of the 2011 4th International Symposium on Advanced Control of Industrial Processes* (Hangzhou, P.R. China), pp. 403–408.

MF.03 "Powstanie Warszawskie" Cz.5. (2015) *Filmoteka Narodowa - Archiwum 1 Chełmska* (Warsaw, Poland), www.fn.org.pl

Michalski, M. W. (1993). *Derivatives of noninteger order and their applications,* Dr Sc. Dissertationes Mathematicae, Inst. Math. Polish Acad. Sci., Warsaw, Poland.

Miller, K. S. and Ross, B. (1993). *An Introduction to the The Fractional Calculus and Fractional Differential Equations,* (A Wiley-Interscience Publications, John Wiley & Sons, Inc., New York).

Mohan, J. J. and Deekshitulu, G. V. (2012). Fractional Order Difference Equations, *International Journal of Differential Equations* **2012**, pp. 1–11.

Monje, C. A., Vinagre, B. M., Chen, Y., Feliu, V., Lanusse, P. and Sabatier, J. (2004). Proposals for fractional $PI^\lambda D^\mu$ tuning, in *Proceedings of Fractional Differentiation and its Applications (FDA'04)* (Bordeaux, France), pp. 156–161.

Monje, C. A., Vinagre, B. M., Feliu, V. and Chen, Y. (2008). Tuning and auto-tuning of fractional order controllers for industry applications, *Control Engineering Practice* **16**, 7, pp. 798–812.

Monje, C. A., Chen, Y., Vinagre, B. M., Xue, D. and Feliu, V. (2010). *Fractional-order Systems and Controls. Fundamentals and Applications (Advances in Industrial Control),* (Springer-Verlag, London).

Morita, T., Sato, K. I. (2014) Detection of Edge with the Aid of Mollification Based on Wavelets, *Applied Mathematics* **5**, pp. 2849–2861. http://dx.doi.org/10.4236/am.2014.518271

Moshrefi-Torbati, M. and Hammond, J. K. (1998). Physical and geometrical interpretation of fractional operators, *J. Franklin Inst* **335B**, 6, pp. 1077–1086.

Nishimoto, K. (1984). *Fractional Calculus. Integration and Differentiation of Arbitrary Order. Vol. 1* (Descartes Press, Koriyama).

Nishimoto, K. (1986). *Fractional Calculus. Integration and Differentiation of Arbitrary Order. Vol. 2* (Descartes Press, Koriyama).

Nishimoto, K. (1989). *Fractional Calculus. Integration and Differentiation of Arbitrary Order. Vol. 3* (Descartes Press, Koriyama).

Nishimoto, K. (1991a). *Fractional Calculus. Integration and Differentiation of Arbitrary Order. Vol. 4* (Descartes Press, Koriyama).

Nishimoto, K. (1991b). *An Essence of Nishimotos Fractional Calculus (calculus in the 21st century) : integrations and differentiations of arbitrary order* (Descartes Press, Koriyama).

Nishimoto, K. (1991c). *Fractional Calculus. Integration and Differentiation of Arbitrary Order. Vol. 5* (Descartes Press, Koriyama).

O'Dwyer, A. (2009). *Handbook of PI and PID Controller Tuning Rules*, 3rd edn. (Imperial College Press, London).

O'Flynn, M. and Moriarty E. (1987). *Linear Systems: Time Domain and Transform Analysis* (John Wiley & Sons, New York).

Ogata, K. (1987). *Discrete-Time Control Systems* (Prentice-Hall International Editions, Englewood Cliffs).

Olderog, E. R. and Dierks, H. (2008). *Real-time systems: formal specification and automatic verification* (Cambridge University Press, Cambridge).

Oldham, K. B. and Spannier, J. (1974). *The Fractional Calculus. Theory and Applications of Differentiation and Integration to Arbitrary Order* (Mathematics in Science and Engineering, Volume 111, Academic Press, New York).

Oliveira, E. C. and Machado, T. J. (2014). A Review of Definitions for Fractional Derivatives and Integral, *Hindavi Publishing Corporation: Mathematical Problems in Engineering* **2014**, pp. 1–6, http://dx.doi.org/10.1155/2014/238459.

Oppenheim, A.— V. and Schaffer, N. W. (1975). *Digital Signal Processing* (Prentice Hall, Englewood Cliffs New York).

Ortigueira, M. D. (2003). On the initial conditions in continuous-time fractional linear systems, *Signal Processing* **83**, pp. 2301–2309.

Ortigueira, M. D. and Coito, F. (2004). From Differences to Derivatives, *Fractional Calculus & Applied Analysis* **7**, pp. 459–471.

Ortigueira M. D. (2011). *Fractional Calculus for Scientists and Engineers* (Springer Science + Business Media B.V., Dodrecht Heidelberg London New York).

Ostalczyk, P. (2000). The non-integer difference of the discrete-time function and its application to the control system synthesis, *Internationl Journal on System Science.* **31**, pp. 1551–1561.

Ostalczyk, P. (2001). *Discrete-Variable Functions. A Series of Monographs No 1018* (The Technical University Press, Łodź).

Ostalczyk, P. (2002). *The one-sided Z-Transform. A Series of Monographs No 1055* (The Technical University Press, Łodź)

Ostalczyk, P. (2003). Fundamental properties of the fractional-order discrete-time integrator, *Signal Processing* **83**, 11, pp. 2367–2376.

Ostalczyk, P. (2004a). Fractional-order backward difference equivalent forms. Part I Horners form, in *Proc. 1st IFAC Workshop on Fractional differentiation and its applications* (Bordeaux, France), pp. 341–346.

Ostalczyk, P. (2004b). Fractional-order backward difference equivalent forms. Part II Polynomial form, in *Proc. 1st IFAC Workshop on Fractional differentiation and its applications* (Bordeaux, France), pp. 347–352.

Ostalczyk, P. (2010a). Stability analysis of a discrete-time system with a variable-fractional-order controller, *Bulletin of the Polish Academy of Sciences* **58**, 4, pp. 613–619.

Ostalczyk, P. (2010b). Graphical Interpretation of the fractional-order backward difference/sum, *Pomiary Automatyka Kontrola* **56**, 5, pp. 418–422.

Ostalczyk, P. (2012). Equivalent description of the discrete-time fractional-order linear system and its stability domains, *International Journal of Applied Mathematics and Computer Science* **22**, 3, pp. 533–538.

Ostalczyk, P. (2014a). Remarks on five equivalent forms of the fractional order backward difference, *Bulletin of the Polish Academy of Sciences, Technical Sciences* **62**, 2, pp. 271–279.

Ostalczyk, P. (2014b). The Fractional Order Backward Difference of a product of Two Discrete Variable Functions (Discrete Fractional Leibnitz Rule), in K. J. Latawiec, M. Lukaniszyn and R. Stanisławski (eds.), *Advances in Modelling and Control of Non integer Order systems* (Springer-Verlag, Berlin Heidelberg), pp. 59–70.

Oustaloup, A. (1983). *Systèmes Asservis Linéaries dOrdre Fractionnaire* (Masson, Paris).

Oustaloup, A., Mathieu, B. and Melchior, P. (1991b). Robust edge detector of non integer order: the CRONE detector, in *Proceedings of the 8th Congres de Cytometrie en Flux et d'Analyse d'Image (ACF'91).* (Mons, Belgium), pp. 2421–2432.

Oustaloup, A., Mathieu, B. and Melchior, P. (1991a). Edge detection using non integer derivation, in *Proceedings of the IEEE European Conference on Circuit Theory and Design (ECCTD'91).* (Copenhagen, Denmark), pp. 130–135.

Oustaloup, A. and Bansard, M. (1993). First generation CRONE control, in *Conference Proceedings of the International Conference on Systems, Man and Cybernetics:* (Le Touquet, France), pp. 130–135.

Oustaloup, A., Mathieu, B. and Lanusse, P. (1993). Second generation CRONE control, in *Conference Proceedings of the International Conference on Systems, Man and Cybernetics:* (Le Touquet, France), pp. 136–142.

Oustaloup, A. (1991). *La Commande CRONE: Commande Robuste dOrdre Non Entier* (Hermès, Paris).

Oustaloup, A. (1995). *La dérivation non entière: theéorie, synthèse et applications* (Hermès, Paris).

Oustaloup, A., Cois, O., Lelay, L. (1995a). *Reprèpsentation et identification par modéle non-entier* (Hermes, Paris).

Oustaloup, A., Mathieu, B. and Lanusse, P. (1995b). The CRONE control of resonant plants: application to a flexible transmission, *European Journal of Control* **1**, 2, pp. 113–121.

Oustaloup, A., Moreau, X. and Nouillant, M. (1996). The CRONE suspension, *Control Engineering Practice* **4**, 8, pp. 1101–1108.

Oustaloup, A., Sabatier, J. and Moreau, X. (1998). From fractal robustness to the CRONE approach, in *ESAIM: Proceedings on Fractional Differential Systems, Models, Methods and Applications* pp. 177–192.

Oustaloup, A. and Mathieu, B. (1999). *La commande CRONE: du scalaire au multivariable* (Hermés Science, Paris).

Oustaloup, A. Levron, F. Nanot, F. and Mathieu, B. (2000). Frequency band complex non integer differentiator: Characterization and synthesis, *IEEE Transactions on Circuits and Systems I* **47**, pp. 25–40.

Oustaloup, A. (2014). *Diversity and Non-integer Differentiation for System Dynamics (Control, Systems and Industrial Engineering)* (Wiley & Sons, Inc., Hoboken).

Padula, F. and Visioli, A. (2011). Tuning rules for optimal PID and fractional-order PID controllers, *Journal of Process Control* **21**, pp. 69–81.

Petráš, I. (1999). The fractional order controllers: methods for their synthesis and appli-
    cation, *Journal of Electrical Engineering,* **50**, 9–10, pp. 284–288,.
Petráš, I., Dorčák, L. and Kostial, I. (2000a). The modelling and analysis of fractional-
    order control systems in discrete domain, in *Proceedings of ICCC'2000* (High Tatras,
    Slovak Republic), pp. 257–260.
Petráš, I., Dorčák, L., O'Leary, P., Vinagre, B. M. and Podlubny, I. (2000b). The modelling
    and analysis of fractional-order control systems in frequency domain, in *Proceedings
    of ICCC'2000* (High Tatras, Slovak Republic), pp. 261–264.
Petráš, I., Dorčák, L., O'Leary, P., Vinagre, B. M. and Podlubny, I. (2000c). The modelling
    and analysis of fractional-order regulated systems in the state space, in *Proceedings
    of ICCC'2000* (High Tatras, Slovak Republic), pp. 185–188.
Petráš, I., Podlubny, I., OLeary, P., Dorčák, L. and Vinagre, B. (2002). *Analogue Re-
    alization of Fractional Order Controller* (FBERG, Technical University of Kosice,
    Kosice).
Petráš, I., Vinagre, B. M., Dorčák, L. and Feliu, V. (1997). Fractional digital control of
    a heat solid: Experimental results, in *Proc. of the International Carpathian Control
    Conference (ICCC 2002)* (Malenovice, Czech Republic), pp. 365–370.
Petráš, I. (2011). *Fractional-Order Nonlinear Systems: Modeling, Analysis and Simulation.
    Series Nonlinear Physical Science* (Springer, Heidelberg).
Petráš, I. (2009a). Fractional Order Feedback Control of a DC Motor, *Journal of Electrical
    Engineering* **60**, 3, pp. 117-128.
Petráš, I. (2009b). Stability of fractional-order systems with rational orders: A survey,
    Fractional Calculus & Applied Analysis **12**, 3, pp. 269-298.
Podlubny. I. (1994a). *Fractional-Order Systems and Fractional-Order Controllers,* (Ústav
    Experimentalnej Fiziky UEF-03-09, Košice).
Podlubny. I. (1994b). *The Laplace transform method for linear differential equations of the
    fractional order,* (UEF-02-94, Slovak Acad. Sci., Kosice).
Podlubny, I., Dorčák, L. and Kostial, I. (1997). On Fractional Derivatives, Fractional-
    Order Dynamic Systems and $PI^\lambda D^\mu$-controllers, in *Proceedings of the 36th Confer-
    ence on Decision and Control* (San Diego, USA), pp. 4985–4990.
Podlubny, I. (1999a). *Fractional Differential Equations* (Academic Press, London).
Podlubny, I. (1999b). Fractional-Order Systems and $PI^\lambda D^\mu$-Controllers, *IEEE Transac-
    tions on Automatic Control* **44**, 1, pp. 208–214.
Podlubny I. (2002). Geometric and Physical Interpretation of Fractional Integration and
    Fractional Differentiation, *Fractional Calculus and Applied Analysis* **2002**, 5, pp.
    367–386.
Podlubny, I., Petráš, I., Vinagre, B. M, Chen, Y., O'Leary, P. and Dorčák, L.
    (2003). Realization of fractional order controllers, *Acta Montanistica Slovaca* **8**, 4,
    pp. 233–235.
Podlubny, I., Despotovic, V., Skovranek, T. and McNaughton, B. H. (2007). Shadows on
    the Walls: Geometric Interpretation of Fractional Integration, *The Journal of On-
    line Mathematics and Its Applications* **7**, $http://www.maa.org/external_archive/
    joma/Volume7/Podlubny/GIFI.html$.
Poinot, Y and Trigeassou, J. C. (2003). A method for modelling and simulation of fractional
    systems, *Signal Processing* **83**, pp. 2319–2333.
Poty, A., Melchior, P. and Oustaloup, A. (2004a). Dynamic path planning for mobile
    robots using fractional potential field, in *First International Symposium on Control,
    Communications and Signal Processing (2004)* (Hammamet, Tunisia), pp. 557–561.
Poty, A., Melchior, P. and Oustaloup, A. (2004b). Dynamic path planning by fractional

potential, in *Second IEEE International Conference on Computational Cybernetics (ICCC2 2004)* (Vienna, Austria), pp. 365–371.

Pratt, W. K. (1991). *Digital Image Processing*, (Wiley, New York).

Prewitt, J., M., S., (1970). Object enhancement and extraction, in B. S. Likin and A. Rosenfeld (eds.), *Picture Processing and Psychopictorics* (Academic Press, London), pp. 75–149.

Proakis, J. G. and Manolakis D. G. (2007). *Digital Signal Processing: Principles, Algorithms, and Applications*, 4th edn. (Upper Saddle River, New Jersey).

Pu, Y., Wang, W., Zhou, J., Wang, Y. and Jia, H. (2008). Fractional differential approach to detecting textural features of digital image and its fractional differential filter implementation. *Science in China Series F: Information Sciences* **51**, 9, pp. 1319–1339.

Pu, Y., Zhou, J., and Yuan, X. (2010). Fractional Differential Mask: A Fractional Differential-Based Approach for Multiscale Texture Enhancement. *IEEE Trans. on Image Processing* **19**, 2, pp. 491–511.

Ralston, A. (1965). *A First Course in Numerical Analysis* (Mc Graw-Hill, Inc., New York).

Raynaud, H. and Zergaïnoh, A. (2000). State - space representation for fractional order controllers, *Automatica* **36**, 7, pp. 1047–1021.

Roberts, L., G. (1963). *Machine Perception Of Three-Dimensional Solids*, (Garland Publishing, New York).

Rosario, J., Dumur, D. and Machado, J. T. (2006). Analysis of fractional - order robot axis dynamics, in *Proceedings of the 2nd Conference on Fractional Differentiation and Its Applications (FDA'06)* (Porto, Portugal), pp. 367–372.

Ross, B. (1974). *Fractional Calculus and its Applications, Lecture Notes in Mathematics 457*, (Springer Verlag, Berlin Heildelberg).

Rubin, B. (1996). *Fractional Integrals and Potentials*, (Addison Wesley Longman Ltd., London).

Sabatier, J., Agrawal, O. P. and Machado T. A. (2007). *Advances in Fractional Calculus. Theoretical Developments and Applications in Physics and Engeneering* (Springer Verlag, Dordrecht).

Salomon, D. (1999). *Computer Graphics and Geometric Modeling* (Springer Science+Business Media, LLC, New York).

Samko, S. G., Kilbas A. A. and Marichev O. I. (1986). *Fractional Integrals and Derivatives and Some of Their Applications* (Nauka i Tekhnika, Minsk) (In Russian).

Samko, S. G., Kilbas A. A. and Marichev O. I. (1993). *Fractional Integrals and Derivatives-Theory and Applications* (Gordon and Breach Science Publishers, London).

Sankowski, D., Mosorov W. and Strzecha K. (2011). *Image processing and analysis in industrial systems* (Wydawnictwo Naukowe PWN, Warszawa).

Sheng, H., Chen, Y. and Qiu, T. (2012). *Fractional Processes and Fractional- Order Signal Processing: Techniques and Applications, Signals and Communication Technology* (Springer, London).

Sierociuk, D. and Dzieliński (2006). Fractional Kalman Filter algorithm for states, parameters and order of fractional system estimation, *International Journal of Applied Mathematics and Computer Science* **16**, 1, pp. 101–112.

Singhal, R., Padhee, S. and Kaur, G. (2012). Design of Fractional Order PID Controller for Speed Control of DC Motor, *International Journal of Control and Automation* **3**, 4, pp. 33–46.

Silva, M. F., Machado, J. T. and Lopes, A. M. (2003a). Position/force control of a walking robot, *Machine Intelligence and Robot Control* **5**, pp. 33–44.

Silva, M. F., Machado, J. T. and Lopes, A. M. (2003b). Comparison of fractional and integer order control of an hexapod robot, in *Proceedings of International Design Engineering Technical Conferences and Computers and Information in Engineering Conference, vol. 5 of 19th Biennial Conference on Mechanical Vibration and Noise (ASME)* (Chicago, USA), pp. 667–676.

Silva, M. F., Machado, J. T. and Lopes, A. M. (2004). Fractional Order Control of a Hexapod Robot, *Nonlinear Dynamics* **38**, 1-4, pp. 417–433.

Silva, M. F. and Machado, J. T. (2006a). Fractional Order PD^α Joint Control of Legged Robots, *Journal of Vibration and Control* **12**, 12, pp. 1483–1501.

Silva, M. F., Machado, T. J. and Jesus, I. S. (2006b). Modelling and simulation of walking robots with 3 dof legs, in *Proceedings of the 25th IASTED International Conference on Modelling, Identification and Control (MIC 06)* (Lanzarote, Spain), pp. 271–276.

Silva, M. F., Barbosa, R. S., Jesus, I. S., Reis, C. M., Marcos, M. G. and Galhano, A F. (2010). Some Applications of Fractional Calculus in Engineering, in J. M. Balthazar, P. B. Gonçalves, S. Lenci, and Y. V. Mikhlin (eds.), *Mathematical Problems in Engineering Theory, Methods, and Applications. Article ID 639801,* (Hindawi Publishing Corporation), pp. 1–32.

Singhal, R., Padhee, S. and Kaur, G. (2012). Design of Fractional Order PID Controller for Speed Control of DC Motor, *International Journal of Scientific and Research Publications,* **2**, 6, pp. 1–8.

Smith, S. W. (2003). *The Scientist and Engineer's Guide to Digital Signal Processing* http://www.dspguide.com/editions.htm

Sobel, I. (1978). Neighborhood coding of binary images for fast contour following and general binary array processing, *Computer Graphics and Image Processing* **8**, 1, pp. 127–135.

Sommacal, L., Melchior, P., Oustaloup, A., Cabelguen, J. M. and Ijspeert, A. J. (2008). Fractional multi - models of the frog gastrocnemius muscle, *Journal of Vibration and Control* **14**, 9-10, pp. 1415-1430.

Sparavigna, A. C. (2009). *Using fractional differentiation in astronomy* (arXiv), `http://www.arxiv.org/abs/0910.4243`.

Stanisławski. R. (2013). *Advances in modeling of fractional difference systems-new accuracy, stability and computational results,* Oficyna Wydawnicza Politechniki Opolskiej.

Steiglitz, K. (1974). *An introduction to discrete systems* (John Wiley & Sons, Inc., New York).

Talai, Z., and Talai, A. (1997). A fast edge detection using fuzzy rules, in *2011 International Conference on Communications, Computing and Control Applications (CCCA)* (Hammamet, Tunisia), pp. 1–5.

Tan, K. K., Wang, Q., Hang, C. C. and Hagglund, T. (2000). *Advances in PID controllers (Advances in Industrial Control)* (Springer-Verlag, London,).

Tas, K., Machado, T. J. and Baleanu, D. (2007). *Mathematical Methods in Engineering* (Springer, Dordrecht).

Tavazoei, M. S., Haeri, M., Bolouki S. and Siami, M. (2008). Stability preservation analysis for frequency-based methods in numerical simulation of fractional order systems, *SIAM J. Numerical Analysus* **47**, 1, pp. 321–338.

Tavazoei, M. S. and Haeri, M. (2009). A note on the stability of fractional order systems, *Mathematics and Computers in Simulation* **79**, pp. 1566–1576.

Tian, Dan., Wu, J., and Yang, Y. (2014). A Fractional-order Laplacian Operator for Image Edge Detection, *Applied Mechanics and Materials,* **536-537**, pp. 55–58. doi:1010.4028/www.scientific.net/AMM.536-537.55

Trujillo, J. J., Rivero, M. and Bonilla, B. (1999). On a Riemann-Liouville Generalized

Taylor's Formula, *Journal of Mathematical Analysis and Applications* **231**, pp. 255–265, http://www.idealibrary.com.

Tuan, V. K. and Gorenflo, R. (1995). Extrapolation to the limit for numerical fractional differentiation, *Zeitschrift Angew. Math. Mech* **75**, 8, pp. 646–648.

Valério, D. and da Costa, J. (2006a). Tuning-rules for fractional PID controllers, in *Proceedings of the Second IFAC Symposium on Fractional Differentiation and its Applications* (Porto, Portugal), pp. 1–6.

Valério, D. and da Costa, J. (2005). Time-domain implementation of fractional order controllers, *IEE Proceedings Part-D: Control Theory and Applications,* **152**, 5, pp. 539–552.

Valério, D. and da Costa, J. (2006). Tuning of fractional PID controllers with Ziegler-Nichols-type rules, *Signal Processing* **86**, pp. 2771–2784.

Valério, D. and da Costa, J. (2013a). *An Introduction to Fractional Control* (The Institution of Engineering and Technology, London, United Kingdom).

Valério, D., Trujillo J. J., Rivero, M., Machado T. J. and Balenau, D. (2013b). Fractional Calculus: A Survey of useful Formulas, *European Physical Journal Special Topics* **222**, pp. 1827–1846.

Vich, R. (1987). *Z - Transform Theory and Applications* (D. Reidel Publishing Company, Dodrecht).

Vilanova, R., and Visioli, A. (2012). *PID Control in the Third Millennium. Lessons Learned and New Approaches* (Springer-Verlag, London).

Vinagre, B. M., Podlubny, I., Hernández, A. and Feliu, V. (2000). Some approximations of fractional order operators used in control theory and applications, *Fractional calculus and applied analysis* **3**, 3, pp. 231–248.

Vinagre, B. M., Petras, I., Podlubny, I. and Chen, Y. (2002). Using fractional order adjustment rules and fractional order reference models in model reference adaptive control, *Nonlinear Dynamics* **29**, pp. 269–279.

Vinagre, B. M., Monje, C. A. and Tejado, I. (2007). Reset and fractional integrators in control, in *Proceedings of the International Carpathian Control Conference (ICCC),* (Strbske Pleso, Slovak Republic,), pp. 24–27.

Wang, J. and Li, Y. (2006). Frequency domain stability criteria for fractional-order control systems, *Journal Chongqing Univ.* **5**, 1, pp. 30–35, http://www.url.com/triality.html.

Watkins, C. C., Sadun, A. and Marenka, S. (1993). *Modern Image Processing: Warping, Morphing, and Classical Techniques,* (Academic Press, Inc.).

Weeks, M. (2011). *Digital Signal Processing. Using Matlab® and Wavelets.* 2nd edn. (Jones and Bartlett Publishers, Sudbury).

Xue, D. and Chen, Y. (2002). A Comparative Introduction of Four Fractional Order Controllers, in *Proceedings of the 4th World Congress on Intelligent Control and Automation* (Shanghai, P.R. China), pp. 3228–3235.

Xue, D., Zhao, C. and Chen, Y. (2006). Fractional Order PID Control of A DC-Motor with Elastic Shaft: A Case Study, in *Proceedings of the 2006 American Control Conference* (Minneapolis, USA), pp. 3182-3187.

Yang, H., Ye, Y., Wang, D. and Jiang B. (2010). A Novel Fractional-order Signal Processing Based Edge Detection Method, in *Proc. 11th Conf. Control, Automation, robotics and Vision (ICARCV2010)* (Singapore), pp. 1122–1127.

Yang, Z., Lang, F., Yu, X. and Zhang, Yu. (2011). The Construction of Fractional Differential Gradient Operator, *Journal of Computational Information Systems* **7**, 12, pp. 4328-4342.

Yeroglu, C. and Tan, N. (2011). Classical controller design techniques for fractional order case, *ISA Transactions* **50**, pp. 461-472.

Ying Luo, Y. C. and Chen, Y. (2012). *Fractional Order Motion Controls.* (John Wiley & Sons, New York).

Yitzhaky, Y. and Peli, E. (2003). A Method for Objective Edge Detection Evaluation and Detector Parameter Selection, *IEEE Transactions on Pattern Analysis and Machine Intelligence* **25**, 8, pp. 1027–1033.

Yu, Q., Liu, F., Turner, I. and Burrage, K. (2013). Numerical simulation of the fractional Bloch equations, *Journal of Computational and Applied Mathematics* **255**, pp. 635–651.

Zhang, Z., Ma, S., Liu, H. and Gong, Y. (2009). An edge detection approach based on directional wavelet transform, *Computers & Mathematics with Applications* **57**, 8, pp. 1265–1271.

Zhang, Y. and Ding, H. (2010). Finite Difference Method for Solving the Time Fractional Diffusion Equation, in *Proceedings of the Asia Simulation Conference Part III (AsiaSim 2012)* (Shanghai, China), pp. 115-123.

Zhao, C., Xue, D. and Chen, Y. (2005). A fractional order PID tuning algorithm for a class of fractional order plants, in *Proceedings of the IEEE International Conference on Mechatronics and Automation (ICMA)* (Niagara, Canada), pp. 216–221.

Zhao, C. and Zhang, X. (2008). The Application of Fractional Order PID Controller to Position Servomechanism, in *Proceedings of the 7th World Congress on Intelligent Control and Automation* (Chongqing, China), pp. 3380-3383.

Zhen, C. (2009). Research on image edge detection based on wavelet transform algorithm, in *Proc. International Conference on Test and Measurement (ICTM'09)* (Hong Kong), pp. 423–436.

Index

Call for Book Proposals

Direct your submissions to *cchen@umassd.edu*

S e r i e s   i n

Computer
Vision

Series Editor

C.H. CHEN

Professor Emeritus
Department of Electrical and Computer Engineering
University of Massachusetts Dartmouth, USA

In recent years, there has been significant progress in computer vision in theory and methodology and enormous advancement on the application front, accompanied by rapid progress in vision systems and technology.

It is hoped that this book series in computer vision can capture most of the important and recent progress and results in computer vision. The target audiences cover researchers, engineers, scientists and professionals in many disciplines including computer science and engineering, mathematics, physics, biology, and medical areas, etc.

Topics include (but not limited to)

- Computer vision theory and methodology (algorithms)

- Computer vision applications in biometrics, biomedicine, etc biometrics, biomedicine, etc

- Robotic vision

- New vision sensors, software and hardware systems and technology

World Scientific
www.worldscientific.com

ICP Imperial College Press
www.icpress.co.uk

Printed in the United States
By Bookmasters